17th International Summer School on Vacuum, Electron, and Ion Technologies

(VEIT 2011)

Journal of Physics: Conference Series Volume 356

Sunny Beach, Bulgaria
19-23 September 2011

Editors:

M. C. M. van de Sanden
Miglena Dimitrova
Chavdar Ghelev

ISBN: 978-1-62276-773-1
ISSN: 1742-6588

Printed from e-media with permission by:

Curran Associates, Inc.
57 Morehouse Lane
Red Hook, NY 12571

Some format issues inherent in the e-media version may also appear in this print version.

Printed by Curran Associates, Inc. (2013)

Institute of Physics
Dirac House, Temple Back
Bristol BS1 6BE UK

Phone: 44 1 17 929 7481
Fax: 44 1 17 920 0979

techtracking@iop.org

Additional copies of this publication are available from:

Curran Associates, Inc.
57 Morehouse Lane
Red Hook, NY 12571 USA
Phone: 845-758-0400
Fax: 845-758-2634
Email: curran@proceedings.com
Web: www.proceedings.com

TABLE OF CONTENTS

ELECTRON, ION BEAM, LASER AND PLASMA ASSISTED PROCESSING

012001 Biomedical Applications and Diagnostics of Atmospheric Pressure Plasma 1
Z L Petrovic, N Puac, S Lazovic, D Maletic, K Spasic, G Malovic

012002 Structural and Photoluminescent Properties of Ag/Zno Nanocomposite Heterostructures 9
M E Koleva, A O Dikovska, N N Nedyalkov, P A Atanasov

012003 Laser Technology for Synthesis of AlN Films: Influence of the Incident Laser Fluence on the Films Microstructure 13
A Szekeres, A Cziraki, G Huhn, K Havancsak, E Vlaikova, G Socol, C Ristoscu, I N Mihailescu

012004 Auger Analysis of High-dose Ion-implantation of Arsenic in Silicon 18
G Spasov

012005 Formation of Si Nanocrystals in Ion Implanted Si-SiO_2 Structures by MeV Electron Irradiation 23
S Kaschieva, A Gushterov, C Angelov, S N Dmitriev

012006 Modification of Polyethylene Terephthalate by Atmospheric Pressure Dielectric Barrier Discharge (DBD) in View of Improving the Polymer Wetting Properties 27
K G Kostov, A L R Dos Santos, P A P Nascente, M E Kayama, R P Mota

012007 Evaluation of the Plasma Parameters in COMPASS Tokamak Divertor Area 32
M Dimitrova, P Ivanova, I Kotseva, T K Popov, E Benova, T Bogdanov, J Stöckel, R Dejarnac

012008 Electronic System for Langmuir Probe Measurements 37
M Mitov, A Bankova, M Dimitrova, P Ivanova, K Tutulkov, N Djermanova, R Dejarnac, J Stöckel, T K Popov

012009 Microwave Discharge in a Finite Length Vessel 41
Z Kiss'Ovski, V Vachkov, S Iordanova, I Koleva

012010 Plasma Parameters of a Small Microwave Discharge at Atmospheric Pressure Obtained by Probe Diagnostics 45
Z Kiss'Ovski, A Ivanov, S Kolev

012011 Analytical Calculation of the Gas Temperature and Time-resolved Measurement of the Electron Temperature of a Gas Discharge in He and Ne-He Mixtures 50
K A Temelkov, S I Slaveeva, N K Vuchkov

012012 2d Numerical Modeling of the Gas Temperature in a Large-volume High-temperature Nanosecond Pulsed Longitudinal Discharge in Helium with Small Admixtures of Neon, Strontium and Bromine 54
T P Chernogorova, K A Temelkov, N K Koleva, N K Vuchkov

012013 Plasma Deposited Polymers as Gas Sensitive Films 58
E Radeva, V Georgieva, J Lazarov, L Vergov, N Donkov

THIN FILMS AND COATINGS: DEPOSITION, ANALYSIS AND PROPERTIES

012014 Optical Properties of Chalcogenide Ge-Te-In Thin Films 62
A Zaidan, V Ivanova, P Petkov

012015 Influence of Burst Pulses on the Film Topography in Picosecond Pulsed Laser Deposition of $LaAlO_3$ 66
E Thelander, B Rauschenbach

012016 Optical Characterization of WO_3-VO_x Thin Films for Application in Electrochromic Devices—'Smart Windows' 71
G Bodurov, T Ivanova, M Aleksandrova, K A Gesheva

012017 Optical Characterization of Zn Coatings Deposited on Low Carbon Steel Substrates 75
G Bodurov, N Boshkov, L Lutov, T Ivanova, K A Gesheva

012018 Hybrid Sol-gel Silica Films with (TiO_2-CeO_2) Binary Nanopowders 79
M Zaharescu, M Nicolescu, M Gartner, A Barau, L Predoana, M Anastasescu, M Stoica, A Szekeres

012019 Surface Topography and Optical Properties of Ge-Sb(As)-S-Te Thin Films by Atomic-force Microscopy and Variable Angle Spectroscopic Ellipsometry 83
N Dulgheru (Nedelcu), M Anastasescu, M Nicolescu, M Stoica, M Gartner, V Pamukchieva, A Szekeres, K Todorova

012020 Planar Homogeneity of the Electrical Properties of $YBa_2Cu_3O_7$/$La_{0.7}Sr_{0.3}MnO_3$ Bi-layers 87
T Nurgaliev, B Blagoev, E Mateev, V Štrbík, Š Benacka, V Šmatko, Š Gaži, Š Chromik

012021 Properties of SFS Heterostructures Prepared by a Focused-ion-beam Technique 91
V Štrbík, Š Benacka, V Šmatko, Š Gaži, Š Chromik, E Mateev, B Blagoev, T Nurgaliev
012022 Impedance Measurements of Epitaxial and Polycrystalline LSMO Thin Films 95
E Mateev, B Blagoev, T Nurgaliev
012023 Vacuum Deposited GeSbSe Thin Films for Photonic Applications 98
J Dikova, R Todorov, T Babeva
012024 Structural Properties of ZnO Layers Deposited on Glass Substrates by PECVD 102
S Kitova, R Kazakov, G Danev
012025 Correlation Between the Haze Ratio and the Surface Roughness of Electrochemically Deposited Nanostructured ZnO Films 106
K Lovchinov, O Angelov, H Nichev, V Mikli, D Dimova-Malinovska
012026 Electrical and Optical Properties of ZnO Thin Films Doped with Nb Deposited by Rf Magnetron Sputtering 110
K Lovchinov, O Angelov, D Dimova-Malinovska
012027 NO_2 Sensing Properties of Amorphous Silicon Films 114
V Georgieva, N Donkov, P Stefanov, M Sendova-Vassileva, A Grechnikov, V Gadjanova
012028 Compositional Dependence of the Optical Properties of Silver Containing As_2Se_3 Thin Films 118
V Ilcheva, P Petkov, T Petkova, V Boev, B Monchev
012029 Preparation and Protection of Iron and Iron Compounds 122
N Koprinarov, M Konstantinova, G Avdeev, T Ruskov, T Tzacheva
012030 Transformation of the Coherent Population Trapping Resonance Into Enhanced Fluorescence in a Paraffin Coated Rb Vacuum Cell 126
E Taskova, E Alipieva, G Todorov
012031 Annealing of Si Surface Region Modified by Plasma Immersion Implantation of Nitrogen 130
S Alexandrova, E Halova, A Szekeres, E Vlaikova, M Gartner, M Anastasescu
012032 Investigation of the Optical Properties of InAs/InGaAs/GaAs Quantum Dot in Quantum Well Multilayer Structures for Infrared Photodetectors 134
T Ivanov, V Donchev, K Germanova, T Tellaleva, K Borissov, V Hongpinyo, P Vines, J P R David, B S Ooi
012033 Room Temperature Humidity Sensors Based on Co-evaporated TeO_2 and Sn 138
B Georgieva, I Podolesheva, G Spasov
012034 Study of Reticulated Vitreous Carbon Surface Treated by Plasma Immersion Ion Implantation for Electrodes Production 142
L L G Silva, D A S Conceição, S S Oishi, A Toth, M Ueda
012035 Surface Modification of Aluminium Alloys Using Hybrid Treatment Techniques 146
P Petrov
012036 XPS Study of ZnO Nanostructures Prepared by Laser Ablation 150
G Atanasova, A O Dikovska, M Stankova, P Stefanov, P A Atanasov
012037 Thin Dye Layers – Vacuum Deposition and Structure 156
D Dimov, J Assa, A Georgiev, R Kazakov, D Karashanova, G Danev, E Spassova
012038 Study of the Charge Transport Mechanism in Pulsed Laser Deposited AlN:Si Films 161
I P Minkov, S Simeonov, A Szekeres, A Cziraki, G Socol, C Ristoscu, I N Mihailescu
012039 Properties of Thin Silicon Carbide Films Prepared by Rapid Thermal Annealing 166
M Beshkova, K Grigorov, I Nedkov, M Massi, B Sismanoglu, H Maciel, N B Velchev
012040 Characterization of Electrophoretic Suspension for Thin Polymer Film Deposition 169
D Mladenova, M Weiter, P Stepanek, I Ouzzane, M Vala, V Sinigersky, I Zhivkov
012041 Structural and Dielectric Properties of TiO_2 Thin Films Deposited by the Sol-gel Method on Si Substrates 173
P Vitanov, A Harizanova, T Ivanova

MODIFICATION OF MATERIALS, DEVICES, AND PROCESSES: MODELLING AND COMPUTER SIMULATION

012042 Investigation of Magnetic Properties of Thin Films Using Computer Simulation 177
N A Balakirev, V A Zhikharev, G G Gumarov
012043 Monte Carlo Simulation of Electron Kinetics in a Hollow Cathode Discharge 184
S Karatodorov, D Mihailova, J Van Dijk, J Van Der Mullen, M Grozeva
012044 Simulation Tools for Computer-aided Design and Numerical Investigations of High-power Gyrotrons 188
M Damyanova, E Balabanova, S Kern, S Illy, S Sabchevski, M Thumm, E Vasileva, I Zhelyazkov

012045 The Role of Plasma Radius As a Condition for Sustaining a Coaxial Discharge at Various Wave Modes 192
K Ivanov, T Bogdanov, E Benova
012046 Nanoscale Biomaterial Interface Modification for Advanced Tissue Engineering Applications 197
V Safonov, A Zykova, J Smolik, R Rogovska, N Donkov, A Goltsev, T Dubrava, I Rassokha, V Georgieva
012047 IR Studies of Impurities in Chalcogenide Glasses and Thin Films of the Ge-Sb-S-Te System 201
V Pamukchieva, K Todorova, O C Mocioiu, M Zaharescu, A Szekeres, M Gartner
012048 Magnetic Properties of Nanosized $MgFe_2O_4$ Powders Prepared by Auto-combustion 207
C Ghelev, T Koutzarova, S Kolev, I Nedkov, K Krezhov, D Kovacheva, B Blagoev, B Vertruyen, C Henrist, R Cloots, A Zaleski, V Nizhankovskii
012049 Investigation of Optical Properties of Thin Polymer Films 212
N G Sultanova, S N Kasarova, I D Nikolov
Author Index

Editorial

The International Summer School on Vacuum, Electron and Ion Technologies (VEIT) has been organized biennially since 1977, when the VEIT Summer School series was launched by the Institute of Electronics, Bulgarian Academy of Sciences. The aim was to act as a forum for the exchange and dissemination of knowledge and ideas on the latest developments in electron-, ion- and plasma-assisted technologies. The organizers of the 2011 edition of the event were the Institute of Electronics, Bulgarian Academy of Sciences, Sofia, Bulgaria and the Department of Applied Physics, Eindhoven University of Technology, Eindhoven, The Netherlands. Whilst the school initially provided a meeting place for researchers mainly from Eastern and Central European countries, its importance has grown issue by issue. The school is now a major scientific event and a meeting place for young scientists from Eastern and Western Europe involved in research and development associated with high-tech industries. Many former school participants have gone on to become leading scientists in research establishments and companies throughout the world. Leading international companies, such as High Voltage Engineering, Balzers, Varian, and Hauzer have used the VEIT forum to present their products through oral presentations, poster contributions and exhibits. The School Proceedings have been published in special issues of the international journals *Vacuum, Plasma Processes and Polymers* and *Journal of Physics: Conference Series.*

The Seventeenth edition of VEIT was held in the Black Sea resort of Sunny Beach, Bulgaria on 19–23 September 2011. It was attended by 96 participants from 18 countries: Belgium, Brazil, Bulgaria, Czech Republic, Denmark, France, Germany, Greece, The Netherlands, Romania, Russia, Serbia, Sweden, Switzerland, Turkey, Ukraine, UK and USA.

Following the tradition of publishing the VEIT Proceedings, a selection of papers presented at the event is published in this special issue of *Journal of Physics: Conference Series*, under the originality and quality criteria of acceptance by the journal, including peer reviewing.

The school comprised thirteen plenary and three poster sessions. At the plenary sessions, 21 invited talks of general interest were presented by highly-renowned international experts in the field, as well as thirteen progress reports by young scientists. In total, 60 contributed papers were presented during the poster sessions. There were several scientific highlights, covering the fundamentals of interaction of fast particles with solids, and challenging practical applications. These ranged from novel techniques for creating hard coatings, optical/protective layers and biocompatible materials, to nanosized structures produced by evaporation, sputtering or external irradiation. The latest results were presented on ion-beam synthesis and modification in both low-energy (deposition and film growth) and high-energy (sputtering, implantation) regimes, the processing of solid materials aimed at surface patterning, and the creation of nanophase systems for electronic or tribological/wear resistant applications.

Despite the busy scientific program, the atmosphere was relaxed and informal. The early afternoons of most conference days were free to stimulate both scientific and social interaction between participants, which often took place on the beach. The social program included a welcome party, an official dinner, and an outing to historical landmarks in the vicinity of Nessebar.

VEIT 2011 owes its success to many people. The International Advisory Committee shaped the scientific program and ensured high-quality plenary presentations by their careful selection of the invited speakers. The Local Committee bore the brunt of the organization, both at the conference site and in dealing with correspondence, abstracts, and manuscripts for these proceedings. We are grateful to our sponsor, the Department of Applied Physics, Eindhoven University of Technology, Eindhoven, The Netherlands, for its generosity that enabled us to support the attendance of students, and provided support dealing with mailing, printing, renting the conference site, etc. We are also grateful to the members of the International Advisory Committee: G Dinescu, A Ehiasarian, N Guerassimov, D Mataras, W Möller, I Petrov, Z Petrovic, B Rauschenbach, M C M (Richard) van de Sanden and M Ürgen, for their continuous concern for the high scientific level and the future of the event, and for their contribution to the development of a new generation of scientists and experts in the field. And, last but not least, we would also like to thank all authors for their valuable contributions to these proceedings and to the school, and all reviewers for their hard and tedious, but very important, work.

The next conference in the series will be held in September 2013.

M C M van de Sanden, Miglena Dimitrova, Chavdar Ghelev
Guest Editors

All papers published in this volume of *Journal of Physics: Conference Series* have been peer reviewed through processes administered by the proceedings Editors. Reviews were conducted by expert referees to the professional and scientific standards expected of a proceedings journal published by IOP Publishing.

Biomedical applications and diagnostics of atmospheric pressure plasma

Z Lj Petrović[1], N Puač, S Lazović, D Maletić, K Spasić and G Malović

Institute of Physics, University of Belgrade, Pregrevica 118, 11080 Belgrade, Serbia

E-mail: zoran@ipb.ac.rs

Abstract. Numerous applications of non-equilibrium (cold, low temperature) plasmas require those plasmas to operate at atmospheric pressure. Achieving non-equilibrium at atmospheric pressure is difficult since the ionization growth is very fast at such a high pressure. High degree of ionization on the other hand enables transfer of energy between electrons and ions and further heating of the background neutral gas through collisions between ions and neutrals. Thus, all schemes to produce non-equilibrium plasmas revolve around some form of control of ionization growth. Diagnostics of atmospheric pressure plasmas is difficult and some of the techniques cannot be employed at all. The difficulties stem mostly from the small size. Optical emission spectroscopy and laser absorption spectroscopy require very high resolution in order to resolve the anatomy of the discharges. Mass analysis is not normally applicable for atmospheric pressure plasmas, but recently systems with triple differential pumping have been developed that allow analysis of plasma chemistry at atmospheric pressures which is essential for numerous applications. Application of such systems is, however, not free from problems. Applications in biomedicine require minimum heating of the ambient air. The gas temperature should not exceed 40 °C to avoid thermal damage to the living tissues. Thus, plasmas should operate at very low powers and power control is essential. We developed unique derivative probes that allow control of power well below 1 W and studied four different sources, including dielectric barrier discharges, plasma needle, atmospheric pressure jet and micro atmospheric pressure jet. The jet operates in plasma bullet regime if proper conditions are met. Finally, we cover results on treatment of bacteria and human cells as well as treatment of plants by plasmas. Localized delivery of active species by plasmas may lead to a number of medical procedures that may also involve removal of bacteria, fungi and spores.

1. Introduction

The choice of the plasma system used for treatment is usually guided by the kind of samples that are treated and the effect these plasmas are intended to have on the samples. The desire to use plasma for in-vivo treatments have made it necessary that several requirements for plasma sources be met. Necessarily, plasmas have to operate at atmospheric pressure if they are to be used for medical treatment of living organisms. At the same time, one needs non-equilibrium plasmas in order to achieve separation of electrons on the one side, and ions and neutrals, on the other. It is an advantage that no expensive vacuum systems are needed, while, on the other hand, it is much more difficult to achieve non-equilibrium (non-thermal) mode of operation at higher pressures.

[1] To whom any correspondence shoul be addressed.

Published under licence by IOP Publishing Ltd

The sensitivity to heat of biomedical samples narrows the choice of non-thermal plasmas. There are many types of plasmas that can be generated under ambient pressure and temperature conditions suitable for treatment of sensitive samples [1, 2, 3]. The motivation is to develop new medical techniques, as plasma offers some possibilities for inducing desired processes with minimum damage to the living tissue [1, 2, 3, 4]. While the first results seemed quite impressive, including effects on tumor cells and even active tumors [5], tooth decay treatment and tooth cleaning [6], wound healing [5], treatment of fungi and spores and even treatment of ulcers and blood vessels, one can still not rule out negative effects that have not been studied over a sufficiently long time scale. Th preliminary results, however, show a large degree of selectivity.

Some of the well-known small-size atmospheric-pressure plasma sources are: plasma needle [7, 4], μAPPJ [8], plasma bullet [9], plasma torch [10] and floating electrode dielectric barrier discharge plasma [11]. Their electrode configuration, voltages and excitation frequencies are very different; some of them work at microwave frequencies, some at 13.56 MHz and others at 5-120 kHz in sine or pulse regime. Yet, all is not understood about their physics and while models are being developed mainly based on low pressure plasmas, the reliable experimental data are limited due to the limited availability of diagnostic techniques that are suited for such plasmas.

Here we will present several diagnostics techniques suited for atmospheric pressure plasmas and the operation of several different plasma systems working at atmospheric pressure. We will also summarize our results in treating living organisms and give examples of results mainly on sterilization of bacteria and biofilms.

2. Atmospheric pressure discharges – different experimental set-ups

Achieving non-equilibrium at atmospheric pressure is difficult since the ionization growth is very fast at such a high pressure. The high degree of ionization on the other hand enables transfer of energy between electrons and ions through Coulomb collisions. Furthermore, heating of the background neutral gas is achieved through collisions between ions and neutrals. Thus, all schemes to produce non-equilibrium plasmas revolve around some form of control of the ionization growth. It can be achieved either by an inhomogeneous field as in corona or by employment of a dielectric barrier which turns the field off after a space charge is deposited on the dielectric. Ionization growth limiting may also be achieved by a time-varying field. Another approach is to operate at the pd value corresponding to the Paschen minimum, i.e. at microscopic dimensions and high pressures. In that case, the breakdown voltage is below the threshold for streamer development and thus a glow discharge may be achieved. If the electronegative nature of the gas is increasing the breakdown and operating voltages, one may mix in an inert gas. The discharge is thus initiated in the inert gas and then the atmospheric gas is mixed to produce chemically active radicals. In most cases, however, atmospheric pressure plasmas have small dimensions making it very difficult to perform standard diagnostics.

Some of the plasma devices designed for in-vivo treatments are the μ-APPJ and the plasma needle, which operate at 13.56 MHz at atmospheric pressure. A micro-atmospheric plasma jet was developed by Schultz van der Gathen and coworkers [12]; this plasma source is interesting both for applications as well as for the study of its basic properties.

The micro-atmospheric pressure plasma jet [μ-APPJ] consists of two symmetrical electrodes of equal length (34 mm) made of stainless steel. The distance between the electrodes can be adjusted with good precision from a few mm up to several hundred micrometers. In all our experiments, the distance between the powered and the grounded electrode was 1 mm. One of the electrodes was powered by a signal generator at 13.56 MHz while the other electrode was grounded. The measurements were made at powers of 40-80 W fed by a RF power supply.

Plasma is ignited along the entire length of the electrodes (figure 1); for certain combinations of power/gas-flow parameters, effluent of plasma coming out of the cuvette can be formed. The main advantage of this design is that both the discharge volume (plasma core) and effluent region are accessible for diagnostics, such as optical emission spectroscopy (OES) and two-photon absorption laser-induced fluorescence (TALIF) [8]. Also, the plane parallel geometry of the electrodes adds to the simplicity when it comes to modeling this type of discharge.

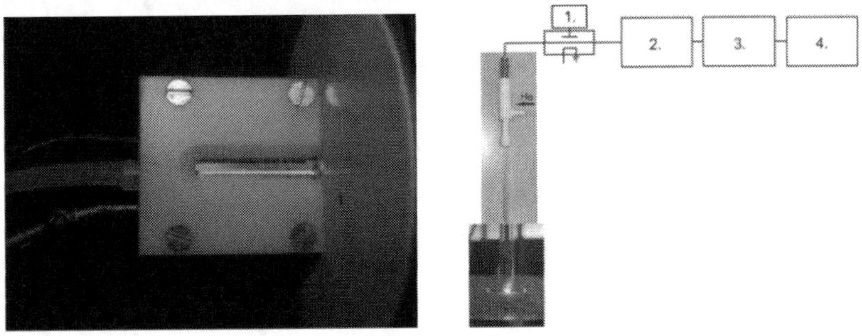

Figure 1. μ-APPJ with formed plasma and plasma needle system set-up.

Another plasma source that meets all the necessary conditions for treatment of organic materials and living tissues is the plasma needle (figure 1, picture on the right-hand side). Most importantly, in such a discharge gas the heating is minimized, while th effects on the tissue and bacteria have been clearly shown to be significant. The needle consists of a central electrode made of tungsten insulated almost to the tip by a slightly larger ceramic tube, both being placed inside a glass tube. The needle body is made of Teflon. We used helium as a buffer gas at several different flow rates. The central electrode is powered by a 13.56 MHz signal generator through an amplifier and a matching network. Both for the plasma needle and for thee μ-APPJ, we have derivative probes and a Hiden HPR60 mass/energy analyzer in order to determine the power applied to plasma and the composition of the discharge, respectively. In both of these systems, inert gas is used to reduce the breakdown voltage and achieve stable non-equilibrium plasma formation. Yet these plasmas have shown several modes of operation and further studies are required to fully understand their operation and make further optimizations.

Another type of atmospheric pressure plasma relies on mixing the inert gas carrying the plasma created by external electrodes with atmospheric gas mixture. It is the so-called plasma jet. The operating frequencies of plasma jets are in the region of several tens of kHz which are much lower frequencies then those used for the plasma needle and μ-APPJ (in MHz). It has been shown that micro jet plasma is not always continuous but often is formed by a train of fast travelling bullets which only appear to be continuous to the human eye. The atmospheric pressure plasma jet/bullet that we constructed was made of a Pyrex glass tube with the electrodes made of a thin copper foil wrapped around the glass tube. The distance between the powered and the grounded electrode was 13.5 mm and their width was 13 mm. One of the electrodes (the left electrode, see figure 2) was grounded. The other electrode, closer to the end of the glass tube, was the powered one (see figure 2). In all experiments the buffer gas was helium. We used a signal generator connected to the custom-made amplifier to power the micro jet. The highest voltages that we could obtain from the amplifier were up to 1 kV, which was not enough to ignite the plasma. In order to the increase the applied voltages to more than 5-6 kV, we had to use an additional homemade transformer. The operating frequency was 80 kHz and the applied voltage was sinusoidal in the range of 6-10 $kV_{peak\text{-}to\text{-}peak}$. Micro jets with plasma bullets have been constructed with a range of different

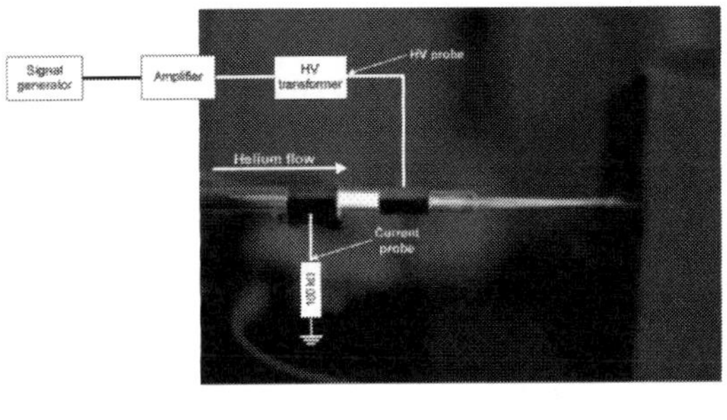

Figure 2. Atmospheric plasma bullet.

Journal of Physics: Conference Series **356** (2012) 012001 doi:10.1088/1742-6596/356/1/012001

geometries, frequencies and shapes of applied voltage. The effect of bullets seems to be quite universal for an optimum range of electrode sizes and flows, while the frequencies and voltage shapes vary a lot. Plasma bullets, however, still need to be fully understood and properly modelled. We will focus here on the plasma jet diagnostics, while at the same time showing some results on the sterilization achieved by a plasma needle.

3. Plasma diagnostics

It has been reported only quite recently that the plasma jets formed by the source operating at low excitation frequency is not continuous. Instead it consists of small plasma packages that are formed in positive and/or negative half cycle of the period [13]. Amazingly these little bullets are formed and travel outside the plasma jet where there is no applied electric field. The velocity of these packages are larger than the speed of the flowing feed gas by several orders of magnitude. Several theories of bullet formation have been proposed [13, 14, 15, 16, 17, 18] but to date a definite explanation of the physical mechanisms involved in creation and propagation of plasma bullet are still not fully understood.

For the current and voltage measurements, we used two commercial probes. The current and voltage waveforms when the plasma is formed and without discharge are shown in figure 3. When the plasma is off, the phase difference between the current and voltage is close to 90°. In this case, we have a capacitive impedance of several MΩ, corresponding to a capacitance of about 0.5 pF. On the other hand, when the plasma is formed, the current signal is larger, deformed and shifted in phase overlapping more with the voltage signal. The plasma ignition changes the slopes of the V_{RMS}–I_{RMS} curves (see figure 4). The mean power calculated increases with the increase of the applied voltage; it was in the range from 1 to 8 W in all measurements.

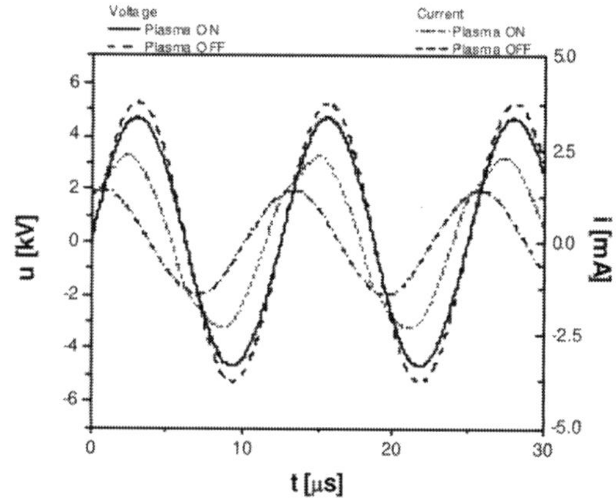

Figure 3. Current and voltage waveforms for helium flow rate of 3 slm. The dashed lines represent the case when discharge is OFF, the solid lines, when discharge is ignited [19].

Integral and time-resolved images of the plasma jet system were obtained by an ICCD camera. For exposure times larger than the cycle period (12.5 μs), the plasma appears to be continuous, like a plume (see figure 1 LHS picture). The length of the plasma plume is up to five centimeters, depending of the flow rate and the voltage applied. For the time-resolved images, we had to use integration on the chip because the light emission in a single shot is not sufficient to obtain clear images with gate widths less than 50 ns. This was facilitated by the high reproducibility of the pulses and the small jitter. Figure 5 shows the plasma bullet images obtained for several different flows of working gas. We can see that with the decrease in the He flow, the plasma bullet starts to be elongated, deformed and its intensity is much smaller. Eventually, bullets are not formed at the very small flows.

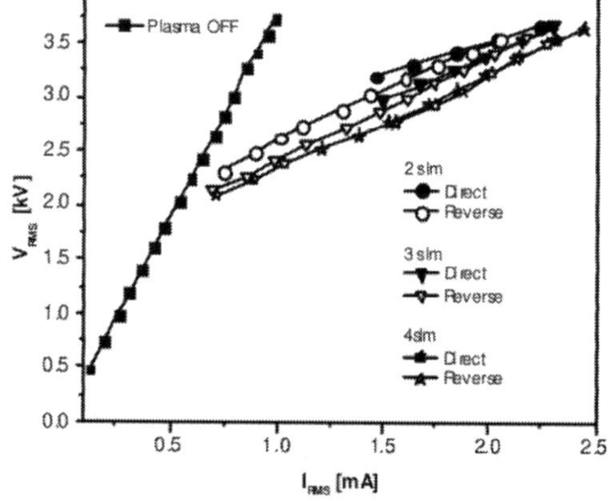

Figure 4. Current-voltage characteristics for three different flows of helium [19].

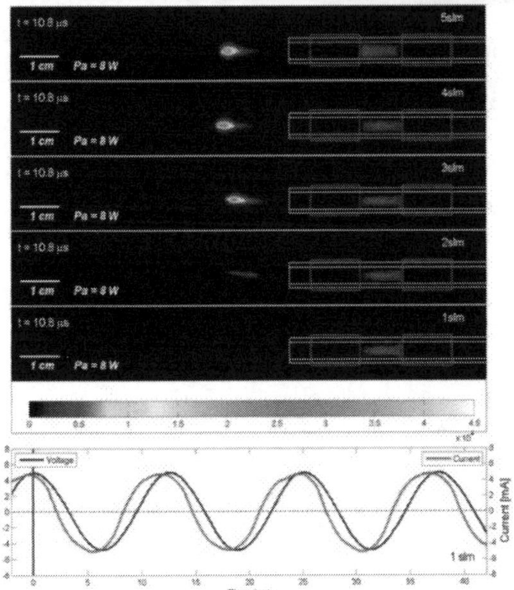

Figure 5. Plasma jet for 1, 2, 3, 4 and 5 slm of He flow. Exposure time 2 ms, gate width 25 ns, gate delay 10.8 μs [19].

Figure 6 shows the development of the plasma over the entire period of applied voltage (12.5 μs). All images are scaled to the same maximum intensity and thus can be compared. One can see that when the current and voltage signals are close to zero, the plasma is not visible. In the negative part of the current and voltage waveforms, the plasma is confined between the electrodes. During the positive part of the waveforms, the plasma is first confined between the electrodes (rising slope) and then, near the maximum of the curves, it leaves the glass tube in the form of a bullet. The dimensions of the bullet are of the order of a few millimeters. We calculated the speed of the bullet at ~20 km/s, depending on the position away from the end of the glass tube. The plasma bullet is much faster than the speed of the buffer gas flow (1 to 7 m/s). Thus we can conclude that our plasma source was not continuous, it consisted of very small plasma packages that traveled at a high speed. By varying the plasma parameters, the length and intensity of the plasma coming out of the tube can be adjusted.

For the two plasma devices operating at a much

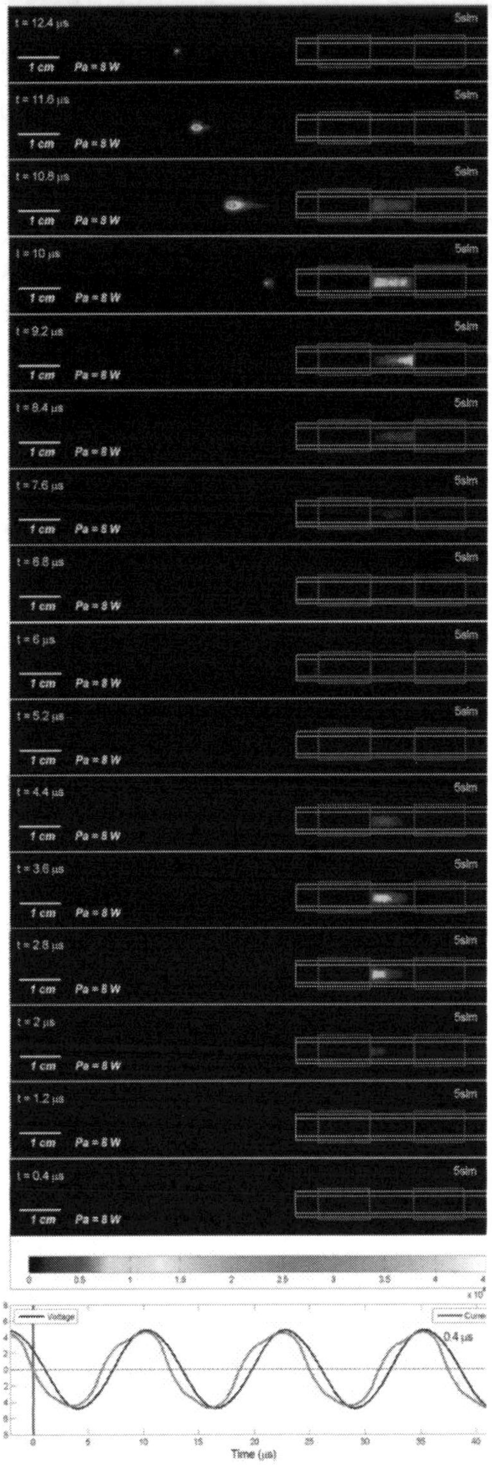

Figure 6. Plasma jet at 5 slm, exposure time 2 ms, gate width 25 ns and gate delay from 0.4 to 12.4 μs [19].

higher frequency (13.56 MHz), the diagnostics was made by using homemade derivative probes in order to determine the power transmitted to the plasma and the operation mode of the discharge. The derivative probes were very sensitive; a numerical procedure for subtracting the displacement current based on accurate calibration of the system was performed so that it was possible to measure powers of the order of 0.1 W or less even with displacement current a couple of orders of magnitude larger than the plasma current.

Besides derivative probes, we used mass spectrometry to analyze the plasma products formed by a μ-APPJ [19] and by a plasma needle [20]. Several problems occurred during the setting up of the experiment; they are described in detail elsewhere [19, 20].

The analysis of the composition of neutrals and ions was motivated by the need to check which species are formed in the discharge. These results may be used as a test of plasma chemical models, to identify radicals and ions (that may be used after acceleration to induce damage to the tissue). The performance of the mass analyzer was tested and techniques were developed to produce data without the uncertainty induced by a contribution of the ionizer to possible dissociation. It was found that the predominant ions created by the plasma are O_2^+, O^+, H_3O^+, N_2^+, N^+, NO^+, OH^+ [19]. When it comes to plasma treatment of samples of biological origin, the chemically active species that are of interest are O, metastables O and O_2, OH, N, H_2O_2 and NO.

4. Plasma sterilization

The entry point for most groups dealing with plasma medicine is a study of sterilization. The effects on bacteria may be shown quickly, although it requires expertise in biomedicine. The benefit is that direct potential applications may be developed outside the realms of strict medical regulations. Yet, in situ sterilization, for example, treatment of wounds to prevent infection, would be a much more important goal. Following preliminary work on sterilization in microwave plasma, albeit at low pressure, we reinitiated the studies of plasma sterilization as a part of our plasma medical project. So far, a plasma needle has been used to induce killing of Streptococcus mutans and Escherichia coli bacteria in the form of planktonic samples. Also, we have the plasma interaction with normal, living cells; for these experiments we used human peripheral blood mesenchymal stem cells (hPB-MSC) as a model system to predict the degree of possible damage to the cell responses [21]. Many factors are responsible for bacterial inactivation. Direct exposure of the bacterial samples to the plasma appears to be more effective than remote exposure. Another factor that determines the efficiency of the specific treatment is the type [22] of bacteria, gram positive or gram negative. Very importantly, we studied the sterilization of bacteria in planktonic samples, where bacteria are dissolved in a small amount of liquid that would otherwise give it some protection from other agents. We showed that efficient sterilization of planktonic samples is not only possible, but may be efficient depending on the initial population [21].

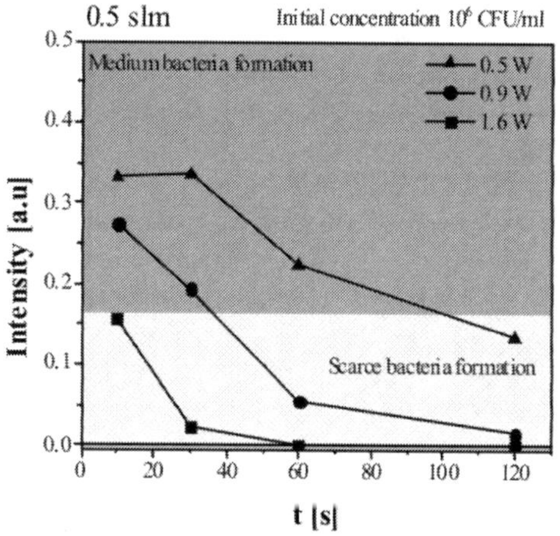

Figure 7. Treatment of MRSA biofilms of Staphylococcus aureus (ATCC 25923) by using plasma needle. Untreated sample showed STRONG bacteria formation (control intensity 0.65 a.u). Initial concentration of bacteria used was 10^6 CFU/ml.

One of the most serious problems in the hospital environment is bacterial contamination of surfaces with methicillin-resistant Staphylococcus aureus (MRSA) responsible for significant nosocomial infections. The pathogenic contaminants form biofilms, which are difficult to treat with routine biocides. The biofilm is not just a secured shelter but a defense mechanism and a nutrition depot for pathogens. We show below the preliminary results of plasma treatment of the MRSA bacteria samples in the form of a biofilm.

In figure 7, we show the optical density of bacteria samples after plasma treatment for several different treatment times (10, 30, 60 and 120 s). The initial concentration of the samples was 10^6 CFU/ml, which corresponds to a measured optical density of 0.65 a.u. The buffer gas flow was 0.5 slm, but studies were also made as a function of the flow rate. The treatment efficiency increases with the increase of the treatment time and the mean power deposited to the plasma. For the highest power and only for the shortest treatment time of 10 s, there was scarcely bacteria formation; for the longer treatment, no bacteria formation was observed after the plasma treatment and yet there was very little or no heating of the gas.

Conclusions

We reviewed shortly our studies of atmospheric pressure plasmas and their application in biomedicine. In particular, we covered new results obtained with a plasma jet showing formation of plasma bullets and their properties as a function of geometry of electrodes and gas flow. In addition, we showd some practical results of sterilization using a plasma needle. The treatment of biofilms is essential, as are the studies of treatment of fungi, spores, prions and viruses. At the same time, one needs to extend the studies to specific medical problems associated with treatment of living organisms, including humans.

As far as plasma goes, some further optimization may be made for localized accurate treatment of cells or sterilization. With good knowledge of the power deposited into the plasma and control of the radicals that are produced, together with spatial emission profiles indicating changing of the regime of operation, a sufficient control over the reproducibility of the plasma needle operation was achieved. Other sources may be sought for more refined interaction with living cells. Different applications may seek more uniform sources extended over larger areas or even more localized treatment, which is all within the reach of the present day techniques.

Acknowledgements

This research has been supported by the MES, Serbia, under contract numbers ON171037 and III41011.

References

[1] Fridman A and Kennedy L A 2004 *Plasma Phys. Eng.* (New York: Taylor and Francis)
[2] Iza F, Kim G J A, Lee S M A, Lee J K A, Walsh A J, Zhang A Y and Kong M 2008 *Plasma Process. Polym.* **5** 322
[3] Stoffels E, Flikweert A J, Stoffels W W, and Kroesen G M W 2002 *Plasma Sources Sci. Technol.* **11**/4 383
[4] Puač N, Petrović Z Lj, Malović G, Đorđević A, Živković S, Giba Z and Grubišić D 2006 *J. Phys.* D: *Appl. Phys.* **39** 3514-19
[5] Fridman G, Friedman G, Gutsol A, Shekhter A B, Vasilets V N and Fridman A 2008 *Plasma Process. Polym.* **5** 503–33
[6] Sladek R E J, Stoffels E, Walraven R, Tielbeek P J A and Koolhoven R A 2004 *IEEE Trans. Plasma Sci.* **32**/4 1540-3
[7] Kieft I E, v d Laan E P and Stoffels E 2004 *New J. Phys.* **6** 149
[8] v d Gathen S V, Schaper L, Knake N, Reuter S, Niemi K, Gans T and Winter J 2008 Spatially resolved diagnostics on a microscale atmospheric pressure plasma jet *J. Phys.* D: *Appl. Phys.* **41** 194004
[9] Shi J, Zhong F, Zhang J, Liu D W and Kong M G 2008 *Phys. Plasmas* **15** 013504
[10] Yonson S, Coulombe S, Leveille V and Leask R L 2006 *J. Phys.* D: *Appl. Phys.* **39** 3508–13
[11] Fridman G, Shereshevsky A, Jost M M, Brooks A D, Fridman A, Gutsol A, Vasilets V and Friedman G 2007 *Plasma Chem. Plasma Process.* **27** 163–176
[12] v d Gathen S V, Buck V, Gans T, Knake N, Niemi K, Reuter St, Schaper L and Winter J 2007

Contrib. Plasma Phys. **47**/7 510
[13] Walsh J L and Kong M G 2008 *IEEE Trans. Plasma Sci.* **36** 1314
[14] Lu X and Laroussi M 2006 *J. Appl. Phys.* **100** 063302
[15] Walsh J L, Iza F, Janson N B, Law V J, Kong M G and Shi J 2010 *J. Phys.* D: *Appl. Phys.* **43** 075201
[16] Zhong F, Zhang J, Liu D W and Kong M G 2008 *Phys. Plasmas* **15** 013504
[17] Mericam-Bourdet N, Laroussi M, Begum A and Karakas E 2009 *J. Phys. D: Appl. Phys.* **42** 055207
[18] Shashurin A A, Shneider M N, Dogariu A, Miles R B and Keidar M 2009 *Appl. Phys. Lett.* **94** 231504
[19] Maletić D, Puač N, Malović G and Petrović Z Lj *unpublished*
[20] Malović G, Puač N, Lazović S and Petrović Z Lj 2010 Mass analysis of an atmospheric pressure plasma needle discharge *Plasma Sources Sci. Technol.* **19** 034014
[21] Lazović S, Puač N, Miletić M, Pavlica D, Jovanović M, Bugarski D, Mojsilović S, Maletić D, Malović G, Milenković P and Petrović Z Lj 2010 *New J. Phys.* **12** 083037
[22] Stoffels E, Sakiyama Y and Graves D B 2008 *IEEE Trans. Plasma Sci.* **36**/4 1441

Structural and photoluminescent properties of Ag/ZnO nanocomposite heterostructures

M E Koleva[1], A O Dikovska, N N Nedyalkov and P A Atanasov

Emil Djakov Institute of Electronics, Bulgarian Academy of Sciences,
72, Tzarigradsko Chaussee, 1784 Sofia, Bulgaria

E-mail: mihaela_ek@yahoo.com

Abstract. We describe the preparation of Ag nanoparticles covered by ZnO and investigation of the Ag/ZnO nanocomposite heterostructures. The silver nanoparticles were produced by pulsed laser deposition (PLD) on quartz substrates in a vacuum chamber and were post-deposition annealed by a Nd:YAG laser for surface nanostructuring. The morphology and optical properties of Ag nanoparticles were studied and the mean particles diameter and size distribution were estimated. Laser ablation was used for the subsequent deposition of ZnO. In order to study the Ag nanoparticles influence on ZnO, X-ray diffraction (XRD) and photoluminescent (PL) spectra were obtained and analyzed. It was found that the band-gap emission of ZnO grown on Ag nanoparticles can be enhanced by about a factor of five compared to the emission of ZnO film without Ag undercoating. The silver nanoparticles are also responsible for the UV shift in the PL emission.

1. Introduction

It is well-known that an electric dipole placed in the proximity of a metal surface may transfer energy to surface plasmons (SP), whch results in modification of its spontaneous emission decay rate [1]. This effect has been used to enhance the photoluminescence (PL) of light emitters, further aiming to raise the emission efficiency of light emission diodes (LEDs) [2, 3]. The metal films morphology has been found to be of paramount importance for SP-mediated emission enhancement [2], because after radiation energy is transferred to SPs, the coupled SP energy can only be transferred to free space radiation by scattering which bridges the momentum gap. Thus, metal films with periodic grating or random corrugation have been typically used to realize light emission enhancement.

Several methods have been applied in view of improving the light emitting efficiency of the optoelectronic devices based on ZnO, such as hydrogen doping, annealing or metal doping [4-7]. In all these methods, the underlying mechanisms for the improvement were attributed to the passivation of defects by doping atoms, or to carrier transfer. Newly devised ZnO/Ag nanocomposites have been based on another mechanism of combined effect of surface modification, namely, band alignment and charge transfer between Ag and ZnO in the nanostructure.

[1] To whom any correspondence should be addressed.

2. Experimental

Silver films with thickness of 180 nm were deposited on 1×1 cm quartz substrates by laser ablation in a vacuum chamber at a base pressure of 3.9×10^{-4} Torr and fluence of 2.6 J/cm^2. The substrate to target distance was 3.5 cm. Laser annealing in air at a fluence of 1 J/cm^2 was applied for nanostructuring the silver layers by a fixed number of laser pulses $N_p = 10$. The as-prepared silver nanoparticles were covered by a ZnO thin film using pulsed laser deposition at room temperature. A reference sample without Ag nanoparticles was also prepared at the same deposition conditions for PL comparison. Since a good quality crystalline and stoichiometric ZnO films were produced also at low oxygen pressures [8, 9], our depositions were carried out at 3 Pa of oxygen flow for 10 min. The laser deposition of Ag and ZnO films and the laser annealing were performed by using a third harmonic Nd:YAG laser (Lotis LS-2147, $\lambda = 355$ nm, pulse duration $\tau = 18$ nm) operating at a repetition rate of 10 Hz for the laser deposition. The morphology of the thin films, nanoparticles and nanocomposites was characterized by scanning electron microscopy (SEM) (Dual beam system SEM/FIB, Lyra\Tescan) and the mean size of the nanoparticles was estimated. The crystalline nature of the films was confirmed by XRD with Cu-Kα radiation (1.54 Å). The transmission spectrum of the Ag nanoparticles was analyzed using a VIS OTS spectrometer (Ocean Optics Inc.) in the range of 380 – 780 nm. The PL measurements were made using a Fluorolog spectrofluorometer (HORIBA Jobin Yvon) under excitation at 320 nm.

3. Results

Laser annealing of silver films at 1 J/cm^2 leads to melting and fragmentation of the material into nanosized droplets. Except on the laser fluence, the particles density and size distribution depend also on the film thickness and the other deposition conditions [10]. The morphology of the Ag nanostructured film and nanocomposite heterostructures of ZnO/Ag are presented in figure 1 (a and b). A mean particles diameter of 70 nm was estimated for the silver nanoparticles in the range of 30 – 110 nm (figure 1a). Figure 1b shows the ZnO/Ag nanocluster formation, where the nanoparticles formed are also spherical and have a mean diameter of 90 nm in the range of 40 – 130 nm. Therefore, the silver nanoparticles are covered by approximately 20-nm ZnO shell. The deposition of ZnO on a quartz substrate at room temperature and oxygen ambient leads to the formation of a continuous film having a flat surface, as presented on figure 1c.

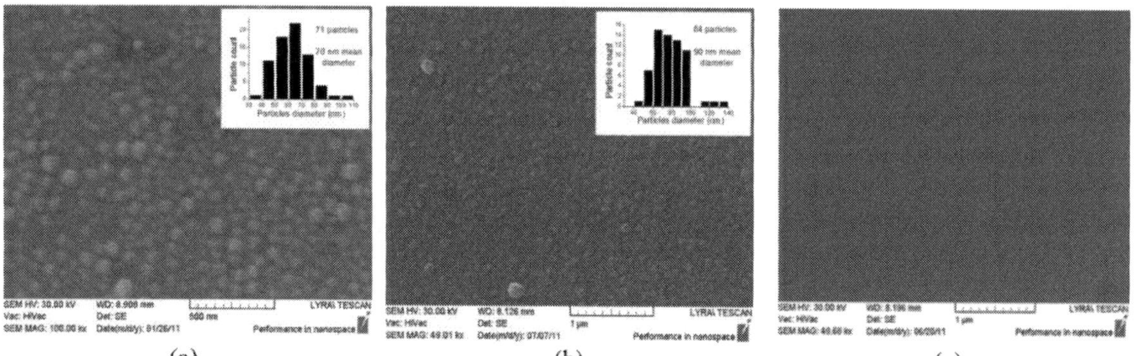

(a) (b) (c)

Figure 1. SEM images and size histograms of (a) Ag nanoparticles (b) ZnO/Ag nanocomposite heterostructures and (c) ZnO thin film.

The transmission spectrum of Ag nanoparticles is given in figure 2. A transmission minimum is observed at 450 nm, which is due to the excitation of surface plasmons by narrow-size-controlled Ag nanoparticles. The small interparticle distances, visible on figure 1a, lead to interaction between the particles, such as plasmon coupling and multiscattering effects.

Crystalline ZnO films were successfully deposited at room temperature in oxygen ambient on quartz substrates. Figure 3 shows XRD patterns of ZnO thin films fabricated with and without a Ag nanostructured layer. The results confirm the single-phase structure of ZnO with a (002) preferred orientation. The peaks obtained are indexed to wurtzite ZnO with high crystallinity. The ZnO film and the ZnO/Ag nanocomposite film exhibit the same XRD spectra, the only difference being the diffraction intensity which depends on the metal undercoating. The presence of Ag nanoparticles is found to induce a reduction in the diffraction intensity due to the Ag nanoparticles role of impurities induced defects in the ZnO matrix. No characteristic peaks of other phases were observed in the samples.

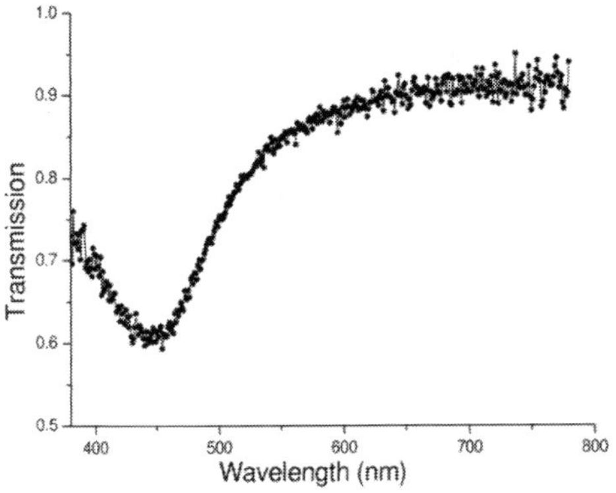

Figure 2. Transmission spectra of silver nanoparticles produced from the film with thickness of 180 nm annealed at 1 J/cm^2.

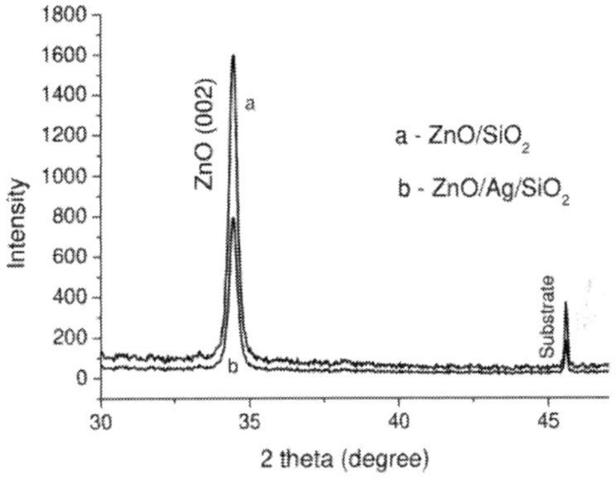

Figure 3. XRD patterns of (a) ZnO and (b) ZnO/Ag on SiO_2 substrates.

A broad photoluminescent peak centered at 420 nm is observed for the pure ZnO thin film on quartz. The photoluminescence is strongly enhanced after ZnO is grown onto the silver nanoparticles. The separation from the substrate by a nanostructured plasmonic silver film with thickness of 180 nm also leads to a shift in the PL peak position from the violet towards the near UV spectrum. The PL enhancement in these heterostructures is attributed to hot electron transfer to the conduction band edge [11]. The origin of the hot electrons can be understood in terms of excitation of local surface plasmons (LSPs) in the metal nanoparticles. The LSP resonances depend on the particular geometry of the individual nanoparticles and on the dielectric environment [12].

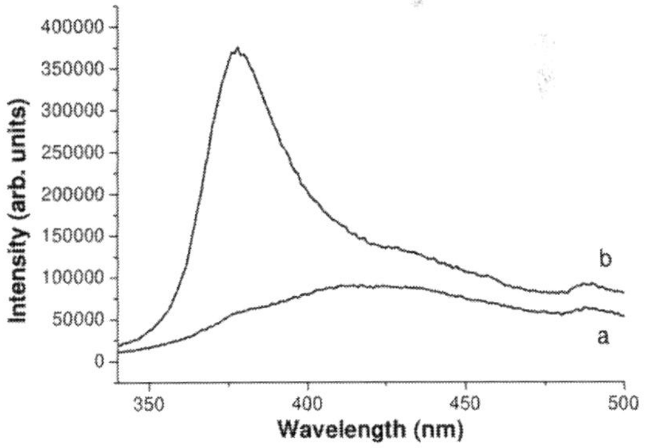

Figure 4. PL spectra of (a) pure ZnO on SiO_2 and (b) ZnO/Ag on SiO_2 substrate.

Conclusions

Silver nanoparticles having a spherical shape with size in the range of 30 – 110 nm and mean diameter about 70 nm were prepared from a 180-nm silver layer by laser annealing in air. Silver films were produced by pulsed laser deposition using a third-harmonic Nd:YAG laser. ZnO films with a flat surface and single crystalline phase structure were deposited on SiO_2 substrates at room temperature. Transformation of the ZnO thin film to a discontinuous structure consisting of small particles was observed when the deposition was performed onto a silver nanostructured surface. Spherical nanocomposite particles with mean diameter of 90 nm and a relatively narrow size distribution were observed. The characteristic silver plasmon minimum was observed in the optical transmission spectrum at 450 nm. A broad peak in the visible band of the PL spectrum was detected for pure ZnO. Utilizing an Ag underlayer, the light emitting efficiency of the ZnO layer was improved considerably. The interfacial charge transfer between metal Ag and ZnO is also responsible for the emission shift from the VIS to the UV spectrum.

Acknowledgments

This work was supported in part by the Bulgarian Ministry of Education, Youth and Science under Contract No. DO02-293/08. The SEM measurements were supported by the Bulgarian Ministry of Education, Youth and Science under Contract No. DO02-56/08. The optical transmission measurements were supported by the Ministry of Education, Youth and Science under the Operational Program "Human Resources Development" of the European Social Fund, operational scheme BG051PO001-3.3.04 - project BG051PO001.3.3.04/54/2009 "Development of interdisciplinary thinking and training of young scientists working in the field of light-matter interactions". The photoluminescence measurements were supported by NSF - DO-02-112/2008 "National Center on Biomedical Photonics".

References

[1] Chance R R, Prock A and Sibley R 1974 *J. Chem. Phys.* **60** 2744-8
[2] Cheng P, Li D, Yang Z, Chen P and Yang D 2008 *Appl. Phys. Lett.* **92** 041119-1-3
[3] Lei D Y, Li J and Ong H C 2007 *Appl. Phys. Lett.* **91** 021112-1-3
[4] Lin H Y, Cheng C L and Chen Y F 2006 *Nanotechnol.* **17** 4391-4
[5] Zhang W, Zhong S and Fu Z 2006 *Appl. Phys. Lett.* **88** 232110-3
[6] Lai C W, An J and Ong H C 2005 *Appl. Phys. Lett.* **86** 251105-3
[7] Lin H Y, Chou Y Y, Cheng C L and Chen Y F 2007 *Opt. Express* **15** 13832-7
[8] Nakata Y, Okada T and Maeda M 2002 *Appl. Surf. Sci.* **197-198** 368-70
[9] Lee M K, Kim T G, Kim W and Sung Y M 2008 *J. Phys. Chem. C* **112**/27 10079
[10] Imamova S, Nedyalkov N, Dikovska A, Atanasov P, Sawczak M, Jendrzewski R, Sliwinski G and Obara M 2010 *Appl. Surf. Sci.* **257** 1075-9
[11] Lin H Y, Cheng C L, Chou Y Y, Huang L L, Chen Y F and Tsen K T 2006 *Opt. Express* **14**/6 2372
[12] Bohren C F and Huffman D R 2004 *Absorption and scattering of light by small particles* (New York: Wiley–Interscience)

Journal of Physics: Conference Series **356** (2012) 012003 doi:10.1088/1742-6596/356/1/012003

Laser technology for synthesis of AlN films: influence of the incident laser fluence on the films microstructure

A Szekeres[1], A Cziraki[2], G Huhn[2], K Havancsak[2], E Vlaikova[1,4], G Socol[3], C Ristoscu[3] and I N Mihailescu[3]

[1] Institute of Solid State Physics, Bulgarian Academy of Sciences, 72, Tzarigradsko Chaussee, 1784 Sofia, Bulgaria
[2] Eotvos Lorand University, 1, Pazmany Peter setany, 1117 Budapest, Hungary
[3] National Institute for Lasers, Plasma and Radiation Physics, PO Box MG-54, RO-77125, Magurele, Ilfov, Romania

E-mail: vlaikova@issp.bas.bg

Abstract. Thin AlN films were synthesized by pulsed laser deposition at 800 °C by a KrF* excimer laser source (λ = 248 nm, τ = 25 ns, 3 Hz) at fluences of 4.8, 8.6 and 10 J/cm^2 in nitrogen ambient at a dynamic pressure varying from 0.1 to 10 Pa. The film microstructurewas studied by X-ray diffractometry, while the film surface morphology of AlN films was examined by AFM imaging. At the low laser fluence (4.8 J/cm^2), a stable hexagonal AlN phase was formed, while at the intermediate laser fluence (8.6 J/cm^2), both hexagonal and metastable cubic crystallites were observed. At the high laser fluence (10 J/cm^2), the films were "XRD amorphous". The surface roughness shows a tendency to increase with the increase in the nitrogen pressure. The root-mean-square roughness values are 0. 2 - 5 nm, depending on N_2 pressure.

1. Introduction

Thin films of AlN with definite microstructure are essential for many applications in surface acoustic wave and UV optical devices. Moreover, AlN films deposited on Si are promising as microelectronic components thanks to their high dielectric constant and thermal conductivity, thermal and chemical stability, and good lattice match with SiC and GaN. However, obtaining AlN films with definite structure and crystalline quality still remains a challenge for most deposition techniques, such as metal organic chemical vapor deposition [1], reactive magnetron sputtering [2,3] and pulsed laser deposition (PLD) [4-7]. Extensive research is still necessary to find out exactly how the crystallographic structure and degree of crystallinity of AlN films depend on the film preparation method and conditions. Vispute *et al.* reported the synthesis of h-AlN with (0001) orientation [6], while Zhu *et al.* obtained cubic AlN films [8]. Metastable c-AlN was found in films synthesized by nitrogen-ion-assisted pulsed laser deposition [7]. Our recent investigations of AlN films prepared by pulsed laser deposition revealed that films deposited at low incident laser fluence on the target surface (3.7 J/cm^2) possess cubic polycrystalline structure [9,10] and the amount of nitrogen in the gas ambient during the

[4] To whom any correspondence shoul be addressed.

deposition plays a decisive role in the crystallization process [10]. Using a higher laser fluence (10 J/cm^2) and varying the N_2 pressure, amorphous films with AlN nanocrystallites could be grown [11].

Extending the findings of our previous investigations, the present work is concerned with synthesis of AlN thin films by pulsed laser deposition in nitrogen varying the incident laser fluence on the target (4.8 - 10 J/cm^2) and the N_2 pressure (0.1 -10 Pa) in the vacuum chamber. The films microstructure was analyzed by X-ray diffractometry (XRD) and the film surface morphology was examined by atomic force microscopy (AFM). The results obtained are discussed in relation to the deposition conditions.

2. Experimental details

AlN films were PLD synthesized on single crystalline Si(100) substrates. The target was bulk polycrystalline stoichiometric AlN with 99% purity and the distance to the substrate was set at 5 cm. After loading the substrate and target, the deposition chamber was pumped down to a residual pressure ~10^{-5} Pa and the substrate was heated up to 800°C for 1 hour, aiming at decompositing the native SiO_2 on the Si surface. The substrate temperature was kept at 800°C during deposition, providing conditions for crystalline growth of AlN. The ablation was performed by a COMPEXPro 205 pulsed KrF* excimer laser source (248 nm, 25 ns) operating at an intensity of ~4x10^8 W/cm^2 and repetition rate of 3 Hz. The incident fluence was changed from 4.8, 8.6 up to 10 J/cm^2. The target was rotated at 0.3 Hz during multi-pulse laser irradiation in order to avoid piercing. For the deposition of one film, 20 000 consecutive laser pulses were applied, which resulted in films thickness ranging from 0.2 - 0.6 μm with a tendency to decrease when increasing the nitrogen pressure and/or incident laser fluence.

To characterize the films structure, X-ray diffraction (XRD) measurements were performed on a large-angle (2Θ = 0–90°) Philips X'Pert equipment. The XRD patterns were identified using ASTM files [12]. We should note that due to the film thickness varying with the deposition conditions, the Bragg peaks of the crystalline phases in the thinner films are very weak and the signal-to-noise ratio is greater.

The surface morphology was studied by recording AFM images on an NT-MDT (Russia) Solver type scanning probe microscope. The surface scanning was performed in a semi-contact mode to avoid altering the sample. Cantilevers with typical force constant of 5.5 N/m, curvature radius of 10 nm and resonant frequency of 130 kHz were used.

3. Results and discussion

Typical X-ray diffraction patterns of the PLD AlN films deposited at different conditions are shown in figure 1. Although the XRD data analysis was difficult due to the small amount of film material and the presence of strong Bragg peaks originating from the Si substrate, peaks with low intensity related to the crystalline AlN phase were detected within the broad band of amorphous AlN phase.

In the case of AlN films deposited at an incident laser fluence of 4.8 J/cm^2 (figure 1a), the XRD patterns show the growth of AlN crystallites with a stable hexagonal phase [12]. From the observed broad peaks one can conclude that the size of the crystallites is on a nano-scale. However, precise estimation of the crystallites size is not possible from the given patterns because, on one hand, the number of the detected Bragg peaks is too small and, on the other, the broadening could also be caused by the highly defective crystalline structure. At a pressure of 1 Pa, the corresponding XRD pattern reveals the same AlN phase, but the ratio of the intensities of H(100) and H(002) Bragg peaks changes to the preferable <001> orientation of AlN crystallites in the films. A further increase of the N_2 pressure up to 10 Pa leads to development of a fully "XRD amorphous" film structure.

Upon increasing the laser fluence to 8.6 J/cm^2, the shape of the XRD patterns (figure 1b) indicates that crystalline phases are also developed in the growing films. For the AlN films deposited at nitrogen pressure of 0.1 Pa, the Bragg peaks reveal that crystallites are formed in both hexagonal and cubic phases. The co-existence of two crystalline phases is in good agreement with our previous observations on AlN films deposited by PLD using a laser source with different parameters [9]. When increasing the pressure to 10 Pa, only metastable cubic AlN crystallites are formed in the amorphous matrix. The size of crystallites cannot be determined due to the reasons discussed above, but it is reasonable to assume that the average size is below 100 nm.

When the deposition is performed with a laser fluence of 10 J/cm^2, the XRD patterns of the AlN films deposited at 0.1 Pa and 10 Pa become featureless (figure 1c), thus revealing that these films are "XRD amorphous". This means that if there is a crystalline phase, the size of the nanocrystallites is too small or their density is too low to be observed with this XRD technique. Indeed, the TEM observations performed on these AlN films [11] have shown that the films deposited at 0.1 Pa are amorphous, but in the films deposited at 10 Pa the TEM micrographs visualize crystallites inside the amorphous film matrix growing in a metastable cubic AlN phase up to a few nanometers.

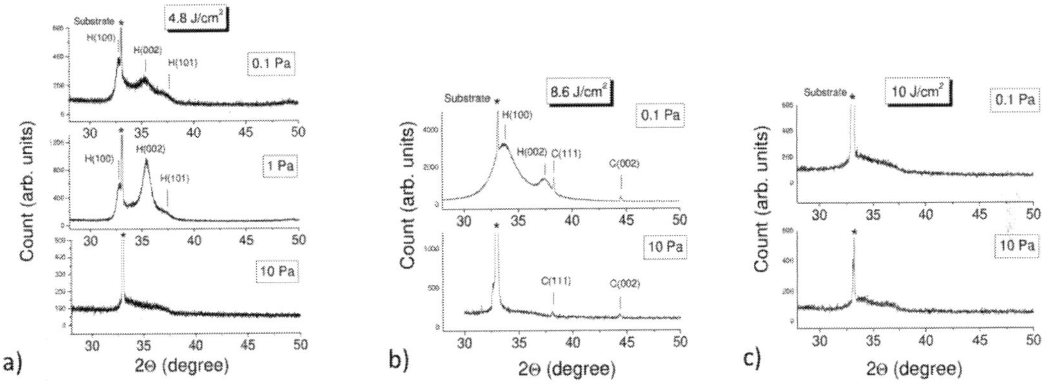

Figure 1. XRD spectra of AlN films deposited in N_2 ambient at different dynamic pressure and at incident laser fluence of 4.8 J/cm^2 (a), 8.6 J/cm^2 (b) and 10 J/cm^2 (c).

At first sight, these XRD results are contradictory to our previous studies on PLD AlN films [9,10] obtained at similar nitrogen pressures but at different laser source parameters. It should be emphasized that in earlier experiments we applied lower laser fluence (3.7 J/cm^2), but the most essential difference in the deposition conditions was the pulse duration, which was more than three times as short (t_{FWHM} = 7 ns) as that (t_{FWHM} = 25 ns) used at the present experiments. Because of these differences, the growing processes and the resulting film microstructures cannot be compared.

The gradual modification of the films structure results also in a change of the surface morphology. Figure 2 represents typical AFM images from 1×1 μm^2 area surface scans of AlN films deposited at the conditions described. They illustrate well the columnar growth of PLD AlN films. At low and intermediate N_2 pressures, the films have smooth and uniform surface morphology, very low values of the root-mean-square (rms) roughness S_q indicating the amorphous nature of the films studied. At the high pressure of 10 Pa, the films still have a homogeneous surface, but the roughness sharply increases and small droplets of different size appear. In the case when 10 J/cm^2 laser fluence is used, this effect is stronger and it correlates well with the reduced growth rate at ahigh N_2 pressure.

The present XRD observations reveal that stable hexagonal phase of AlN can be formed in the case when the films are deposited at a considerably low laser fluence and a low nitrogen pressure. A possible explanation for this could be that, as the laser fluence is increased the species evaporated from the target acquire a much higher kinetic energy and, as they reach the surface of the growing film, this excessive energy can be transformed to adatoms obstructing the ordering in a crystalline network. This phenomenon is reflected well in the XRD results (figure 1), in particular for deposition at low nitrogen pressure (0.1 Pa). As the N_2 pressure is raised, the collisions between particles in a gaseous phase most probably reduce considerably the kinetic energy and, hence, the adatoms' mobility

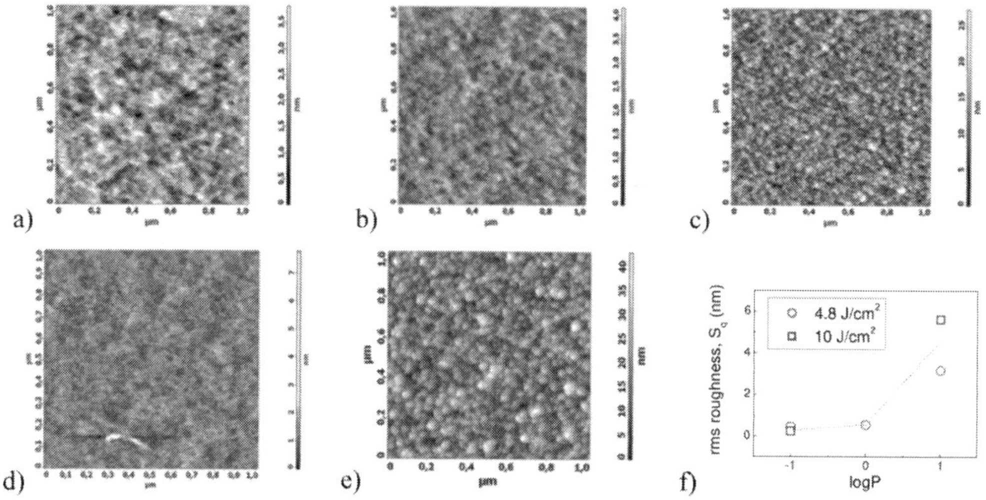

Figure 2. 2D AFM images from a 1x1 μm^2 surface area of AlN films (a-e) deposited with laser fluence of 4.8 (a,b,c) and 10 J/cm^2 (d,e) at a N_2 pressure of 0.1 Pa (a,d), 1 Pa (b) and 10 Pa (c,e). The rms surface roughness S_q values are also given (f).

on the growth surface. This would facilitate the formation of crystallites. However, increasing the nitrogen pressure further leading to an excess amount of nitrogen atoms adsorbed on the surface of the growing film, larger amounts of nitrogen atoms are build in the growing film volume, thus retarding the crystallization process. According to our observations, this effect becomes dominant for films deposited at 10 Pa, where the film structure is "XRD amorphous" independently from the laser fluence (figure 1). In this case, the excess nitrogen atoms most probably form clusters leading to an irregular growth of film with columnar structure and to an increased surface roughness, as is observed (figure 2f).

Conclusions

The XRD and AFM studies of AlN films synthesized by PLD at the conditions described revealed that by varying the laser fluence incident on the target and the nitrogen pressure during deposition, nano-crystallites with different AlN phases and orientations can be formed in the amorphous AlN matrix. The tendency was found that stable hexagonal AlN phase can be formed only when the film deposition proceeds at low incident laser fluence and low nitrogen pressure. At high laser fluence (10 J/cm^2) the AlN film structure is "XRD amorphous". The PLD AlN films studied exhibit homogeneous surface morphology and considerably low values of the surface roughness, being dependent only on the N_2 pressure. A possible explanation for the formation of the film microctructure is given.

References

[1] Kung P, Saxler A, Zhang X, Walker D, Wang T C, Furguson I and Razeghi M 1995 *Appl. Phys. Lett.* **66** 2958
[2] Liu W-J, Wu S-J, Chen C M, Lai Y-C and Chuang C-H 2005 *J. Cryst. Growth* **276** 525
[3] Stan G E, Pasuk I, Galca A C and Dinescu A 2010 *Digest J. Nanomater.Biostruct.* **5/4** 1041
[4] Ristoscu C, Ducu C, Socol G, Craciunoiu F and Mihailescu I N 2005 *Appl. Surf. Sci.* **248** 411
[5] Shukla G and Khare A 2008 *Appl. Surf. Sci.* **255** 2057
[6] Vispute R D, Narayan J, Wu H and Jagannadham K 1995 *J. Appl. Phys.* **77** 4724
[7] Ren Z M, Lu Y F, Ni H Q, Liew T Y F, Cheong B A, Chow S K, Ng M L and Wang J P 2000 *J. Appl. Phys.* **88** 7346
[8] Zhu J, Zhao D, Luo WB, Zhang Y and Li YR 2008 *J. Crystal Growth* **310** 731

[9] Bakalova S, Szekeres A, Cziraki A, Lungu C P, Grigorescu S, Socol G, Axente E and Mihailescu I N 2007 *Appl. Surf. Sci.* **253** (19) 8215

[10] Bakalova S, Szekeres A, Cziraki A, Huhn G, Havancsak K, Grigorescu S, Socol G, Axente E, Mihailescu I N and Gavrila R 2009 *J. Optoelectr.&Adv. Mater.* **11**/10 1479

[11] Szekeres A, Fogarassy Zs, Petrik P, Vlaikova E, Cziraki A, Socol G, Ristoscu C, Grigorescu S and Mihailescu I N 2011 *Appl. Surf. Sci.* **257** 5370

[12] *Index to the Powder Diffraction File 2000* (Joint Comm. Powder Diffractions Standards)

Auger analysis of high-dose ion-implantation of arsenic in silicon

G Spasov

Acad. J. Malinowski Institute of Optical Materials and Technologies,
Bulgarian Academy of Sciences, 109 Acad. G. Bonchev Str., 1113 Sofia, Bulgaria

E-mail: gspassov@clf.bas.bg

Abstract. The paper deals with Auger electron spectroscopy (AES) analysis of high-dose ion-implantation of arsenic in silicon. Six profile distributions are obtained with doses covering uniformly the range 2×10^{16} - 2×10^{17} atoms/cm^2. The peak from the matrix (Si) is only monitored; it is assumed that the alteration in its intensity is due to the other element of the binary composition (As). This method of quantification is applicble to ion implantation. The profile depth is calibrated by Rutherford backscattering spectroscopy (RBS). The Ar ion (3 keV) sputtering rate depends weakly on the dose.

1. Introduction

Ion implantation (IIm) is a basic method for modifying surfaces of materials. However, studies by AES of the depth distribution of the introduced implants is limited to high-dose implantation because of the comparatively low detection threshold of the method. But it is precisely in such cases where AES is very useful, as the accompanying processes (ion sputtering, amorphization, recrystallization) make the experimental data critical in modeling the implantation.

About 145 scientific articles have been devoted to the research on IIm and related problems by means of AES. The first ones (for P in Si and F in GaAs) date from 1972; by 1980, their number has become 11, and has monotonically increased afterwards. However, for As implanted in Si, literature is scarce. In [1-3], the distributions of the implant are presented, but the doses do not surpass 1×10^{16} atoms/cm^2. The quantification has always been performed using the elemental sensitivity factors method (ESFM).

This work presents the experimental distributions of As in Si for the dose range 2×10^{16} - 2×10^{17} atoms/cm^2 obtained through AES profiling by argon ion sputtering. Additional Rutherford backscattering spectrometry (RBS) data are used to calibrate the profile depth. The way we apply AES can be of interest in other cases of high-dose IIm.

2. Experimental

2.1. Samples

Arsenic is implanted in silicon at energy 40 keV with six doses (2×10^{16}, 4×10^{16}, 8×10^{16}, 1.2×10^{17}, 1.6×10^{17}, 2×10^{17} atoms/cm^2), well covering the range (2×10^{16} - 2×10^{17}). More details for the samples preparation can be found in [4, 5], which works are also the source of the RBS analysis data.

Journal of Physics: Conference Series **356** (2012) 012004 doi:10.1088/1742-6596/356/1/012004

2.2. Auger analysis and experiment

The AES analysis is performed by a microprobe (including a cylindrical mirror analyzer with a coaxial electron gun) mounted in a stainless-steel bell at a working pressure of 1.3×10^{-8} Pa. The analyzer's energy resolution is about 0.3%; a 0.5-μA 3-keV electron beam with a diameter 10 μm directed normally to the surface is used. The spectra are differential; the modulation voltage is 4 V p.t.p. The profiling is performed by a 3 keV Ar^+-ion sputtering. The ion beam is incident at an angle of 79° to the surface normal. Only the silicon peak is monitored during the continuous sputtering. To guarantee a constant etching rate, the SiO_2 standard is etched before and after the profiling of every implanted sample and the data obtained is used to correct the profile time axis.

2.3. Quantification

The arsenic content (in at. %) is obtained from the Auger data by the formula:

$$c_{As} = 100\,(I_{Si0} - I_{Si})/I_{Si0}\,. \qquad (1)$$

3. Results

3.1. Derivation of equation (1) for AES quantification

The experimentally determined relation of proportionality between the content, c_A, of the element A in the sample and the intensity of its Auger peak, I_A, is a main postulate of Auger quantification. However the relation, obtained as quotient, after applying the aforesaid for two samples,

$$c_A/c_A' = I_A/I_A' \qquad (2)$$

is true only if all the conditions concerning the Auger process are equal for the two samples. Let us assume this is correct and the second sample is a standard of the element A (for a "pure sample" c_0 is 1):

$$c_A = I_A/I_A\,. \qquad (3)$$

Let the sample examined be binary. The concentration of the second element, c_B, is obviously $(1 - c_A)$. This relation and (3) yield (1), in atomic parts. The main requirement for the conclusion's correctness is equality of the Auger conditions for the two samples.

3.2. Calibration of the x-axis

We equalized the time axes for all profiles, but the absolute profiling depth is not known. The influence of the high implant doses over the etching rate is not entirely clarified. A priory, different – even opposite – tendencies are possible (for example, the included arsenic, being a heavier element, should slow the etching down, while silicon amorphization should speed it up [6]). Therefore, we resort to the available RBS profiles. Taking the middle of the dose range profile (1.2×10^{17} atoms/cm^2) we fit it over the corresponding RBS. For that purpose, first we approximate it with splines. After that, we change the x-axis scale of this analytic curve, while monitoring (by the least squares method) its fit over the RBS profile. The optimal fitting gives the wanted depth calibration of the AES profile. The same stretching of the x-axis is done for the remaining AES profiles. The superposition of the corresponding AES and RBS profile is shown in figure 1. The fitting is acceptable for all couples, i.e. it does not depend strongly on the dose. The same holds for the ion etching rate of the layers. The profiles for all doses studied are presented together on figure 2.

4. Discussion

4.1. Reasons proving the correctness of using formula (1)

It is known that Auger quantification requires the use of relative intensities (This compensates the intensity difference between the samples, which do not have concentrational origin.). Although only one peak (Si) is monitored (at each etching step) during our analysis, the profiling should be considered as a unified experiment, for which all the Auger conditions are equal (the sample, the apparatus and the analytical mode remain strictly the same). Actually, the quantification is performed using relative intensity (the normalization is carried out by the intensity I_{Si0}, which can be considered

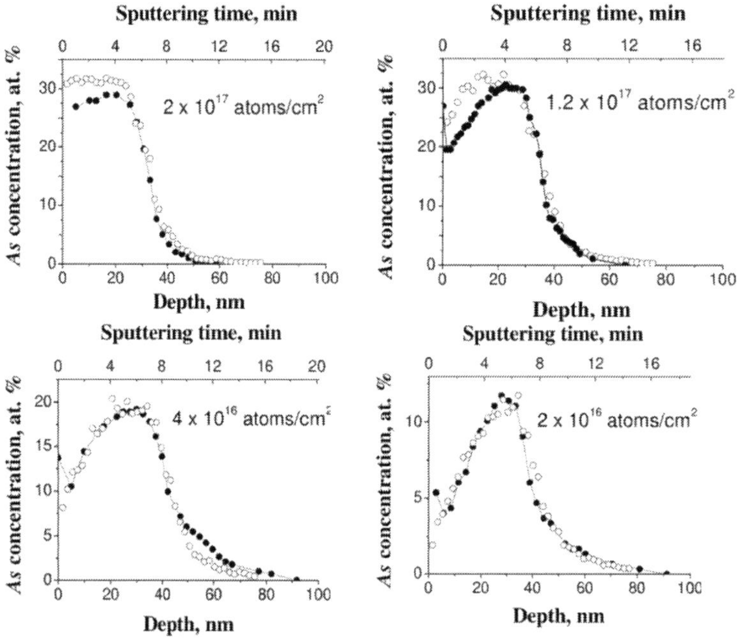

Figure 1. Depth scale calibration, showing the fitting between AES and RBS data.

as obtained in the same experiment). From the analytical conditions, only two points of the profile can be attributed to the matrix effects or changes caused by the etching; however, accounting for such differences is not an important problem of the Auger practice.

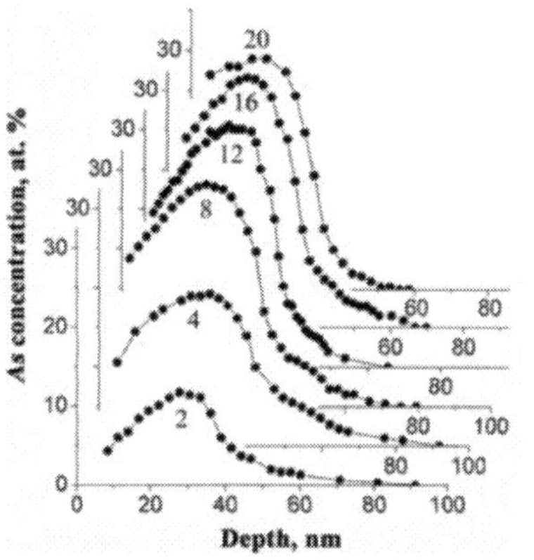

Figure 2. As concentration profiles for all doses.

4.2. About the matrix effects and their corrections

In the quantification variant suggested by (1), the matrix effects are not taken into account. This is preferable to introducing corrections using unclear criteria (for the ion-implanted materials). Here we suggest an evaluation of their order of magnitude. The silicon intensity in (1) is corrected relative to the intensity of the silicon standard. The correction of the backscattering effect (for the plateau in the beginning of the profile), performed according to [7], is shown in table 1. The correction is in the right direction, but exceeds the RBS data. A combined rough correction for a change in the atomic density, N, and the attenuation depth can be expressed as $(N/N_0)^{1/2}$ [8]. The atomic density of As is higher by 10 % than that of crystalline Si. On the contrary, the silicon atomic density should decrease in high-dose IIm; one tentative value is that of amorphous Si (5% less than the value of the crystalline Si [9]). In the presence of voids, however, there exist data showing a decrease by 18 % (or even 25 %). For the calculations, we use the value of 10 % and assume that silicon is fully amorphized and As is built in as a substitute atom. The result from the total correction is also presented in table 1. Despite the close match with the RBS results, we should emphasize that real corrections necessitate a crystallographic characterization of the layer.

4.3. Detection limit of the implanted element and accuracy in its determination

The minimal implant concentration, c_{min}, if the signal exceeds the noise by a factor of 2.5 (with double-amplitude N), is:

$$c_{min}\,(\%) = 250(I_{T0}/N)^{-1} = 250(S/N)_C^{-1}(S_{C0}/S_{T0}). \qquad (4)$$

Table 1. Matrix correcting for the averaged maximum of the As distribution.

Dose, cm^{-2}	As concentration, at. %			
	Without correcting	With a backscattering correction	With a total correction	RBS data
2×10^{17}	28.0	33.6	30.0	31.3
1.6×10^{17}	31.4	37.3	33.9	–
1.2×10^{17}	29.9	36.6	32.2	31.5
8×10^{16}	27.9	33.5	29.9	–
4×10^{16}	18.9	23.6	19.4	19.4
2×10^{16}	11.3	14.5	9.8	10.9

There, the signal from a pure target element, I_{T0}, determines the ratio "signal-noise" for the apparatus, I_{T0}/N. If the latter is known as an apparatus parameter, $(S/N)_C$ (for a standard element C), the second record can be used; S_{C0} and S_{T0} are the standard and the target sensitivities. The implant detection limit is determined by the element of the target (improved by a factor of S_{T0}/S_{C0} compared to the ESFM calculated).

The relative error of the implant concentration (from the intensities), $\Delta c/c$, is determined from (1). The expression, written as an explicit function of c, decreases with the decrease of c:

$$\Delta c/c = (\Delta I_{Si}/I_{Si0}).(2 - c)/(1 - c). \qquad (5)$$

For example, for a dose 2×10^{17} atoms/cm^2, its value in the beginning of the profile (c is about 0.3) is around $2.4N/I_{Si0}$ and decreases to $2N/I_{Si0}$ in the profile "tail". This is opposite to the error behavior in the ESFM quantification, in which case the error increases when c decreases, reaching 40 % at the detection limit (20 times as high as in our method). This feature leads to "smoother" profile „tails".

Conclusions

I. The profile distributions of As implanted in silicon at energy 40 keV in the dose range 2×10^{16}-2×10^{17} atoms/cm^2 are obtained using AES.

IIa. In what concerns quantification, the silicon peak is only monitored during the profiling and the arsenic content is calculated by the decrease in Si intensity relative to the Si elemental standard.

IIb. The profile depth scale calibration is carried out using additional (RBS) data.

III. For ion implantation with dose over 2×10^{16} atoms/cm^2 in the system "As in Si", the sputtering rate by Ar (3 keV) ions does not depend strongly on the dose.

Acknowledgement

The author would like to thank Professor H.I. Budinov of Instituto de Física, Porto Alegre, for providing the implanted samples and the RBS data.

Reference

[1] Fujiwara K, Ohtani M, Kanayama K and Ogata H 1976 *Surf. Sci.* **61** 435
[2] Cerofolini G F, Meda L, Queirolo G, Nava F and Ottaviani G 1984 *J. Appl. Phys.* **56** 2981
[3] Skinner D K 1995 *Microsc. Microanal. Microstruct.* **6** 321
[4] Budinov H I 1990 *Thesis* (Institute of Micro- and Optoelectronics, Botevgrad) (*in Bulgarian*)

[5] Budinov H I and Karpuzov D S 1993 *Nucl. Instr. Meth. B* **73** 352
[6] Roosendaal H 1981 *Sputtering by Particle Bombardment* ed R Behrisch (Springer) p 296
[7] Shimizu R 1983 *Jap. J. Appl. Phys.* **22** 1631
[8] Seah M P and Dench W A 1979 *Surf. Interface Anal.* **1**/2
[9] Renner O and Zemek J 1973 *Czech. J. Phys.* **23** 1273

Formation of Si nanocrystals in ion implanted Si-SiO_2 structures by MeV electron irradiation

S Kaschieva[1,5], A Gushterov[2], Ch Angelov[3] and S N Dmitriev[4]

[1] Institute of Solid State Physics, Bulgarian Academy of Sciences,
72, Tsarigradsko Chaussee, 1784 Sofia, Bulgaria
[2] Institute of Nanostructure Technologies and Analytics,
University of Kassel, Germany
[3] Institute for Nuclear Research and Nuclear Energy, Bulgarian Academy of Sciences,
72, Tsarigradsko Chaussee, 1784 Sofia, Bulgaria
[4] Joint Institute for Nuclear Research, Flerov Laboratory of Nuclear Reactions,
Dubna, Moskow region, 141980 Russia

E-mail: kaschiev@issp.bas.bg

Abstract. Si-SiO_2 structures implanted with Si^+ or O^+ ions were studied after high-energy (MeV) electron irradiation. 15 keV silicon or 10 keV oxygen ions with doses of 1.5×10^{12} cm^{-2} were implanted in n-type Si-SiO_2 structures. The ion-beam energy was chosen so that the maximum number of the implanted ions would be deposited close to or at the Si-SiO_2 interface. The reference (non-implanted) and ion implanted samples were simultaneously irradiated by 20 MeV electrons with a flux of about 1×10^{15} cm^{-2}. The MeV electron irradiation effect on the redistribution of the implanted ions in Si-SiO_2 structures was studied using RBS spectroscopy. The SiO_2 surface roughness changes induced by ion implantation and high-energy electron irradiation of the structures were observed by atomic force microscopy (AFM).

1. Introduction

Silicon nanocrystals generated in a SiO_2 matrix have attracted considerable interest due to their excellent optical and electronic properties [1-2]. The most important property of these nanocrystals is the intense photoluminescence in the visible and the red/near-infrared ranges of the optical spectrum [2,3]. This is why Si nanocrystals are potential candidates for Si-based optoelectronic devices. Klimenkov et. al. investigated the influence of high-energy electron irradiation on high-dose ion-implanted Si-SiO_2 structures. They showed that high-energy electron irradiation can cause motion of clusters or atoms on the surface [4] and that, alternatively to the thermal treatment, the cluster formation can be stimulated by high-energy electron irradiation [5].

Using RBS/C spectra, it has previously been shown that the silicon and oxygen concentrations before and after high-energy electron irradiation are different [9]. As after electron irradiation both (oxygen and silicon) peaks of the RBS/C spectra widen in the same way, it can be assumed that the electron irradiation generates additional radiation defects, which are the reason for the radiation-

[5] To whom any correspondence should be addressed.

enhanced diffusion of oxygen and an insignificant increase of the amorphous Si layer depth [9]. We compare here these results with AFM measurement of the same samples.

In this work, silicon nanocrystals generated by MeV electron irradiation are observed in a SiO_2 matrix of Si-SiO_2 structures implanted with silicon or oxygen ions (with a dose of 1.5×10^{12} cm^{-2}). The high-energy electron irradiation decreases insignificantly the oxygen concentration and increases slightly the silicon concentration and the amorphous layer thickness on the surface of the implanted Si-SiO_2 samples. We show that silicon accumulation is observed and silicon nanocrystals are formed at the SiO_2 surface after high-energy electron irradiation. We assume that during MeV electron irradiation (as a result of breaking Si-O bonds) free oxygen is generated and radiation-stimulated motion through defects takes place.

2. Experimental details

n-type CZ Si <100> wafers were cleaned and then oxidized in dry O_2 at 950 °C to oxide thickness of 20 nm. The oxide thickness was determined by ellipsometry. After oxidation, the Si-SiO_2 samples were implanted through the oxide by 15 keV silicon or 10 keV oxygen ions with the same dose of 1.5×10^{12} cm^{-2}. The ion energy was chosen so as to produce maximum ion damage at the Si-SiO_2 interface. The implanted samples and non-implanted samples were simultaneously exposed to 20 MeV electrons perpendicular to the SiO_2 surface. The energy is sufficiently high for the electrons to penetrate through the entire sample (SiO_2 and Si wafer) to create radiation defects. The electron irradiation at a flux of 1.2×10^{15} el/cm^2 was carried out on Microtron MT-25 equipment in Flerov Laboratory of Nuclear Reactions of the Joint Institute of Nuclear Research, (FLNR, JINR) Dubna, Russia. The irradiation was performed in vacuum at a pressure of about 1×10^2 Pa. The sample temperature was controlled during all irradiation and kept close to room temperature. The distance between the Microtron window and the samples was 150 mm.

The microscopic morphology of reference sample SiO_2 surfaces and of implanted or irradiated Si-SiO_2 structures was observed by atomic force microscope (AFM) in the atmosphere. The AFM images were taken over areas of 1 μm×1 μm on the surface of the SiO_2-Si structures before and after MeV electron irradiation of each sample. AFM was used as a measurement technique to study the changes observed in the dynamic properties of a vibrating tip that interacts with the samples surface.

3. Results and discussion

Figure 1 shows AFM images of the same non-implanted sample before (a) and after 20 MeV electron irradiation (b). The first sample does not show any ordered surface structures before electron irradiation (figure 1 a). Accumulation of nanostructures at the SiO_2 surfaces is observed only after electron irradiation. We assume that the MeV electron irradiation breaks Si-O bonds and the free oxygen atomes move through radiation defects, so that Si nanostructures are generated in the SiO_2. This is in good agreement with our earlier results [7-9] where we showed that oxygen accumulates at the SiO_2-Si interface. We also assume that the radiation defects (created by the MeV electron irradiation) act as a source for Si nanostructures generation in the SiO_2.

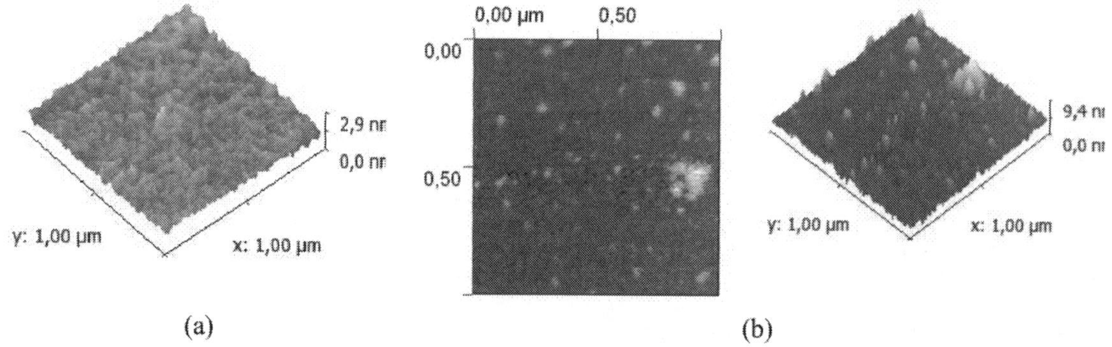

(a) (b)

Figure 1. Surface morphology of Si-SiO_2 sample before and after electron irradiation.

Figures 2 and 3 present AFM images of silicon- and oxygen-implanted non-irradiated samples (a), and such samples irradiated by 20 MeV electrons (b), respectively. It is obvious that a dose of 1.5×10^{12} cm^{-2} for both types of implanted ions is not sufficient to create nanostructures in the oxide. The AFM observation shows structural modifications on the SiO_2 surface after the ion implantation. We assume that the implanted Si atoms (figure 2) migrate and form small precursor precipitations in the SiO_2 matrix, because the implanted ions and the implantation-induced damages are in an activated state. The following 20-MeV electron irradiation (flux of 1.2×10^{15} el/cm^2) generates a new silicon nanostructure in SiO_2 with different appearance and mode. We assume that these nanostructures are silicon nanocrystals generated in SiO_2 by MeV electrons. The AFM images show that the nanocrystal sizes are different depending on the ions implanted in SiO_2. The lateral size and height of these structures are presented in table 1. The silicon nanocrystal size vary with the implantation dose as well [10]. The variation of the nanocrystal size can be related to the reduction of the oxygen ion concentration at the oxide surface. The nanocrystals are larger in silicon implanted samples, but have lower density. The nanocrystals in oxygen implanted samples (figure 3) are narrower and with higher density. This means that the effect of the ion implantation is to rearrange the Si-0-Si groups in a different form. The ion implantation breaks many of the Si-0 bonds and may also knock atoms from their sites. However, most of the atoms, silicon or oxygen, bond again shortly after being dislodged.

Table 1. Structural parameters of the nanocrystals observed on the surface of the electron irradiated samples.

Sample	l_{mean} [nm]	h_{mean} [nm]	l_{max} [nm]	h_{max} [nm]
20 MeV	30	2.2 to 3.2	223	4.2
Si^+10^{12}+20 MeV	50	2.3 to 4.2	50 to 124	4.2
O^+10^{12}+20 MeV	30	2.3 to 3.2	50	6

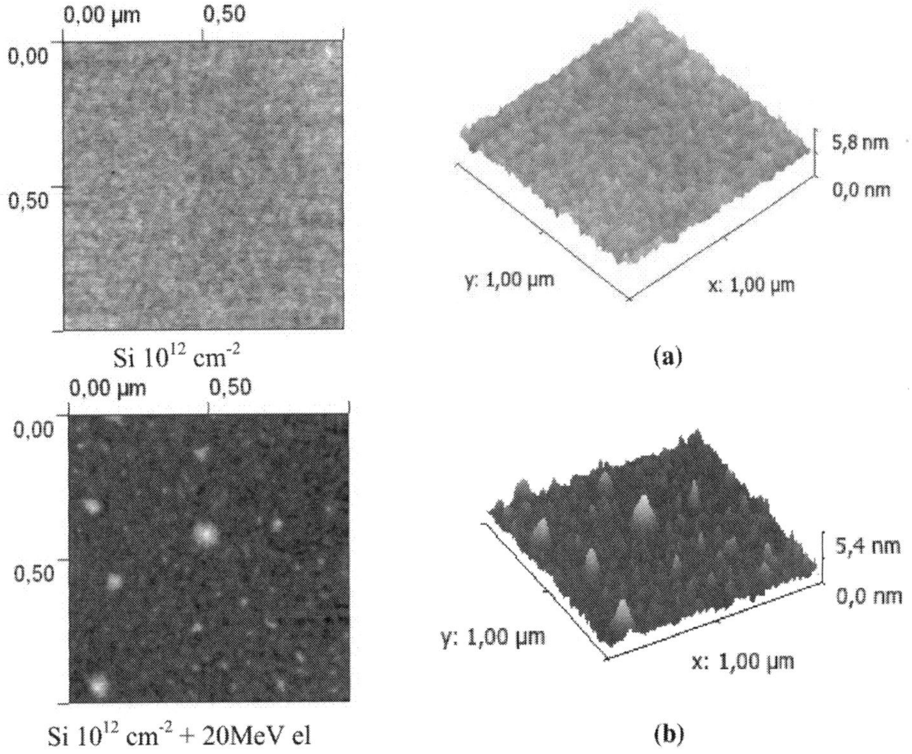

Figure 2. Surface morphology of silicon implanted ions Si-SiO_2 sample (with dose of 10^{12} cm^{-2}) before (a) and after electron irradiation (b).

Journal of Physics: Conference Series **356** (2012) 012005 doi:10.1088/1742-6596/356/1/012005

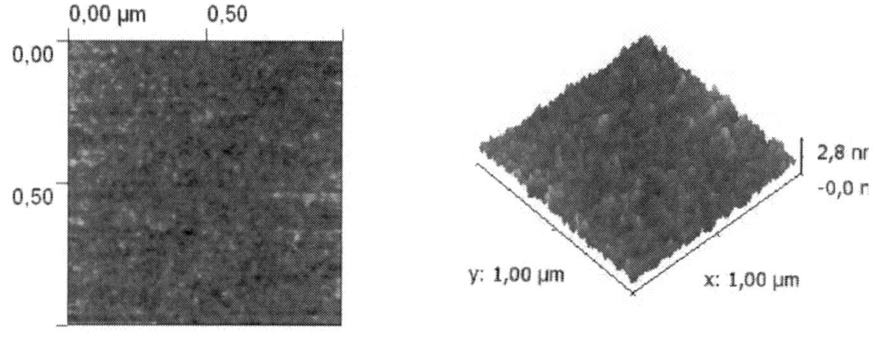

O 10^{12} cm^{-2} **(a)**

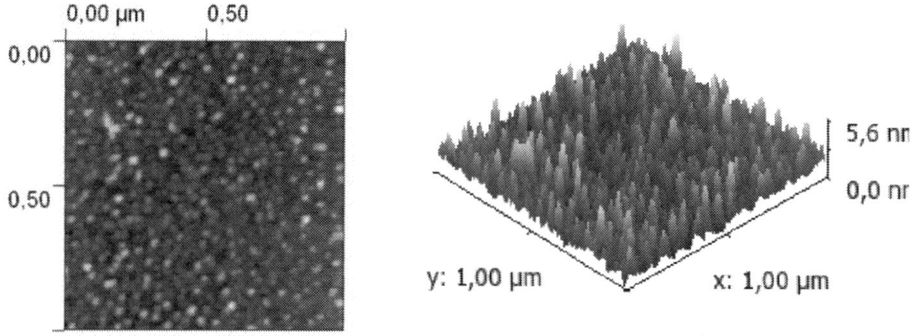

O 10^{12} cm^{-2} + 20MeV el **(b)**

Figure 3. Surface morphology of oxygen implanted ions Si-SiO_2 sample (with dose of 10^{12} cm^{-2}) before (a) and after electron irradiation (b).

Conclusions

Structural changes at the silicon surface of the implanted Si-$Si0_2$ samples are observed after high-energy electron irradiation. The MeV electron irradiation effects on the surface morphology of Si-SiO_2 structures implanted with silicon or oxygen ions (with dose of 1.5×10^{12} cm^{-2}) are by AFM. It is demonstrated that MeV electrons (at a flux of 1×10^{15} cm^{-2}) generate additional radiation defects in the entire Si-SiO_2 structure and radiation-enhanced diffusion of oxygen takes place which can be the reason for the creation of Si nanocrystals in the SiO_2 films in both kinds of implanted samples. The shape and density of these Si nanocrystals depend on type of implanted ions. The SiO_2 surface morphology yields additional evidence that Si nanocrystals are created in SiO_2 films after electron irradiation only.

References

[1] Chen X Y, Lu Y F, Wu Y H, Cho B J, Liu H, Dai D Y and Song W 2003 *J. Appl. Phys.* **93** 6311
[2] Sias U S, Behar M, Boudinov H and Moreira E C 2007 *Nucl.Instr. Meth.B* **257**/
[3] Pavesi L, Mazzoleni C, Tredicucci A and Pellegrini V 1995 *Appl. Phys. Lett.* **67** 3280
[4] Klimenkov M, Matz M and Borany J 2000 *Nucl. Instr. Meth.B* **168** 367
[5] Nepijko S, Klimenkov M, Kuhlenbeck H and Freund H 1999 *Vac. Sci. Technol. A* **17** 577
[6] Nordlund K, Arvi T T J, Meinander K and Samela J 2008 *Appl. Phys. A* **91** 561–6
[7] Kaschieva S 1999 *DrSci Thesis* (Dubna Russia)
[8] Kurmaev E, Shamin S, Galakhov V, Kirilova M, Kurennykh T, Viykhodets V and Kaschieva S 1997 *J.Phys. Cond. Matter.* **9** 6969
[9] Kaschieva S, Angelov Ch, Dmitriev S N and Tsutsumanova G 2006 *Plasma Process. Polym.* **3** 233-6
[10] Kaschieva S, Gushterov A, Gushterova P and Dmitriev S N 2010 *J. Phys.: Conf. Series* **223** 12031

Journal of Physics: Conference Series **356** (2012) 012006 doi:10.1088/1742-6596/356/1/012006

Modification of polyethylene terephthalate by atmospheric pressure dielectric barrier discharge (DBD) in view of improving the polymer wetting properties

K G Kostov[1,3], A L R dos Santos[1], P A P Nascente[2], M E Kayama[1] and R P Mota[1]

[1] State University of São Paulo – UNESP, 12516-410, Guaratinguetá – SP – Brazil
[2] Federal University of São Carlos – UFSCar, 13565-905, São Carlos – SP – Brazil

E-mail: kostov@feg.unesp.br

Abstract. Surface treatment of polymers by discharge plasmas has increasingly found industrial applications due to its capability of modifying uniformly the surface without changing the material bulk properties. This work deals with surface modification of polyethylene terephthalate (PET) by a dielectric barrier discharge (DBD) at atmospheric pressure. The treatments were conducted in air, nitrogen or argon plasma. The polymer surface was characterized by contact angle measurement, X-ray photoelectron spectroscopy (XPS) and atomic force microscopy (AFM). The results show that the plasma treatment introduces oxygen- and nitrogen- related polar groups on the polymer surface and promotes the surface roughening. Both plasma-induced surface modifications contribute to the enhancement of the polymer wettability.

1. Introduction

Polymers possess good mechanical and chemical properties; however, their use in various applications involving composite and biomaterials, packing, printing, dying etc, has been restricted due to their poor adhesion properties. Improvement of the adhesion can be accomplished by chemical methods (adding functional groups on the surface) and/or trough introducing micro- or nano-scale roughness on the surfaces. Recently, the classical chemical technologies have been gradually replaced by the environmentally friendly, versatile and low-cost plasma treatment. The atmospheric pressure plasmas are especially attractive as surface modification tools due to their applicability to a variety of substrates, easy scaling-up, quick processing time, and low cost [1].

This work describes the surface modification of a PET polymer by atmospheric pressure dielectric barrier discharge DBD. This type of discharge occurs between two planar or cylindrical electrodes provided that at least one of them is covered by a dielectric layer. As a consequence, after the gas breakdown takes place, the charge accumulation on the dielectric surface prevents the corona-to-arc transition. The DBD plasmas have many applications ranging from surface functionalization and grafting [2], improvement of adhesion and hydrophylicity [3], sterilization [4], processing of textile and fabrics [5] and, most recently, the so-called plasma medicine [6].

[3] To whom any correspondence should be addressed.

2. Experimental

The experimental arrangement used to create plasma at atmospheric pressure is sketched in figure 1. The DBD discharge was generated between two 9.5-cm-diameter parallel aluminum electrodes fitted in a 150-mm-diameter cylindrical enclosure made of Delrin®. A 2.0-mm-thick glass layer completely covered the bottom reactor electrode. The upper electrode was grounded while the lower one was the high-voltage electrode. For all DBD treatments the gas gap was fixed at 3 mm.

Prior to the DBD treatment, the samples (0.5-mm-thick, 10×15 mm^2) were ultrasonically cleaned in distilled water, washed in isopropyl alcohol and placed on the lower electrode. The gas flux into the reactor chamber was maintained at 2.4 slm using a rotameter. Three different feeding gases (Air Liquid, Brazil) were used: dry air, argon and nitrogen.

The high-voltage power supply consisted of a step-up high-voltage transformer (110/20000 V, 60 Hz) driven by a Variac autotransformer. The AC voltage applied to the reactor lower plate was measured by a high-voltage probe and displayed on an oscilloscope.

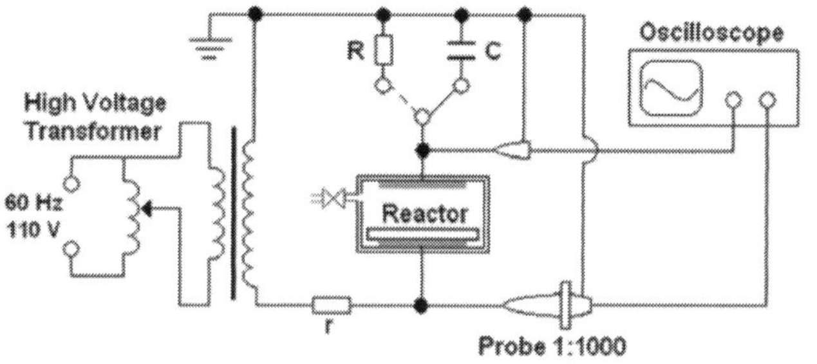

Figure 1. Experimental set-up.

The water contact angle (WCA) was measured by the sessile drop method on a standard Rame-Hart goniometer. The contact angle measurements were made immediately following the plasma treatment and also after some time of storage at controlled ambient conditions (21 °C, and 60 % humidity) to evaluate the stability of the plasma-modified surfaces.

Surface chemical characterization was carried out by X-ray photoelectron spectroscopy using a Kratos XSAM HS system. The spectra obtained were processed by the code provided by the manufacturer using the Shirley method for background subtraction. The value of 284.8 eV of the hydrocarbon C1s core level was used for calibration of the energy scale.

Atomic Force Microscopy–AFM, was used to analyze the surface morphology of the polymer samples. The AFM analysis was performed in air by a Nanoscope–V atomic force microscope. During the analysis, the microscope was operated in tapping mode at a scanning rate of 0.5 Hz.

3. Results and discussion

The DBD discharge operated in filamentary mode and consisted of many tiny streamers distributed over the entire area of the dielectric barrier. The discharge current consisted of a large number of short micro-pulses typical for the DBD discharges. To calculate the charge $Q(t)$ stored on the reactor electrodes, we used a serial capacitor of 0.91 μF. For a DBD driven by an AC power supply, the shape of the Q-V curve (called Lissajous figure) is nearly a parallelogram [7]. The area enclosed by the figure represents the electrical energy consumed for one voltage cycle. The discharge power P is then calculated by multiplying the electrical energy by the frequency of the applied voltage (60 Hz).

Figure 2 shows the Q-V figures of DBD in different gases. As can be seen in figure 2, the shape of the Lissajous figures depends on the feeding gas used. The DBD discharge in Ar was quite unstable and, if operated at voltage with amplitude above 5 kV, turned into arc. Thus, in this case, the DBD voltage was set at 10.0 kVp-p, while the discharge in air and N_2 was operated at 30.0 kVp-p. The mean discharge power measured for the DBD in nitrogen, air and argon gas precursors was 3.3 W, 2.8 W and 0.9 W, respectively. Since the discharge power in different gas precursors is different, to compare the plasma modification effect the treatment intensity was expressed in terms of the energy density deposited. The DBD treatments at lower power (e.g., in Ar) were conducted for a longer period of time in order to achieve the same energy dose [8], which was set at 13 J/cm^2. Increasing further the treatment time did not lead to significant changes in the PET contact angle.

Reactions occurring between the plasma species and the polymer chain introduce polar groups on the surface, which results in enhancement of the surface energy and improves the wettablity [7, 8]. Immediately after the treatment, the PET samples exhibited WCAs around 50-55°. These values are about 37 % lower than those of the WCA of untreated PET sample (82°). The DBD treatment in N_2 seems to be the most efficient, as resulting in a WCA of about 50°. This finding will be further discussed when presenting the XPS analysis results. It is known that the polymer surfaces treated by plasma and left in ambient conditions tend to recover partially their original wetting properties [9]. This process is referred to as hydrophobic recovery and can be observed in Fig. 3 for the three DBD treatments. The samples treated in air, Ar and N_2 plasmas show similar behavior and, after 3-4 days of storage, the WCA of the PET surface for all DBD treatments reached a saturation value, which was well below than that of the untreated specimen.

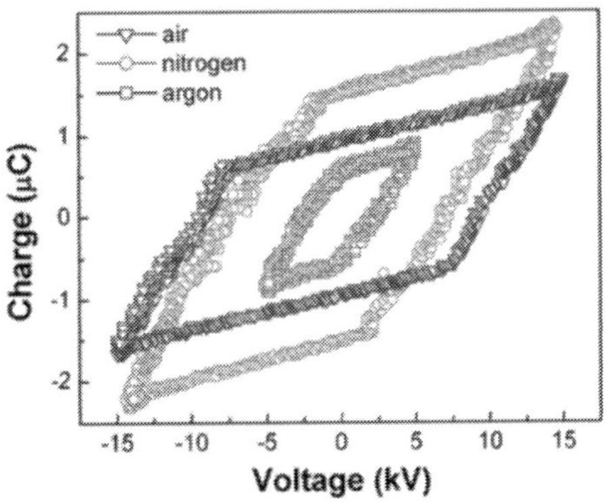

Figure 2. Q-V diagrams for DBD in air, nitrogen and argon.

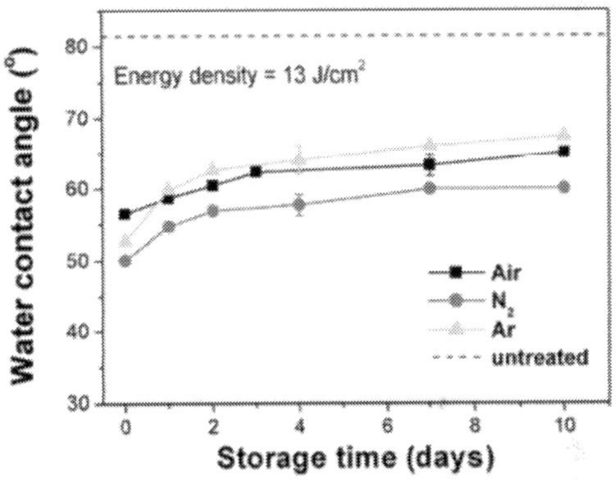

Figure 3. Evolution of water contact angle of plasma-treated PET samples.

The surface atomic composition of the plasma-modified polymers, determined by XPS analysis, is shown in table 1. The measurements were performed 2 weeks after the DBD treatment; therefore, the XPS analysis presents the atomic composition of the saturated polymeric surfaces. As can be seen, all plasma treatments increased the polymer O/C ratio to about the same extent. Since the experiments were conducted in an open environment, it was impossible to ensure a DBD discharge in pure N_2 or Ar atmosphere. Due to residual air trapped in the reactor, oxygen impurities were present in the discharge. This is why the oxygen content on the surface of polymers treated in different discharge gases is about the same. Nevertheless, the O/C ratio is almost the same for the three gases used, while the DBD in N_2 resulted in the highest efficiency in reducing the WCA, as seen in figure 3. This may be explained by the fact that, besides O atoms, the DBD in N_2 also added N atoms to the PET surface (see table 1).

Table 1. Atomic composition of PET samples.

	C (%)	O (%)	N (%)	O/C
Untreated PET	76.1	23.9	-	0.31
Air plasma	70.4	29.6	-	0.42
N_2 plasma	68.2	28.0	3.8	0.41
Ar plasma	70.8	29.2	-	0.41

To assess the PET surface structural modification, the surface morphology was analyzed by AFM; the results are presented in figure 4. The AFM images show some new nodule-like structures on the treated polymers surface. The size of the structures tends to increase with the treatment time. Formation of similar droplet-like structures, whose sizes depend on the energy dose, was also observed by other authors [9, 10].

Besides oxygen incorporation on the surface through hydrogen abstraction, the plasma reactive species (electrons and UV photons) can also break C-C bonds in the polymer chains producing short polymer segments. Thus, as the treatment proceeds, increasingly lighter oligomers are generated on the surface. These highly oxidized species tend to agglomerate into "bumps"- or "mounds"-like structures on the surface [9]. The formation of these globular shapes can be explained in terms of the difference in the surface energy. The polar and mobile oligomers formed during the plasma exposure would rather prefer to form a cohesive drop instead of spreading on the low-energy polymer surface.

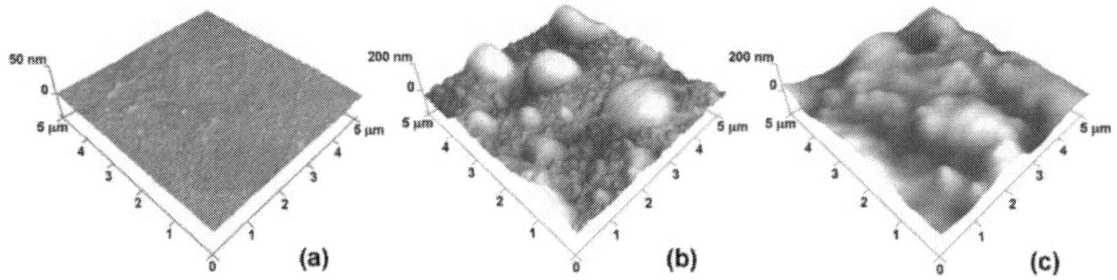

Figure 4. AFM images of PET samples treated in air for a) 0.0 min; b) 5.0 min and c) 10.0 min.

Another factor that influences the material wetting properties is the surface roughness. A hydrophilic surface would have a better wettability as its roughness increases. The roughness values of the plasma-treated PET and a pristine polymer sample are summarized in table 2. As can be seen, the polymers surface roughness increases as a result of the plasma exposure. However, beyond 5 min of treatment the roughness slightly decreased due to excessive surface etching. A similar increase of the polymer roughness was observed in low-pressure plasma treatments of PET and is caused by the formation of nodular structures [10].

Table 2. Surface roughness of PET samples.

	Pristine PET	5-min treated	10-min treated
RMS roughness (nm)	3.68	30.7	25.6

Conclusions

PET samples were subjected to atmospheric pressure DBD plasma in air, argon and nitrogen. The treatments resulted in incorporation of polar oxygen and nitrogen groups on the PET surface, which enhanced the material's wetting properties. The DBD process in nitrogen proved to be more efficient probably due to the higher amount of N atoms added to the surface. The polymer surfaces treated in Ar are more likely to undergo hydrophobic recovery. All polymer samples recovered partially their original wetting properties. The plasma treatment resulted also in surface structural modifications and surface roughness enhancement due to surface etching and polymer surface degradation.

Acknowledgment

This work was supported by FAPESP and CNPq under grants 2006/04585-0 and 471928/2008-9.

References

[1] Kogelschatz U 2003 *Plasma Chem. Plasma Process.* **23** 1
[2] Walther M, Heming M and Spallek M 1996 *Surf. Coat. Technol.* **80** 200

[3] Borcia G, Andeson C A and Brown N M D 2005 *Plasma Sources Sci. Technol.* **14** 259
[4] Birmingham J B 2004 *IEEE Trans. Plasma Sci.* **32** 1526
[5] Gorensek M, Gorjanc M, Bukosek, Kovac J, Petrovic Z and Puac N 2010 *Textile Res. J.* **80** 1633
[6] Bárdos L and Baránková H 2009 *Vacuum* **83** 522
[7] De Geyter N, Morent R, Leys C, Gengembre L, Payen E, van Vlierberghe S and Schacht E 2008 *Surf. Coat. Technol.* **202** 3000
[8] Morent R, De Geyter N, Leys C, Gengembre L and Payen E 2007 *Surf. Coat. Technol.* **201** 7847
[9] Fang Z, Yang H and Qiu Y 2010 *IEEE Trans. Plasma Sci.* **38** 1615
[10] Pandiyaraj K, Selvarajan V, Deshmukh R. and Bousmia M 2008 *Surf. Coat. Technol.* **202** 4218

Evaluation of the plasma parameters in COMPASS tokamak divertor area

M Dimitrova[1], P Ivanova[1], I Kotseva[2], Tsv K Popov[2,4], E Benova[2], T Bogdanov[2], J Stöckel[3] and R Dejarnac[3]

[1] Emil Djakov Institute of Electronics, Bulgarian Academy of Sciences, 72, Tsarigradsko Chaussee, 1784 Sofia, Bulgaria
[2] Faculty of Physics, St. Kliment Ohridski University of Sofia, 5, J. Bourchier Blvd., 1164 Sofia, Bulgaria
[3] Institute of Plasma Physics, Academy of Sciences of the Czech Republic v.v.i., Za Slovankou 3, 182 00 Prague 8, CR

E-mail: tpopov@phys.uni-sofia.bg

Abstract. We report results of the evaluation of the plasma potential, the electron temperature and the electron density during the current shots in the COMPASS tokamak, (IPP.CR, Prague, Czech Republic). The comparisons with results from model calculations based on an extended formula for the electron probe current show a satisfactory agreement.

The results presented demonstrate that the procedure proposed allows one to acquire additional plasma parameters using the electron part of the current-voltage Langmuire probe (LP) characteristics in tokamak edge plasma probe measurements.

1. Introduction

Among the contact methods of plasma diagnostics, the electric probes are the most reliable diagnostic tools allowing one to measure edge plasma parameters with a sufficiently high temporal and spatial resolution. In non-magnetized low-density plasmas, LPs allow local measurements of the plasma potential, the charged particles density and the electron energy distribution functions, $f(\varepsilon)$, (EEDF) [1]-[7].

In magnetized plasma, the interpretation of the electron part of the current-voltage (IV) characteristics above the floating potential, U_{fl}, still remains difficult [8] because the electron part of the IV characteristics is distorted due to the influence of the magnetic field. For this reason, in the strongly magnetized tokamak plasmas, the ion saturation branch of the IV and the part around the floating potential are usually used when retrieving the plasma parameters [9]. The precise evaluation of the plasma potential and the real EEDF is of great importance in understanding the underlying physics of the processes occurring at the plasma edge in tokamaks, such as the formation of transport barriers, plasma-wall interactions, edge plasma turbulence, etc.

In this paper we report results of the evaluation of the plasma potential, the electron temperatures and the electron density during the current shot in the new COMPASS tokamak, (Prague, IPP.CR).

[4] To whom any correspondence should be addressed.

The results obtained from the data measured by divertor probes confirmed the applicability of the technique proposed.

2. First derivative Langmuir probe method

The first-derivative probe method for evaluating the plasma parameters in tokamak edge plasma was published and discussed in detail in [8,10]. The theory for magnetized plasmas was developed for LPs in a non-local approach when the electrons reach the probe in a diffusion regime [11]. The electron current flowing to a cylindrical probe negatively biased by potential U is given by:

$$I_e(U) = -\frac{8\pi e S}{3m^2}\int_{eU}^{\infty}\frac{(W-eU)f(W)dW}{\gamma_0(W)\left[1+\frac{(W-eU)}{W}\psi(W)\right]}, \tag{1}$$

where W is the electron energy; e, m and n are the electron charge, mass and density; S is the probe area; $\psi(W)$ is the diffusion parameter; U is the probe potential with respect to the plasma potential U_{pl} ($U = U_p - U_{pl}$). The geometric factor $\gamma = \gamma(R/\lambda)$ assumes values in the range $0.71 \le \gamma \le 4/3$ (R being the probe radius) [11].

Here $f(W)$ is the isotropic EEDF normalized by:

$$\frac{4\pi\sqrt{2}}{m^{3/2}}\int_0^{\infty} f(W)\sqrt{W}dW = \int_0^{\infty} f(\varepsilon)\sqrt{\varepsilon}d\varepsilon = n. \tag{2}$$

As it was shown in [8,10] for probes oriented parallel to the magnetic field, the diffusion parameter is:

$$\psi_{//}(W) = \frac{\pi L'}{64\gamma_0 R_L(W,B)}. \tag{3}$$

Here L' and $R_L(W,B)$ are the characteristic length of the turbulences and the electron Larmor radius.

It was shown in [8] that at strong magnetic fields, i.e., for high values of the diffusion parameter $\psi(W) = \psi_0/\sqrt{W} >> 1$, the minimum of the first derivative of the IV characteristics is shifted negatively by a value between T_e and $1.5T_e$ in volts with respect to the plasma potential. In this case, and in order to evaluate U_{pl} accurately, we propose the following procedure: the electron temperature is evaluated from the slope in logarithmic scale of the first derivative of the non-distorted part of the experimental IV characteristics. Using this temperature, a model curve of the first derivative (derivative of equation (1)) is calculated. Then the best fit with the experimental first derivative provides the value of the plasma potential.

As it was shown in [8], the EEDF is not represented by the second derivative of the electron probe current (Druyvesteyn formula), but rather by its first derivative. Taking into account equation (3) for the EEDF measured by a cylindrical probe parallel to the magnetic field, we obtain:

$$f(\varepsilon) = -\frac{3\pi\sqrt{2m}L'}{128e^3 S R_L(\varepsilon,B)U}\frac{dI}{dU}. \tag{4}$$

3. Langmuir probe measurements in COMPASS divertor area

The divertor probe system on COMPASS consists of 39 single graphite Langmuir probes embedded parallel to the magnetic field in the divertor target providing profiles with typical spatial resolution in the poloidal direction down to 5 mm. The probe area exposed parallel to the magnetic field in the plasma is $S = 112.5\times10^{-6}$ m^2 (figure 1 (a) and (b)).

Using the main shot parameters and the probe data during the experiment, we found that the plasma is moving in a poloidal (figure 1 (c)) direction. Until the end of COMPASS operation in the middle of October 2010, a regime with stable position of the plasma column could not be achieved – plasma was moving in radial and poloidal directions. The signal varied in time so that we did not observe any quasi-stationary phase. Commissioning of the fast feedback for plasma position is presently ongoing. Full performance is planned to be achieved, including elongated diverted plasma, maximum plasma current and H-mode discharges.

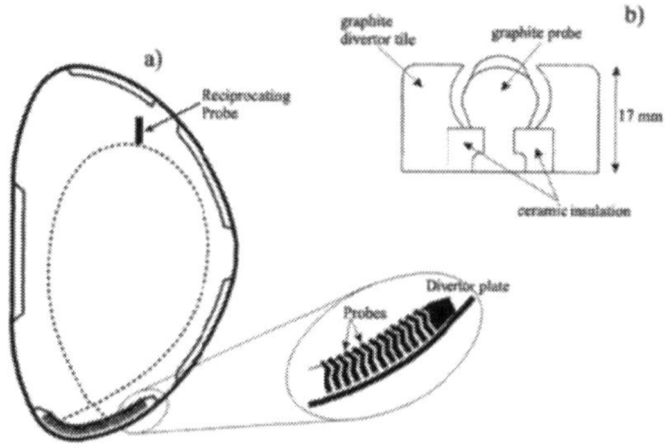

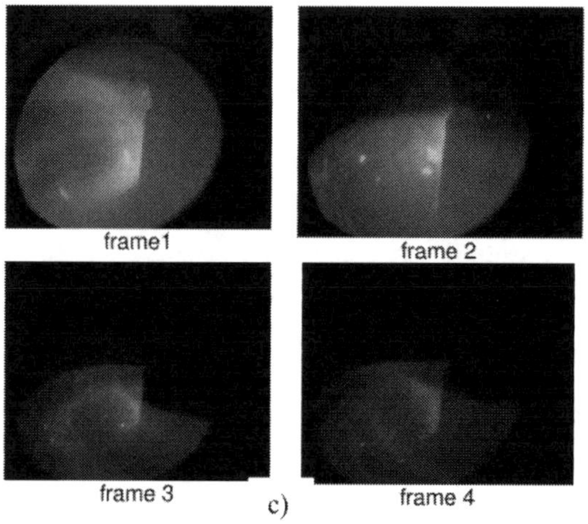

Figure 1. (a) Poloidal cross-section of Compass with the position of the divertor probe system; (b) toroidal cross-section through the divertor target showing a Langmuir probe; (c) circular plasma moves to the divertor.

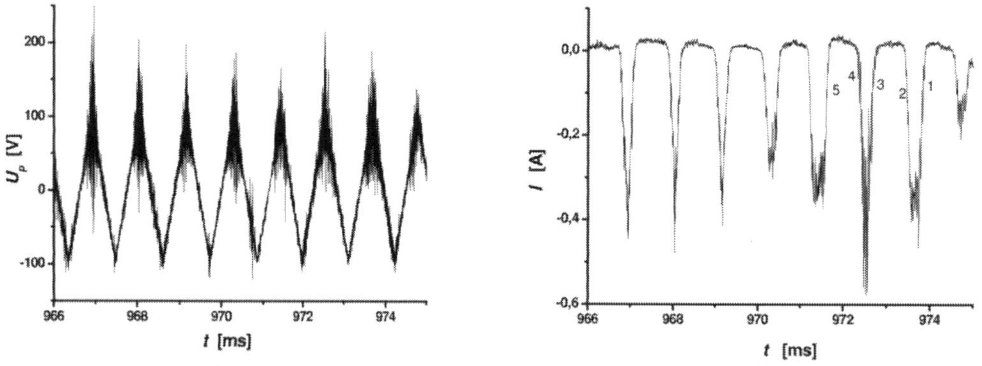

a) b)

Figure 2. a) Probe bias U_p and **b)** probe current $I(U_p)$ as a function of the time t during shot 1654 for probe #38.

Figure 2 presents the record of the sweeping probe bias U_p (a) and the probe current I (b) during shot 1654 in probe 38.

To demonstrate the procedure of evaluating U_{pl}, $f(\varepsilon)$ and n, we will analyze a single IV characteristic, shot 1654 for probe #38, IV5 working gas H_2.

Figure 3 presents the first derivative of the smoothed IV curve and the fit with the model curve (first derivative of equation (1)). Using this comparison one can evaluate the plasma potential $U_{pl} = 39$ V. One can also observe there a more or less pronounced bend in the experimental curve. In practice, even a small increment of $I(U)$ at a probe potential U positive with respect to plasma potential leads to $I'(U)$ deviating from zero at U_{pl}. Additional reasons for this effect are the plasma potential fluctuations due to the plasma turbulence and the smoothing of the experimental IV characteristics.

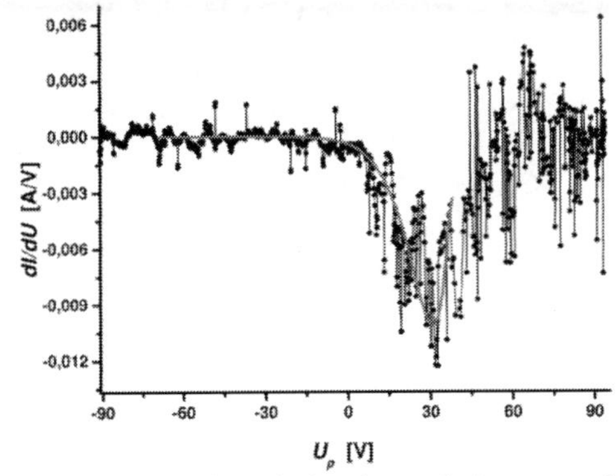

Figure 3. First derivative of the smoothed experimental IV curve (dots), shot 1654 for probe #38, IV5. Model curve – solid line.

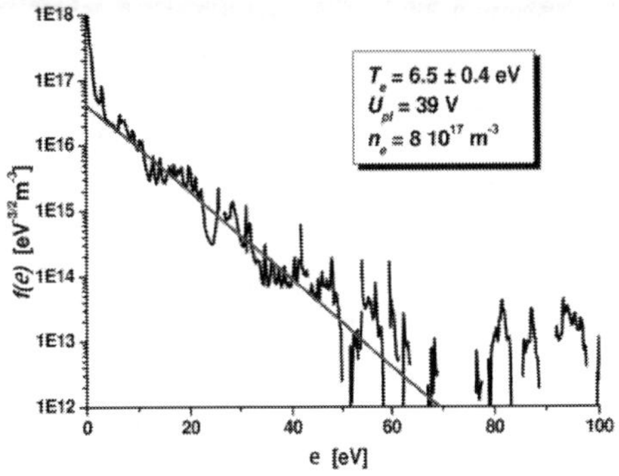

Figure 4. The EEDF obtained by using the first derivative probe method.

Figure 4 presents the EEDF obtained by parallel probe #38, shot 1654, by using equation (4) (dots). The EEDF may be approximated by a Maxwellian with electron temperature $T = 6.5$ eV (solid line). The electron density is calculated by using equation (2) and the estimated value is $n_e = (8 \pm 2)10^{17}\ \mathrm{m}^{-3}$.

Finally, using the values obtained by the first derivative probe method, we performed model calculations for the IV characteristics (eq. (1)). Examples of the comparison with the experimental IV curves are presented in figure 5. As one can see, the agreement is satisfactory.

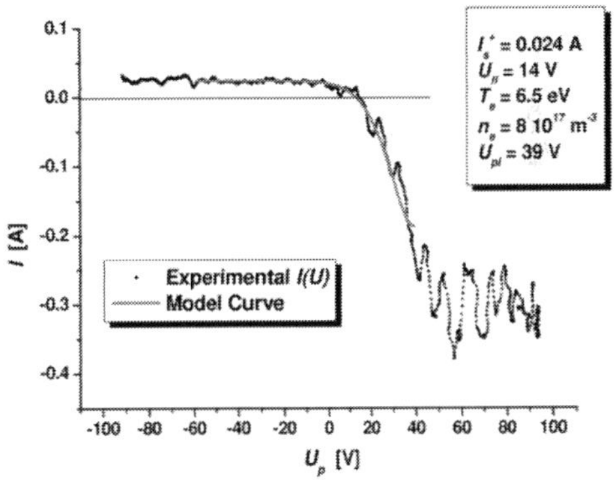

Figure 5. Comparison between model calculations (solid lines) and IV5, probe #38 for H_2 (dots).

The uncertainties in the values are calculated by regression analysis and the electron temperature was evaluated with accuracy of 7%. The best fit for plasma potential evaluation was sought with an accuracy of about 15%.

Conclusions

The first derivative probe method was used for precise evaluation of the plasma potential, the EEDF (or, respectively, the electron temperature) and the electron density during the current shots in the COMPASS tokamak, (IPP.CR, Prague, Czech Republic). The results obtained after processing the data measured by divertor probes in H_2 are reported. The good agreement between the model and experimental IV curves can be seen.

Acknowledgement

This research is in implementation of task P4 of Work Plan 2011 of the Association EURATOM/INRNE.BG in collaboration with the Association EURATOM/IPP.CR, Prague, Czech Republic and JOINT RESEARCH PROJECT between Institute of Electronics, Bulgarian Academy of Science and Institute of Plasma Physics v.v.i., AS CR, 2011.

References

[1] Druyvesteyn M J 1930 *Z. Phys.* **64** 781-98
[2] Kagan Y M and Perel V I 1964 *Sov. Phys. Usp.* **6** 767
[3] Chen F F 1965 *Electric Probes in Plasma Diagnostic Techniques* eds Huddlestone R H and Leonard S L (Academic Press, New York) Chapter 4 pp 113-200

[4] Swift J D and Schwar M J R 1969 *Electrical Probes for Plasma Diagnostic* (Elsevier, New York)
[5] Hershkowitz N 1989 *How Langmuir Probes Work in Plasma Diagnostics* **1** eds Auciello O and Flamm D L Chapter 3 (Academic Press Boston)
[6] Demidov V I, Kolokolov N B and Kudriavtsev A A 1996 *Probe Methods for Low-Temperature Plasma Investigations* (Moscow: Energoatomizdat in Russian)
[7] Popov Tsv K, Dimitrova M, Dias F M, Tsaneva V N, Stelmashenko N A, Blamire M G and Barber Z H 2006 *J. Phys. Conf. Series* **44** 60–91
[8] Popov Tsv K, Ivanova P I, Stockel J and Dejarnac R 2009 *Plasma Phys. Control. Fusion* **51** 065014
[9] Tagle J A, Stangeby P C and Erents S K 1987 *Plasma Phys. Control. Fusion* **29** 297-301
[10] Ivanova P, Popov Tsv K, Dimitrova M, Benova E, Bogdanov T, Stöckel J and Dejarnac R 2011 *Acta Technica* **56** T71-T83
[11] Arslanbekov R R, Khromov N A and Kudryavtsev A A 1994 *Plasma Sources Sci. Technol.* **3** 528-38
[12] Godyak V A and Demidov V I 2011 *J. Phys. D: Appl. Phys.* **44** 233001

Journal of Physics: Conference Series **356** (2012) 012008 doi:10.1088/1742-6596/356/1/012008

Electronic system for Langmuir probe measurements

M Mitov[1], A Bankova[1], M Dimitrova[2], P Ivanova[2], K Tutulkov[1], N Djermanova[1], R Dejarnac[3], J Stöckel[3] and Tsv K Popov[1,4]

[1] Faculty of Physics, St. Kl. Ohridski University of Sofia,
5, J. Bourchier Blvd., 1164 Sofia, Bulgaria
[2] Emil Djakov Institute of Electronics, Bulgarian Academy of Sciences,
72, Tsarigradsko Chaussee, 1784 Sofia, Bulgaria
[3] Institute of Plasma Physics, Academy of Sciences of the Czech Republic v.v.i.,
Za Slovankou 3, 182 00 Prague 8, CR

E-mail: tpopov@phys.uni-sofia.bg

Abstract. A newly developed Langmuir probe system for measurements of current-voltage (IV) characteristics in the tokamak divertor area is presented and discussed. The system is partially controlled by a computer allowing simultaneous and independent feeding and registration of signals. The system is mounted in the COMPASS tokamak, Institute of Plasma Physics, Academy of Sciences of the Czech Republic.

The new electronic circuit boards include also active low-pass filters which smooth the signal before recording by the data acquisition system (DAQ). The signal is thus less noisy and the data processing is much easier. We also designed and built a microcontroller-driven waveform generator with resolution of 1 Ms/s. The power supply is linear and uses a transformer. We avoided the use of a switching power supply because of the noise that it could generate.

Examples of measurements of the IV characteristics by divertor probes in the COMPASS tokamak and evaluation of the EEDF are presented.

1. Introduction

Langmuir probes are known for their ability to provide local measurements of very important plasma parameters, namely, the plasma potential, the electron density and the electron energy distribution function (EEDF) [1-3]. The probes are the least expensive and are still the fastest, most reliable diagnostic tool to acquire local values of the plasma parameters, but their application in strongly magnetized tokamak plasma is still under discussion [4-6].

Since we use the new advanced first-derivative method to evaluate the plasma parameters, the main problem to be solved was eliminating the noise originating from vacuum pumps, magnetic field coils, etc, i.e. noises of non-plasma origin.

In this work we present the new electronic system for measurements of the current-voltage (IV) characteristics in the tokamak divertor area on the COMPASS tokamak.

[4] To whom any correspondence should be addressed.

2. Probe circuit

The device described is the product of two years of research and development after all divertor probes characteristics were taken into account. The differential amplifier, made by Tsv. Popov's research group (Mladen Mitov, Ana Bankova and collective) was designed for current voltage (IV) characteristic measurements in COMPASS tokamak (IPP, CR) edge plasma. In the previous version of the amplifier, the signal was measured in a fully differential manner. This is why the common mode signal was high (~100 V). Although the amplifier had high common-mode rejection ratio (~77 dB) the signal at the output of the amplifier caused by the ion saturation current and the signal caused by the common-mode signal were comparable. To avoid this problem, we chose to use common ground measuring (while keeping the possibility of fully differential measurements). The common ground measurement has the advantage of absence of common mode signal, and the disadvantage of necessity of a power amplifier in each channel.

The probe circuit used to measure the signal of the divertor probes is presented schematically in figure 1 a, b.

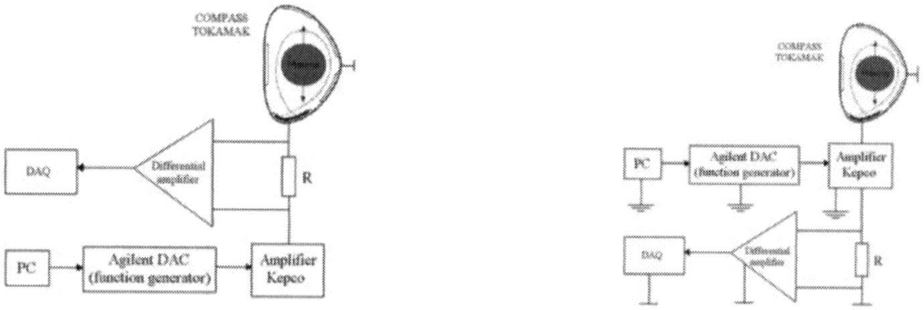

a) b)

Figure 1. Schematic diagram of the probe circuit used to measure the IV divertor probes characteristics: **a)** fully differential measurement, **b)** common ground measurement.

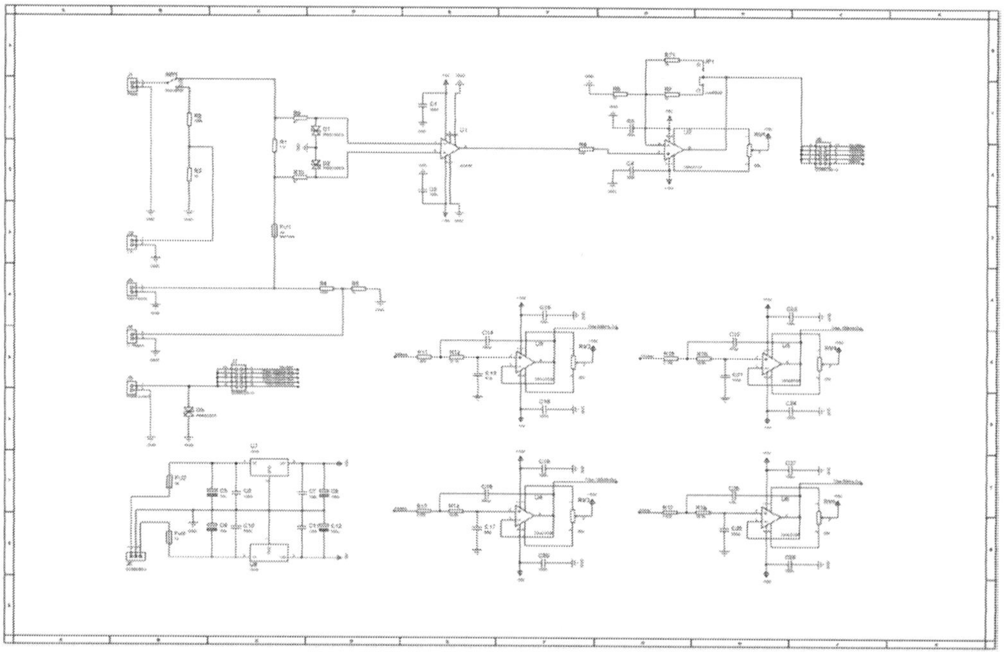

Figure 2. Electrical circuit for measuring voltage and/or current by the divertor probes.

The electrical circuit for measuring the voltage and/or current by the divertor probes is presented in figure 2. The amplifiers have characteristics superior to the ones that are currently used. They avoid the main problems previously encountered – the inability to suppress sufficiently the common-mode signal, so that the true signal and the noise generated by the common mode signal were comparable. This necessitated complicated and time consuming data processing. In addition, a set of low-pass filters was added to the amplifier circuit in order to (when necessary) suppress the high-frequency noise generated by the plasma itself. The new cards are second generation electronics, made by Tsv. Popov's research group in Bulgaria, and are shown in figure 2.

3. Langmuir probe measurements in COMPASS divertor area

The divertor probe system in COMPASS consists of 39 single graphite Langmuir probes embedded in parallel to the magnetic field in the divertor target providing profiles with typical spatial resolution in the poloidal direction down to 5 mm. A sweeping voltage (typically in the range −100 V – +40 V) is applied to generate the current-voltage characteristics. The sweeping probe bias U_p (a) and the probe current I (b) during shot 1597 recorded by probe 5 are presented in figure 3.

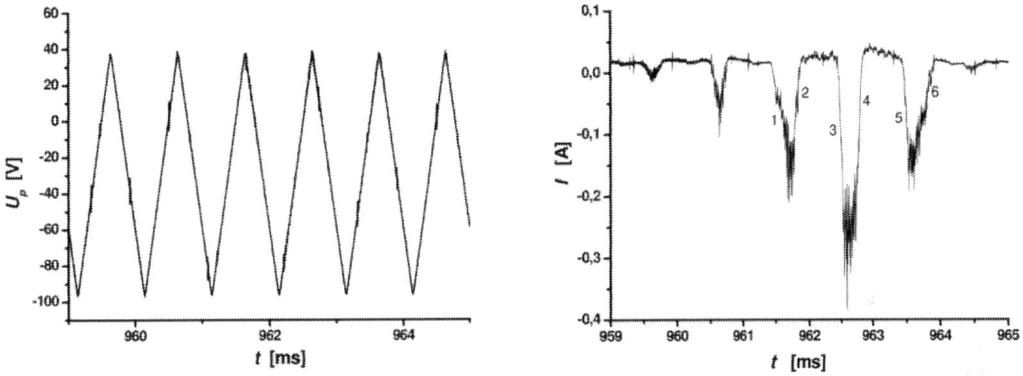

(a) (b)

Figure 3. a) Probe bias U_p and **b)** probe current $I(U_p)$ as a function of the time t during shot 1597 for probe #5.

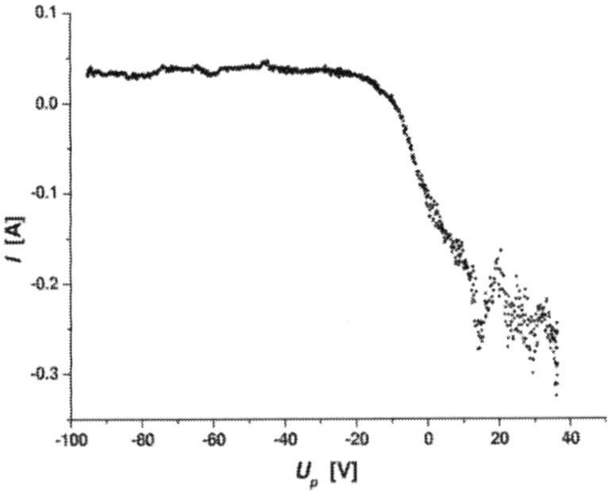

Figure 4. Single IV characteristic, shot #1597 for parallel probe #5, IV 4 (dots).

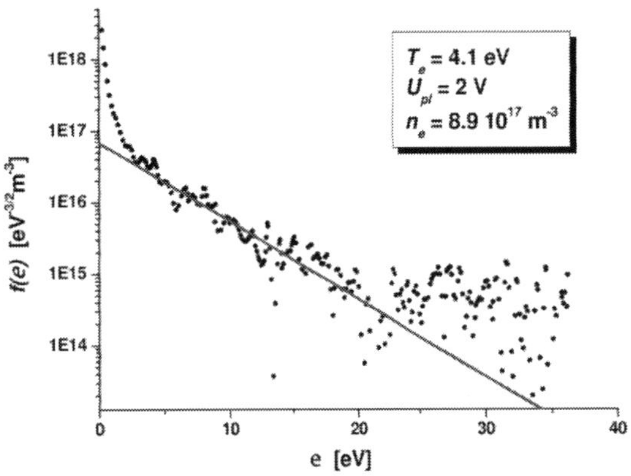

Figure 5. The EEDF obtained by using the first derivative probe method.

Journal of Physics: Conference Series **356** (2012) 012008 doi:10.1088/1742-6596/356/1/012008

The method of evaluating plasma parameters, such as the EEDF, respectively the electron temperature and density, and the plasma potential, have been presented and widely discussed in our previous works [7,8]. Using the data obtained, we can construct the single IV characteristics (figure 4). Figure 5 presents the EEDF obtained by parallel probe #5, shot 1597 (dots). The EEDF can be approximated by a Maxwellian with electron temperature $T_e = (4.1 \pm 0.3)$ eV (solid line). The uncertainties in the values are calculated by regression analysis and the electron temperature was evaluated with accuracy of 7%. The best fit for plasma potential evaluation was sought with an accuracy of about 15%.

Conclusions

A newly developed Langmuir probe system for measurement of the current-voltage (IV) characteristics in the tokamak divertor area is presented and discussed. The system is partially controlled by a computer allowing simultaneous and independent feeding and registration of signals.

The system is mounted in the COMPASS tokamak, Institute of Plasma Physics, Academy of Sciences of the Czech Republic and can also be used for other probes installed in the COMPASS tokamak.

Examples of measurements of the IV characteristics by the divertor probes in COMPASS tokamak and evaluation of the EEDF are presented.

Acknowledgement

This research is in implementation of task P4 of Work Plan 2011 of the Association EURATOM/INRNE.BG in collaboration with the Association EURATOM/IPP.CR, Prague, Czech Republic and JOINT RESEARCH PROJECT between Institute of Electronics, Bulgarian Academy of Science and Institute of Plasma Physics v.v.i., AS CR, 2011.

References

[1] Chen F F 1965 *Electric Probes in Plasma Diagnostic Techniques* eds Huddlestone R H and Leonard S L (Academic Press New York) Chap 4 pp 113-200

[2] Swift J D and Schwar M J R 1969 *Electrical Probes for Plasma Diagnostic.* (Elsevier, New York)

[3] Hershkowitz N 1989 *How Langmuir Probes Work in Plasma Diagnostics* **1** eds Auciello O and Flamm D L (Academic Press Boston) Chap. 3

[4] Demidov V I, Kolokolov N B and Kudriavtsev A A 1996 *Probe Methods for Low-Temperature Plasma Investigations* (Moscow: Energoatomizdat, in Russian)

[5] Arslanbekov R R, Khromov N A and Kudryavtsev A A 1994 *Plasma Sources Sci. Technol.* **3** 528-538

[6] Godyak V A and Demidov V I 2011 *J. Phys. D: Appl. Phys.* **44** 233001

[7] Popov Tsv K, Ivanova P I, Stockel J and Dejarnac R 2009 *Plasma Phys. Control. Fusion* **51** 065014

[8] Ivanova P, Popov Tsv K, Dimitrova M, Benova E, Bogdanov T, Stöckel J and Dejarnac R 2011 *Acta Technica* **56** T71-T83

Microwave discharge in a finite length vessel

Zh Kiss'ovski[1], V Vachkov, S Iordanova and I Koleva

Faculty of Physics, St. Kliment Ohridski University of Sofia,
5, J. Bourchier Blvd., 1164 Sofia, Bulgaria

E-mail: kissov@phys.uni-sofia.bg

Abstract. A microwave surface-wave discharge at low pressure in a finite length vessel is experimentally investigated. Argon plasma is created in a dielectric capillary with length of 15 mm, the tube being an extension of an open-ended coaxial structure. Microwave power at frequency 2.45 GHz is coupled into the source applicator at power levels 6-15 W. The plasma column increases with the power applied and standing wave mode is realized when its length is equal to the capillary length. The electron temperature and plasma density are obtained simultaneously by passive optical emission spectroscopy using the line-ratio method. The surface waves phase-diagram shows that their wavelengths are much lower than the free-space wavelength. The input impedance of the plasma column - modelled as a lossy transmission line - has resonance behaviour when its length is equal to $\lambda_g/2$ or λ_g (λ_g being the surface waves wavelength). The results obtained for the plasma parameters, the surface waves wavelength and the vessel length show correlation to the power reflected, absorbed and radiated from the plasma column. The radiation pattern of the column shows that the main lobe is nearly perpendicular to the capillary axis.

1. Introduction

Microwave discharges operated at low and atmospheric pressure, especially surface-wave (sw) sustained discharges, have been lately intensively investigated. Recent studies on sw discharges pointed out the role of the plasma boundaries on the wave-plasma power coupling and on the distribution of the plasma parameters [1]. The propagation of surface waves in a finite length vessel with insulated closed ends can be modelled as a propagation of EM waves in a lossy transmission line coupled at the microwave launcher. The results obtained [2] show that if the tube is ¼ or ½ wavelength long, the input impedance of the line is low or very high, respectively.

In this study, a microwave discharge was investigated in argon gas (0.026 Torr) in a finite length vessel (quartz capillary with insulated closed ends). Signal with frequency f = 2.45 GHz from a microwave generator was fed to the surface-wave coaxial exciter [3]. In our source, the plasma column was a complex load [4]. Its value varied depending on the power applied and the discharge conditions. In this case, the reflected power also varied and part of the applied power was radiated in the free space surrounding the source. A line-ratio spectroscopy method was used for rapid simultaneous measurement of the two main plasma parameters in our source – the electron temperature and the plasma density. The collisional-radiative model for low pressure argon plasma [5, 6] was used for the calculation of these parameters. The axial wave number and attenuation

[1] To whom any correspondence should be addressed.

coefficient of the surface waves were obtained from the dispersion relation. The results obtained for the input impedance and radiation pattern of the plasma column showed a correlation between the sw wavelength, discharge tube length and the values of the reflected and radiated power.

2. Experimental set-up and diagnostics

Microwave signal at frequency f = 2.45 GHz from a generator was fed to the source through a Pasternack-PE2219-30 double directional coupler and a triple stub (figure 1). The exciter of the surface waves was the end of an open coaxial line [3] with length of $l \sim \lambda_0/4\sqrt{\varepsilon_d}$ (λ_0 - wavelength in the free space, ε_d - dielectric constant of line filling). The discharge tube was a quartz capillary (ε_d=3.8) with inner radius of r = 1 mm, outer radius of 1.5 mm and length of 15 mm. The capillary had two insulated ends and was filled with argon gas at pressure p =0.026 Torr.

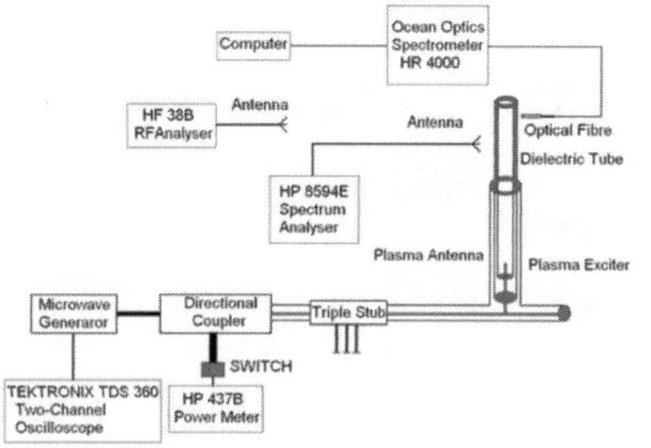

Figure 1. Experimental set-up.

The initial precise tuning of the source at frequency 2.45 GHz by the triple stub resulted in good matching with the microwave generator. The forward and reflected power were measured by a HP 437B power meter. The radial component of the surface waves electric field was registered by one radial microwave probe positioned outside the tube and connected to the spectrum analyzer HP 8594E. The radiated microwave power from the plasma column was measured in the far-field region at distance 2.7 m by a HF-analyzer HF 38B. The equipment for optical emission spectroscopy included a collimating lens, an optical fibre, an Ocean Optics HR4000 spectrometer and an infrared thermometer. High-sensitivity measurements were performed by an infrared thermometer to determine the wall temperature of the quartz capillary. The gas temperature in the plasma column was obtained by theoretical estimation [3] from the wall temperature and the microwave power applied.

3. Results and discussion

The plasma column in the capillary increased with the increase of the absorbed power until reaching the end of the tube. At power levels above 10 W, the discharge was in a standing-wave mode. The reflected power could be reduced to 0.1 W at forward power of 12 W. The calculated average gas

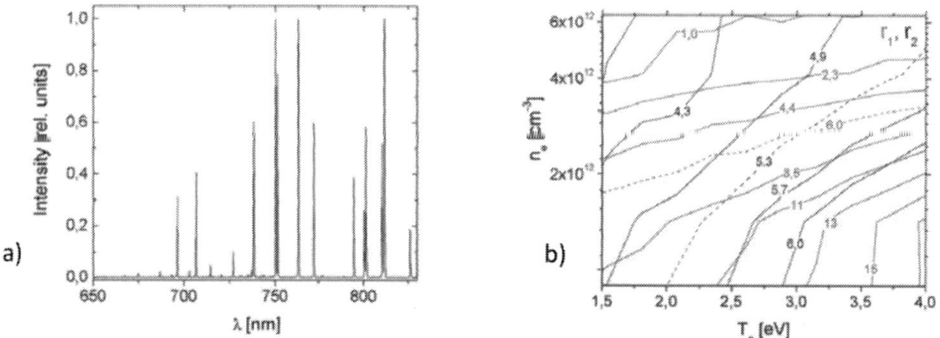

Figure 2. a) Argon spectrum in relative units; **b)** two-parametric (T_e, n_e) dependence of the line intensity ratios $r_1 = I_{738.4}/I_{727.3}$ (grey isolines) and $r_2 = I_{811.5}/I_{826.5}$ (black isolines).

temperature in the plasma column was T_g = 600 K. At these conditions, spectra of microwave discharge were recorded in the wavelength range 650–830 nm (figure 2a). They were mainly composed of neutrals lines in the spectral region 690-830 nm, belonging to transitions from the $3p^5 4p$ to $3p^5 4s$ configuration.

The spectral diagnostic method applied in the study was proposed initially in [5] and extended further in [6]. It combines measurements of certain argon line-intensities with collisional-radiative modeling (CRM) of the excitation kinetics at low pressures. The model includes kinetic processing determination of the population densities of the first fourteen excited levels. In the simulation, the gas pressure, the discharge tube radius and the gas temperature are fixed input parameters. The results generated by the model are the excited level population densities, the line-intensities and the line-intensity ratios as a function of the electron density n_e and the temperature T_e. The line-intensity ratios are imposed on application of *cross-point* method for simultaneous determination of n_e and T_e, described in detail in [5, 6]. The plasma parameters, $n_e = 2.6\times10^{12}$ cm^{-3} and T_e = 3.1 eV, in a continuous mode of discharge operation, were obtained by averaging the results of 5 line-intensity ratio combinations ($I_{794.8}/I_{826.5}$, $I_{706.7}/I_{727.3}$, $I_{738.4}/I_{696.5}$, $I_{738.4}/I_{727.3}$, $I_{751.5}/I_{727.3}$ each one of them with $I_{811.5}/I_{826.5}$). The lines were selected so as to be close in wavelength, which ensures the same spectral response of the optical detecting system and sensitivity to temperature and electron density variations. Figure 2b illustrates the grid of isolines for line-intensity ratio $r_1 = I_{738.4}/I_{727.3}$ (grey isolines) and $r_2 = I_{811.5}/I_{826.5}$ (black isolines). The dashed isolines correspond to the values of the experimentally measured line intensity ratios $r_1 = 6$ and $r_2 = 5.3$.

The dispersion relation of sw waves in a symmetrical cylindrical system plasma-dielectric-vacuum [7] was used for calculating the space damping rate α and axial wave number β (phase constant).

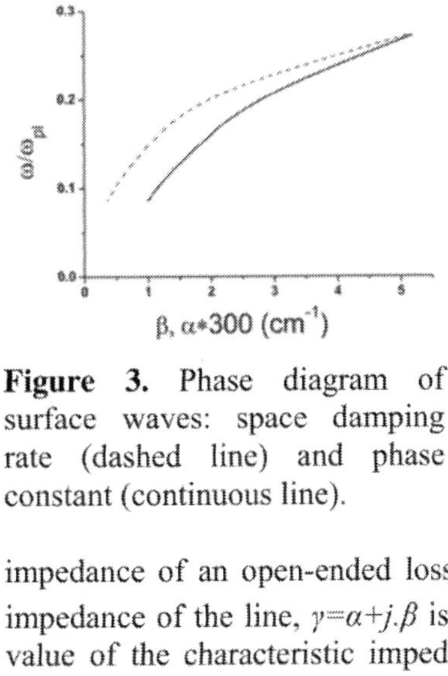

Figure 3. Phase diagram of surface waves: space damping rate (dashed line) and phase constant (continuous line).

In the calculations, $k=\beta+j.\alpha$ is the complex wave number, r=1 mm is the discharge tube inner radius and d=0.5 mm is the tube wall thickness. The media are characterized by the dielectric constants ε, ε_d, ε_v for plasma, dielectric and vacuum, respectively. The dispersion behavior of the waves was investigated in the region of plasma density close to the experimentally obtained $n_e=(1\text{-}10)\times10^{12}$ cm^{-3}. The theoretical results (figure 3) show that the surface waves in the plasma column are weakly damped waves with wavelengths much shorter (λ_g ~1.2-6.5 cm) than the wavelength in the free space (λ=12.24 cm). The column length l of 15 mm is approximately ½ λ_g of the wavelength of surface waves at plasma density n_e = 2.6 x10^{12} cm^{-3}. According to [2], the input impedance of the plasma column in a finite length vessel with insulated end plates can be calculated as the impedance of an open-ended lossy transmission line $Z_{in} = \eta/\tanh(\gamma.l)$, where η is the characteristic impedance of the line, $\gamma=\alpha+j.\beta$ is the propagation constant and l is the vessel length. The calculated value of the characteristic impedance of this line is η= 45 Ω at our experimental conditions. The dependence of the input impedance of the plasma column ($Z_{in} = R_a + jX_a = \mathrm{Re}[Z] + j.\mathrm{Im}[Z]$) on the values of α and β is presented in figure 4. The results show that the input impedance of the plasma column in a finite length vessel has resonance behavior when its length is equal to $\lambda_g/2$ or λ_g. The real part has a maximum while the imaginary part is equal to zero. The reactive resistance changes its sign in these regions from negative (capacitance) to positive (inductance) and vice versa. In this mode, the plasma column can be considered as an antenna with real part of the impedance $R_a=R_L+R_R$, i.e., the sum of the antenna losses and the radiation resistance, respectively. The microwave power radiated from the plasma column with a small ground plate increases up to 1.5 W in the standing-wave mode and works as a monopole plasma antenna. This result is confirmed by radiation pattern measurements and simulations of the electric field of the real column (vertical capillary) in the far-field region by EM simulator HFSS (figure 5).

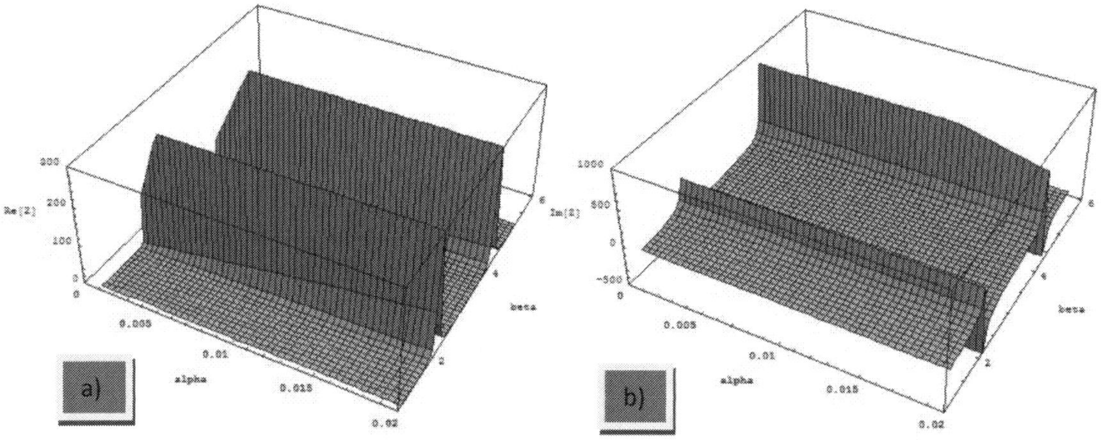

Figure 4. Input impedance of the plasma column: a) real part and b) imaginary part.

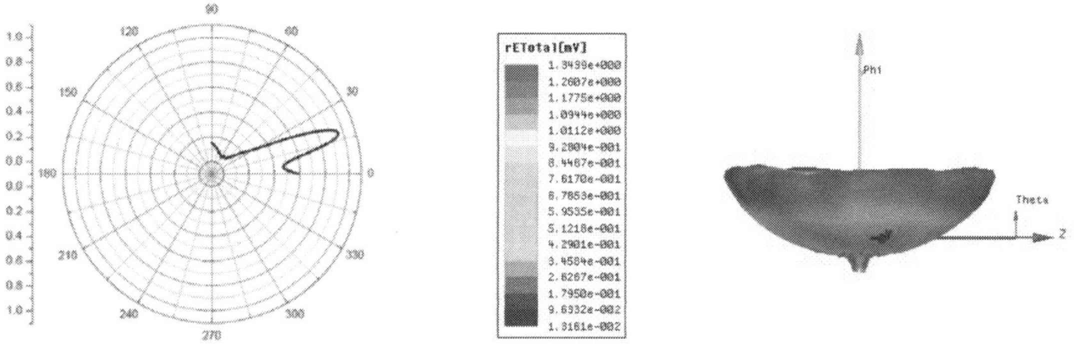

Figure 5. Radiation pattern of plasma column: a) measured E-plane characteristic and b) simulated 3D radiation pattern.

Conclusions

The input impedance of the plasma column in a standing-wave mode in a finite-length dielectric vessel with insulated ends shows resonant behavior when the column length is equal to $\lambda_g/2$ or λ_g of the surface waves sustaining the discharge. The real part of the impedance has a maximum and the imaginary part is equal to zero. The plasma column with a ground plate radiates microwaves as a monopole plasma antenna and the main lobe in the radiation pattern is close to being perpendicular to the capillary axis.

Acknowledgements

This work is financially supported by the Science Fund of University of Sofia, project N26.

References

[1] Alvarez R and Alves L L 2007 *J. Appl. Phys.* **101** 103303
[2] Rayner J and Cheetham A 2010 *IEEE Trans. Plasma Sci.* **38** 62
[3] Kiss'ovski Zh, Kolev M, Ivanov A, Lishev St and Koleva I 2009 *J Phys. D: Appl. Phys.* **42** 182 004
[4] Vachkov V, Kiss'ovski Zhand Ivanov A 2010 *Annual J. of Electronics* **4**/2 70
[5] Iordanova S and Koleva I 2007 *Spectrochim. Acta* B **62** 344
[6] Crintea D L, Czarnetzki U, Iordanova S, Koleva I and Luggenhölscher D 2009 *J. Phys. D: Appl. Phys.* **42** 045208
[7] Aliev Yu M, Schlüter H and Shivarova A 2000 *Guided-wave-produced plasmas* (Berlin: Springer)

Plasma parameters of a small microwave discharge at atmospheric pressure obtained by probe diagnostics

Zh Kiss'ovski[1], A Ivanov and St Kolev

Faculty of Physics, St. Kliment Ohridski University of Sofia,
5, J. Bourchier Blvd., 1164 Sofia, Bulgaria

E-mail: kissov@phys.uni-sofia.bg

Abstract. We present the design of a system for probe diagnostics and the results obtained for the plasma parameters in a small portable microwave plasma source operating at atmospheric pressure. The average electron temperature and the argon plasma density in the discharge are estimated from the current-voltage (I-V) characteristics of an asymmetric double probe. The calculated value of the electron temperature is T_e ~1.7±0.3 eV. Three models for the probe ion saturation current in a flowing plasma at atmospheric pressure are applied to calculating the plasma density. The calculated value of the plasma density varies with the input power P = 15 - 25 W and gas flow rate 250 - 350 sccm. The plasma density estimations obtained by the models lie within a wide range and are compared with results from optical diagnostics.

1. Introduction

The environmental and industrial applications of the microwave discharges at atmospheric pressure, such as detoxification of hazardous gases, surface cleaning and modification and air composition control, require the development of simple techniques for rapid control of the plasma parameters. The double probe diagnostics is a method which allows rapid and simultaneous determination of the main plasma parameters - electron temperature and density.

This work presents the design of a system for probe diagnostics and the results obtained for the plasma parameters in a small portable microwave plasma source [1]. This novel surface-wave plasma source creates dense plasma with stable parameters in a ceramic capillary at atmospheric pressure both in continuous and pulsed regimes. The electron temperature and density of the argon plasma in the discharge are estimated from the asymmetric double probe characteristics. A system was developed for precise measurements of the I-V probe characteristics in continuous and pulsed regimes of the discharge. The acquisition system is based on averaging over many periods of the ramp generator or a fixed number of probe characteristics in a single pulse with controlled time delay with respect to the start. Smoothing and differentiation procedures are applied to the averaged probe characteristics. The electron temperature is obtained by the standard formula, while the plasma density is estimated using the ion saturation current of a small cylindrical probe positioned in the middle of the plasma column and applying the three theories for a probe in a high pressure flowing plasma. These are the "sheath-convection" regime [2], collisional plasmas with high Peclet number [3], and Chung theory fora probe in flowing plasmas [4]. The results obtained for the plasma density value are in a wide range; we

[1] To whom any correspondence should be addressed.

Journal of Physics: Conference Series **356** (2012) 012010 doi:10.1088/1742-6596/356/1/012010

compared them with results from optical emission spectroscopy. The single probe at recombination regime [5] can also be applied for determining the plasma parameters in a microwave atmospheric plasma jet if a reference electrode is available.

2. Experimental set-up and diagnostics

The experimental set-up is presented in figure 1. Microwave power at frequency f = 2.45 GHz supplied by a generator MPG-4M (0-120 W) is fed to the source by a flexible coaxial line through a double directional coupler Pasternack PE2219-30 and a triple stub Maury 1878C. The forward and reflected power are measured by a HP 437B power meter. This low power source is built by pieces of semi-rigid coaxial lines. The exciter of the surface waves is an open coaxial line with outer copper conductor with a diameter of 2 mm and inner conductor – a steel capillary with a diameter of 1 mm. The dielectric of the line is a ceramic tube with an inner diameter of D=2.R = 1mm and an outer of 2 mm of alumina with ε_d = 9.3. This thin tube is also used as a discharge tube, because the ceramics can with-stand high gas temperatures and has high thermal conductivity and small losses. This material reduces significantly the exciter dimensions l to the length of 10 mm ($l \sim \lambda_0/4\sqrt{\varepsilon_d}$, λ_0 – wavelength in free space). A discharge ceramic tube longer by 11 mm than the exciter is used. The initial tuning of the source at frequency 2.45 GHz is carried ou by a network analyzer HP8757A (S_{11} - measurements). The neutral gas is fed to the discharge tube through the steel capillary. The gas flow (150 - 350 sccm) is measured by a mass-flow controller APEX-AX-MC. The working gas is argon.

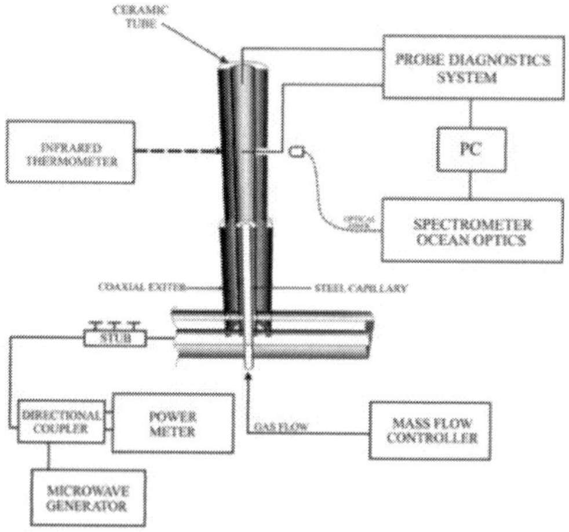

Figure 1. Experimental set-up and diagnostics.

The efficient working regime of the source is ensured by additional matching to the generator by using a triple stub. Thus, the plasma source is self-ignited in continuous and pulsed regimes at a specific threshold value of the input power.

The electron temperature and density are estimated from the asymmetric double probe characteristics measured. The first probe is a tungsten cylindrical wire with a length of 0.6 mm and radius of 0.1 mm positioned perpendicularly to the column in the middle of the discharge through a small hole (~0.5 mm) in the tube wall. The second probe is a wire with length of 4 mm and radius of 0.1 mm, positioned inside the discharge tube at its top.

The equipment for optical emission spectroscopy includes a collimating lens, an optical fiber waveguide, Ocean Optics spectrometers and an infrared thermometer.

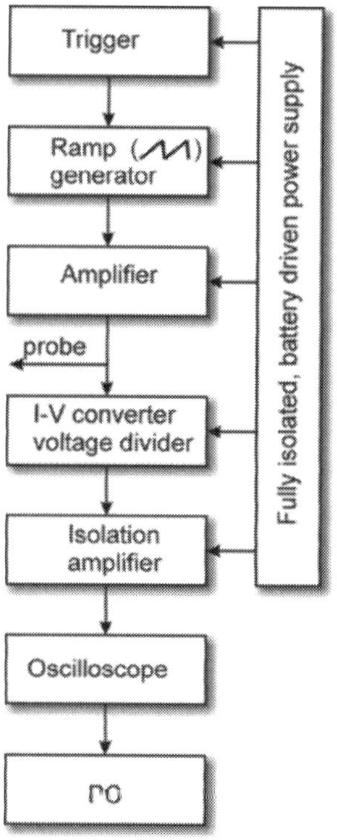

Figure 2. Block-diagram of the probe diagnostics system.

The probe system used during the experiments consists of several modules presented in figure 2 as a block-diagram. The system applies a ramp-shaped signal to the probe. The "trigger" allows the ramp signal generation to be synchronized (or unsynchronized) with an external source. This option is needed for diagnostics of pulsed discharges. The "ramp generator" produces the desired voltage shape which is further amplified by the HV amplifier based on an APEX PA78EU operational amplifier. The system allows a large variation of the applied probe voltage parameters: frequency range: 0.1 – 10000 Hz; voltage range: –200 V – +200V; AC peak-to-peak amplitude: from 1 to 200 V; DC level: –200 – +200 V; maximum probe current: 150 mA. The probe current is converted to a voltage signal by the voltage drop over a 30 Ω resistor, while the probe voltage is reduced by a RC voltage divider. The two voltage signals are further transferred to an oscilloscope through isolation amplifiers (AD215). The data records in the oscilloscope are then send to a PC for calculation of the plasma parameters. The data acquisition is based on averaging over several periods of the ramp generator, i.e. one period corresponds to a whole probe characteristic and many probe (up to 256) characteristics are averaged to produce the final I-V curve. The system is designed to operate in two modes: continuous and pulsed mode. In continuous mode, it is assumed that the discharge parameters do not change during the probe characteristic acquisition and the final (averaged) data is a result of averaging of up to 256 consecutive probe characteristics. The pulsed mode of the system is intended for use in pulsed discharges. In this case, the ramp generator is triggered by an external signal corresponding to the pulses from the discharge power supply. In this way the probe voltage signal starts with the same phase at the beginning of every discharge pulse and if the frequency of the ramp is high enough, one can record a fixed number of probe characteristics within a single pulse. Consequently, with proper adjustment of the delay and the time span of the oscilloscope, we can average over probe characteristics (in many consecutive pulses), having the same time delay with respect to the pulse start.

3. Evaluation procedure and results

The I-V probe characteristics recorded by an asymmetric double probe at three values of the input power are presented in figure 3. The negative probe current is collected by the small electrode of the probe positioned in the middle of the tube, while the positive current is collected by the electrode positioned at the top of the tube. The distance between the tips of the electrodes of the double probe is about 1.5 -2.5 mm. The probe can measure the average plasma parameters in the region of the plasma column between the electrodes. The asymmetry of the double probe characteristic (figure 3a) depends on the length of the plasma column (absorbed microwave power, gas flow). The probe characteristics obtained are subjected to a smoothing procedure by a Savitzky-Golay filter (figure 3b) and three-point differentiation [5] for calculation of the first derivative of the current with respect to the applied voltage.

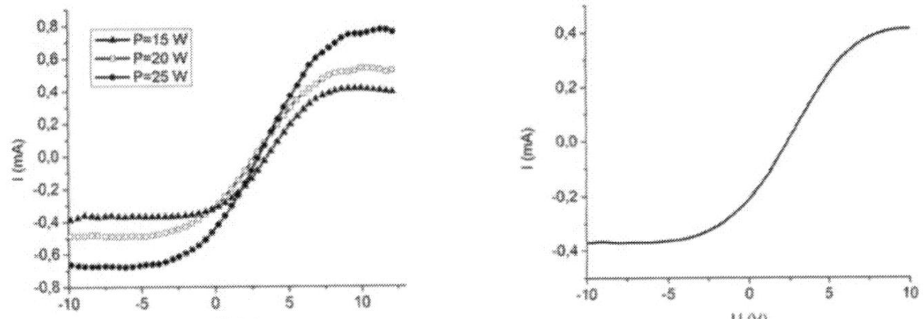

a) b)

Figure 3. I-V double probe characteristics: a) at 15, 20, 25 W b) smoothed characteristic at 15 W.

The theoretical model [6] of this discharge show that the EEDF is close to Maxwellian at our experimental conditions. The value of electron temperature is obtained from the formula:

$$\frac{kT_e}{e} = \frac{I_{1sat} \cdot I_{2sat}}{I_{1sat} + I_{2sat}} \left(\frac{dI}{dU} \right)^{-1} \Big|_{I=0} ,$$

Journal of Physics: Conference Series **356** (2012) 012010 doi:10.1088/1742-6596/356/1/012010

where I_{1sat} and I_{2sat} are the ion saturation currents of the electrodes and dI/dU is the first derivative of the probe current at $I = 0$. The calculated value of the electron temperature is $T_e \sim 1.7 \pm 0.3$ eV. Its accuracy depends on many factors, such as the electrode currents fitting curves, the probe characteristics smoothing, the effect of the microwave field, the heating of the probe, etc. The microwave field can cause a distortion in the single probe characteristics and, consequently, overestimation of the electron temperature, but it has a weaker effect on the double probe characteristics. The opposite case of underestimation of T_e due to plasma cooling by the probe was not observed in our measurements.

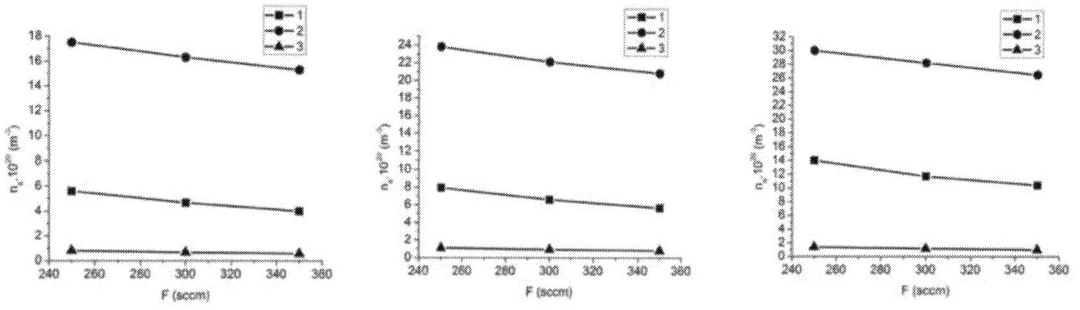

a) b) c)

Figure 4. Plasma density obtained by the models 1, 2 and 3: a) at 15 W b) at 20 W and c) 25 W.

Taking into account that the plasma is not isothermal ($T_e > T_g$, T_i) the density was estimated for three power values (figures 4a, 4b, 4c) by the models of collisional plasmas with high Peclet number (model 1), "sheath-convection" regime of ions (model 2) and Chung theory for probe in flowing plasmas (model 3):

$$I_{sat} = -\left(\frac{6.3}{Sc^{0.03}}\right)Pe^{0.4}.e.n_e.D_i.l, \quad 0.5 \le Pe \le 20 \quad Pe = v_f \cdot r / D_a, Sc = v_{vis} / D_a \,, \tag{1}$$

$$I_{sat} = 5.3(\varepsilon_0.\mu_i.r_p)^{1/4}(n_e.e.v_f)^{3/4} l.V^{1/2}, \alpha = \lambda_D/2r_p < 1, \chi = eV/kT_e > 1, \mathrm{Re} = \frac{v_f.2r_p}{\mu_i(kT_i/e)} > 1, \mathrm{Re}.\alpha^2 < 1, \mathrm{Re}.\alpha^2\chi^2 > 1 \,, \tag{2}$$

$$I_{sat} = n_e.e.S.v_f.\sqrt{\frac{l}{\mathrm{Re}}}.(1 + \frac{T_e}{T_i})\frac{0.293 + 0.181.Sci}{\sqrt{2}Sci}, Sci = v_{vis} / D_a / n_e \,, \tag{3}$$

where Pe is the Peclet number, Sc, Schmitd number, D_a and D_i, the ambipolar and ion diffusion coefficients, Re, Reynold's number, l and r_p are the length and radius of the probe, S, the probe surface collecting area, μ_i, the mobility of ions, v_f, the plasma flow velocity and V, the probe voltage. The results show that the plasma density increases with the power applied and decreases with the increase of the plasma flow. The results obtained by optical spectroscopy measurements [7] for $F = 250$ sccm and $P = 20$ W are $T_e = 1.6$ eV and $n_e = 4.5 \times 10^{20}$ m^{-3}. This demonstrates that the model of "sheath-convection" regime for the probe saturation current gives plasma density values consistent with optical diagnostics.

Conclusions

A probe diagnostics system was developed for measurement of the plasma parameters of a small portable microwave source in continuous and pulsed regimes of operation. The asymmetric double probe technique is applied for determining the average plasma parameters in the middle of the plasma column. The electron temperature was calculated by the standard formula taking into account the factors determining the accuracy. The plasma density was estimated from the ion saturation current by an appropriate model for the specific conditions of the experiment and after calibration with another diagnostics method.

Acknowledgements
This work was supported by Science Fund of University of Sofia, project N26.

References

[1] Kiss'ovski Zh, Kolev M, Ivanov A, Lishev St and Koleva I 2009 *J Phys. D: Appl. Phys*. **42** 182004
[2] Clements R M and Smy P R 1970 *J. Appl. Phys.* **41** 3745
[3] Benilov M S and Rogov B V 1991 *J. Appl. Phys.* **70** 6726
[4] Chung P M, Talbot L and Touryan K 1975 *Electrical Probes in Stationary and Flowing Plasmas: Theory and Application* (Berlin, Springer)
[5] Porteanu H, Kühn S, Gesche R 2010 *J. Appl. Phys.* **108**, 013301
[6] Henriques J, Ivanov A, Tatarova E, Kiss'ovski Zh and Ferreira C M 2011 *Bulletin APS* **56** 35
[7] Kiss'ovski Zh, Ivanov A, Iordanova S and Koleva I 2011 *J Phys. D: Appl. Phys*. **44** 205203

17th International Summer School on Vacuum, Electron, and Ion Technologies (VEIT 2011) IOP Publishing
Journal of Physics: Conference Series **356** (2012) 012011 doi:10.1088/1742-6596/356/1/012011

Analytical calculation of the gas temperature and time-resolved measurement of the electron temperature of a gas discharge in He and Ne-He mixtures

K A Temelkov[1], S I Slaveeva and N K Vuchkov

Metal Vapor Lasers Laboratory, Institute of Solid State Physics,
Bulgarian Academy of Sciences, 72, Tsarigradsko Chaussee, 1784 Sofia, Bulgaria

E-mail: temelkov@issp.bas.bg

Abstract. The gas and electron temperatures were determined in a nanosecond pulsed longitudinal discharge in a new high-temperature discharge tube design developed for a high-power large-volume middle-infrared He-$SrBr_2$ laser. Assuming that the gas temperature varies only in the radial direction, analytical solution of the steady-state heat conduction equation was obtained at uniform power input and, for the first time, for each zone of the discharge tube, namely, discharge zone, ceramic tube, discharge-free zone incompactly filled with zirconia fibers insulation, and quartz tube. The line-ratio optical emission spectroscopy method was used to determine experimentally for the first time the time-resolved electron temperature in the discharge afterglow, namely, measurement of the relative intensities of some He and Ne spectral lines originating from different upper levels. The average values of the electron temperature were also found by averaging the time-resolved electron temperature over the time.

1. Introduction

One of the main fundamental and relevant problems in plasma physics is determining the main plasma parameters, such as the electron density, the gas and electron temperatures. This is particularly important for the gas-discharge laser physics, gaseous discharges, plasma technologies, gas-discharge mass spectroscopy, absorption and emission spectroscopy and plasma in general. As was demonstrated in [1], the characteristic constants for the heavy-particle interaction depend on the gas temperature. In what concerns the metal or metal-halide vapor lasers, the thermal mode, as well as the radial temperature distribution, are of great importance for the stability of the laser operation and for achieving high output characteristics, because they not only control the laser-levels kinetics, i.e. the creation of population inversion, but also the concentration of the active particles, i.e. particles on whose transition laser oscillation is obtained. The experimental technique for gas temperature measurement relying on measurements of the Doppler broadening of spectral lines and thermal lens focal distance are definitely imprecise. The electron temperature determines the characteristic constants for elastic and inelastic electron-atom and electron-ion collisions, as well as the three-body electron-ion recombination [2, 3]. For gas-discharge lasers in particular, the electron temperature influences the creation of population inversion via the abovementioned processes, which populate or

[1] To whom any correspondence should be addressed.

depopulate the upper and lower laser levels. Typical methods to measure the electron temperature include the use of a Langmuir probe and laser Thomson scattering (LTS), which are inapplicable to the gas discharge used in our experiments. Several models have been described which predict, among other parameters, the electron temperature values with considerable discrepancies and, moreover, without overlap [2, 3].

Two methods have been proposed and investigated to increase the operating temperature (up to 1300 °C) of the gas-discharge tube typical for middle-infrared (MIR) He-$SrBr_2$ laser pumped by a nanosecond pulsed longitudinal discharge [4]. This laser has enormous prospects for application in laser medicine and as an alternative laser source of free electron lasers. Briefly, the methods consist in adding Ne to the He buffer-gas without additional active-volume insulation [5] and, the one presented in this paper, development of a new discharge tube design with ZrO_2 active-volume insulation. For this discharge tube, we present here calculations of the gas temperature distribution and experimental time-resolved measurements of the electron temperature applying for the first time the line-ratio method of OES to the nanosecond pulsed longitudinal discharge. The average values of the electron temperature were also found by averaging the time-resolved electron temperature over the time.

2. Experimental setup

The new high-temperature large-volume discharge tube is shown schematically in figure 1. The basic tube with a 71.5-mm inside diameter and a 76-mm outside diameter is made of fused quartz. A ceramic insert, confining the discharge zone, with a 30.5-mm inside diameter, a 38.5-mm outside diameter and a length of 98 cm is coaxially sleeved in the basic tube. The zone between the ceramic insert, basic quartz tube and the holders of the ceramic insert is incompactly filled with ZrO_2 fibrous insulation. The discharge operates in a self-heating regime. The gas discharge tube is not loaded with $SrBr_2$, as is shown in figure 1. The temperature at the quartz tube surface is measured by a

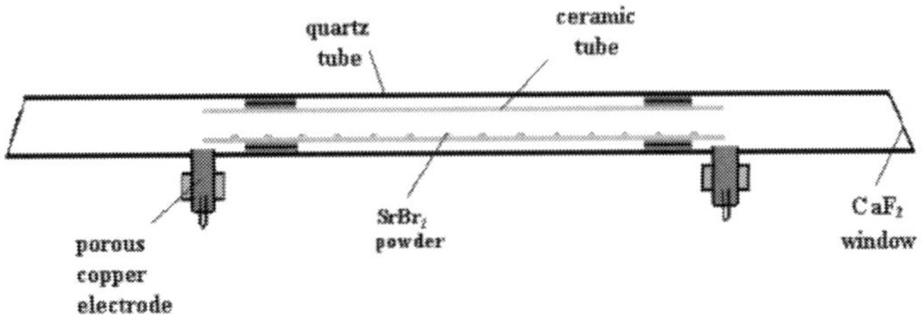

Figure 1. Discharge tube design.

thermocouple. The electrodes of the discharge tube are made of porous copper with a special design. CaF_2 windows are glued to the ends of the discharge tube. The nanosecond pulsed longitudinal discharges investigated are excited by an electrical scheme with interacting circuits (IC scheme). The spectral studies were performed using a Bentham M300 spectrometer equipped with one 1800 groove/mm holographic grating and a DH-2 photo-multiplier. The spontaneous emission pulses were displayed on a Tektronix 2455A oscilloscope with a 20 MHz cutoff filter.

3. Theoretical and experimental results

3.1. Analytical solution of the steady-state heat-conduction equation

Assuming that the gas temperature varies only in the radial direction, the steady-state heat conduction equation is solved at uniform power input as in [5]. For the first time, the well-known solution of this equation is obtained for the case of one discharge zone with radius R_1 and three discharge-free zones, namely, a ceramic tube within $R_1 \le r \le R_2$, a compact or incompact fill of ZrO_2 insulation within $R_2 \le r \le R_3$, and a basic quartz tube within $R_3 \le r \le R_4$. The dependence of the thermal conductivity k of gases and gas mixtures has the form $k = B.T_g^{\,a}$, where B and a are constants (within a certain temperature range) specific for each gaseous or solid medium. The constants B and a, which determine the thermal conductivity, could be obtained through fitting the existing experimental data taken from [6]. The thermal conductivities of He, the ceramic Al_2O_3 tube, the compact fill of ZrO_2 insulation, the incompact fill of He-ZrO_2, and the basic quartz tube are presented in table 1.

Table 1. $k = B.T_g^a$ – thermal conductivity (B and a are for k in W m^{-1} K^{-1}).

	He	Al_2O_3	ZrO_2	He-ZrO_2	quartz
B	34.9x10^{-4}	44323.1	7326.2x10^{-4}	655.9x10^{-4}	705.9x10^{-4}
a	0.670	-1.227	0.130	0.366	0.487

In figure 2, a comparison is shown between the gas temperature distributions for two discharge tube designs, namely, without and with zirconia insulation in the discharge-free zone for a gas discharge in pure helium. The following cases are considered: 1 and 2 are gas temperature profiles neglecting and taking into account, respectively, the influence of the ceramic tube radiance on the gas temperature distribution, which were already presented in [5], for a gas discharge tube without zirconia insulation; 3 is the case experimentally studied with compact fill of zirconia insulation; 4 and 5 are the cases experimentally investigated with incompact fill of zirconia insulation for two different average input powers of 2.1 kW and 2.6 kW, respectively.

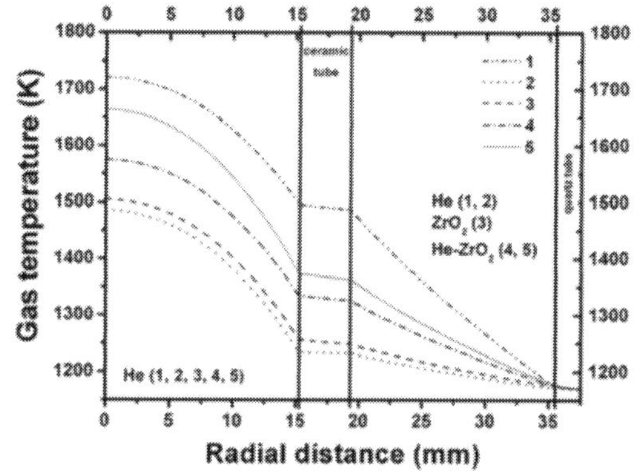

Figure 2. Radial distribution of the gas temperature for gas a discharge in helium for different types of active-volume thermal insulation.

3.2. Experimental determination of time-resolved and average electron temperature

Under local thermodynamic equilibrium (LTE) conditions, measuring the relative intensity ratio of two lines originating from different upper levels i and j would be sufficient for determining the electron temperature by using the relative intensity ratio below, the Bolzmann relation for the relative equilibrium populations of the two upper levels and the expression derived from them:

$$\frac{I_i}{I_j} = \frac{\lambda_j n_i A_i}{\lambda_i n_j A_j}, \qquad \frac{n_i}{n_j} = \frac{g_i}{g_j} e^{\frac{E_i - E_j}{k_b T_e}},$$

$$T_e = \frac{\left(E_j - E_i\right)}{k_b} \cdot \left[\ln\left(\frac{I_i \lambda_i g_j A_j}{I_j \lambda_j g_i A_i} \right) \right]^{-1}, \quad (1)$$

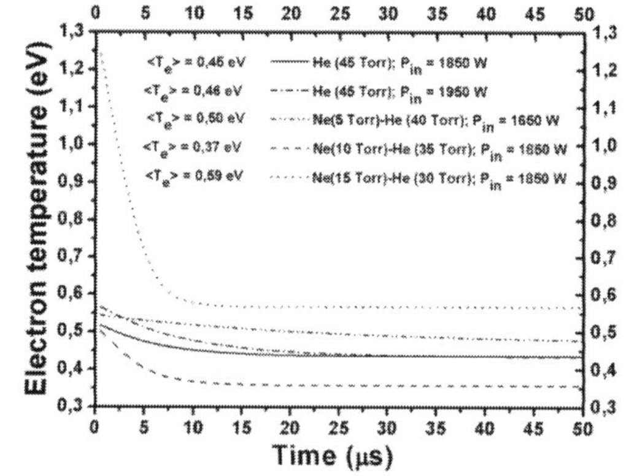

Figure 3. Time dependence of the electron temperature in the discharge afterglow of a nanosecond pulsed longitudinal discharge in pure helium and Ne-He mixtures.

where λ_i and λ_j are the wavelength of the transitions measured from levels i and j; A_i and A_j are the transition probabilities, i.e. the spontaneous rate coefficients for the transitions from measured levels i and j; n_i and n_j are the populations in levels i and j; g_i and g_j are the statistical weights, i.e. the degeneracies of levels i and j; E_i and E_j are the energies of levels i and j; and k_b is the Boltzmann constant. Using the data required for experimental determination of the electron temperature T_e presented in [5] for each He and Ne transition, the time dependence of the electron temperature is shown in figure 3 for the afterglow of a nanosecond pulsed longitudinal discharge in pure helium (45 Torr) and with addition of a small amount of neon (5, 10 and 15 Torr, maintaining the total Ne-He mixture pressure of 45 Torr constant) to the helium buffer gas. The average input power used is also given in the figure, as well as the average values of the electron temperature found by averaging the electron temperature over the time. As one can see, the dependence of the time-resolved and average electron temperatures on the added neon pressure has a pronounced minimum at 10 Torr.

Conclusions

By solving analytically the steady-state heat-conduction equation for uniform power input and under the assumption that the gas temperature varies only in the radial direction, it is shown that the crucial problem impeding the further development and application of high-temperature large-volume He-$SrBr_2$ laser is solved through the development of a new discharge tube design with incompact ZrO_2 active-volume insulation. Th line-ratio OES method of remote measurement is the only method applicable to nanosecond pulsed longitudinal discharges. This method is used for the first time to determine experimentally the time-resolved electron temperature in the nanosecond pulsed longitudinal discharge afterglow.

References

[1] Temelkov K A, Vuchkov N K, Ekov R P and Sabotinov N V 2008 *J. Phys. D: Appl. Phys* **41** 105203
[2] Carman R J 1990 *IEEE J. Quantum Electr.* **26**/9 1588-608
[3] Carman R J, Brown J W and Piper J A 1994 *IEEE J. Quantum Electr.* **30**/8 1876-94
[4] Temelkov K A, Vuchkov N K, Mao B, Atanassov E P, Lyutov L and Sabotinov N V 2009 *IEEE J. Quantum Electr.* **45**/3 278-81
[5] Temelkov K A, Slaveeva S I and Vuchkov N K 2011 *IEEE Trans. Plasma Sci.* **39**/3 831-5
[6] Handbook 1976 *Tables of physical constants*, ed I K Kikoin (Moscow: Atomizdat) pp 257-61

2D numerical modeling of the gas temperature in a large-volume high-temperature nanosecond pulsed longitudinal discharge in helium with small admixtures of neon, strontium and bromine

T P Chernogorova[1], K A Temelkov[2,3], N K Koleva[2] and N K Vuchkov[2]

[1] Faculty of Mathematics and Informatics, St. Kl. Ohridski University of Sofia, 5, James Bourchier Blvd., 1164 Sofia, Bulgaria
[2] Metal Vapor Lasers Laboratory, Institute of Solid State Physics, Bulgarian Academy of Sciences, 72, Tzarigradsko Chaussee, 1784 Sofia, Bulgaria

E-mail: temelkov@issp.bas.bg

Abstract. A numerical 2D model (r, z) of the gas temperature was developed for the case of axial symmetry and uniform power input. The model determines the gas temperature of a nanosecond pulsed longitudinal discharge in helium with small additives of neon, strontium and bromine. The gas discharge is excited in the newly designed large-volume high-temperature discharge tube.

1. Introduction

It is well known that the gas temperature is one of the basic gas-discharge plasma parameters, which determines the heavy particles interaction, heavy particles concentration, gas discharge stability, etc., The experimental or theoretical determination of the gas temperature is of fundamental importance and is, therefore, applied to problems in various fields, such as gas-discharge laser physics, gaseous discharges, plasma technologies, gas-discharge mass spectroscopy, absorption and emission spectroscopy and plasma in general. It is also well known that the techniques widely used for gas temperature measurement based on measurements of spectral lines Doppler broadening and thermal lens focal distance are definitely imprecise, i. e. with inadmissible experimental error.

Assuming that the gas temperature varies only in the radial direction and considering uniform and non-uniform power input, we calculated gas temperature distributions through analytical solution of the steady-state heat-conduction equation for nanosecond pulsed longitudinal discharge in a series of gas-discharge tube designs [1, 2]. Unfortunately, despite the tremendous efforts, solving analytically the abovementioned steady-state heat conduction equation for 2D (r, z) and 3D (r, φ, z) cases encountered some insuperable obstacles.

2. Experimental setup

A schematic diagram of the new high-temperature large-volume discharge tube is shown in figure 1. The basic tube with a 71.5-mm inside diameter and a 76-mm outside diameter is made of fused quartz.

[3] To whom any correspondence should be addressed.

A ceramic insert, confining the discharge zone, with a 30.5-mm inside diameter and a 38.5-mm outside diameter and a length of 98 cm is coaxially sleeved in the basic tube. The discharge operates in a self-heating regime. The temperature at the quartz tube surface is measured by a thermocouple.

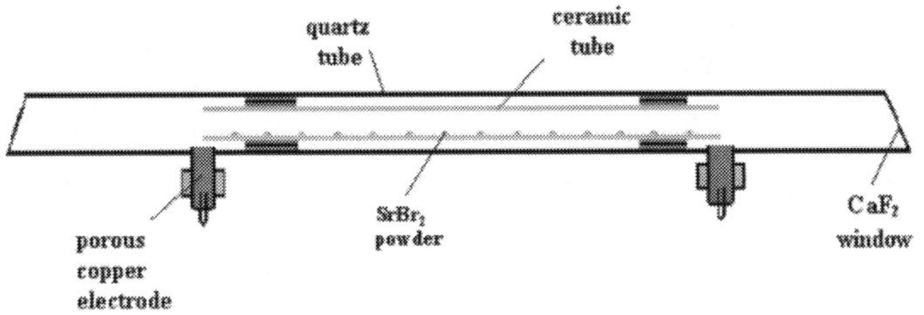

Figure 1. Discharge tube design.

The electrodes of the discharge tubes are made of porous copper with a special design. CaF_2 windows are glued to the ends of the discharge tubes. The nanosecond pulsed longitudinal discharges investigated are excited by an electrical circuit based on a high-voltage rectifier and a high-voltage pulsed excitation circuit with interacting circuits (IC circuit).

3. Results and discussion

A 2D numerical solution of the steady-state heat-conduction equation was derived for the case of one discharge zone with radius R_1, and three discharge-free zones, namely a ceramic tube within $R_1 \le r \le R_2$, a gaseous discharge-free zone within $R_2 \le r \le R_3$, and a basic tube made of quartz within $R_3 \le r \le R_4$. The dependence of the thermal conductivity k of gases and gas mixtures has the form $k = B.T_g^{\,a}$, where B and a are constants (within a certain temperature range), which are specific for each gaseous or solid medium. The constants B and a, which determine the thermal conductivity, were obtained through fitting the existing experimental data taken from [3]. The thermal conductivities of He and Ne-He mixtures, a ceramic tube made of Al_2O_3, and a basic tube made of quartz are presented in table 1.

Table 1. $k = B.T_g^{\,a}$ – thermal conductivity.

	He (45 Torr)	Ne (45 Torr)	Ne-He (5-40 Torr)	Ne-He (10-35 Torr)	Ne-He (15-30 Torr)	He-Sr	He-Br	Al_2O_3	quartz
B	34.9×10^{-4}	9.7×10^{-4}	30.5×10^{-4}	26.4×10^{-4}	22.9×10^{-4}	26.8×10^{-4}	28.9×10^{-4}	44323.1	705.9×10^{-4}
a	0.670	0.685	0.672	0.673	0.675	0.680	0.675	-1.227	0.487

A brief description of the 2D numerical model is given in the appendix. Figure 2 shows two-dimensional gas temperature distributions for a nanosecond pulsed longitudinal discharge in He (a), Ne (b), Ne-He mixtures (c), (d) and (e), and in He with small admixtures of strontium (f) and bromine (g). The partial pressure of Ne in the Ne-He mixtures is 5 (c), 10 (d) and 15 (e) Torr, maintaining the total Ne-He mixture pressure of 45 Torr constant, while the partial pressures of strontium and bromine are 0.6 Torr and 1.2 Torr, respectively.

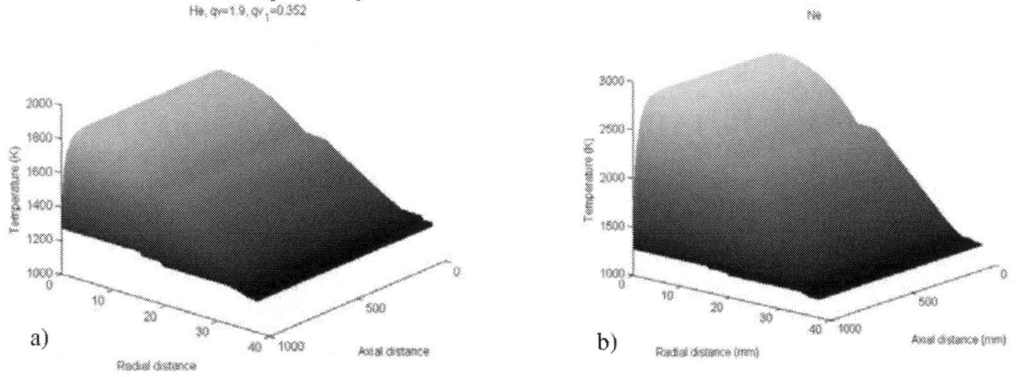

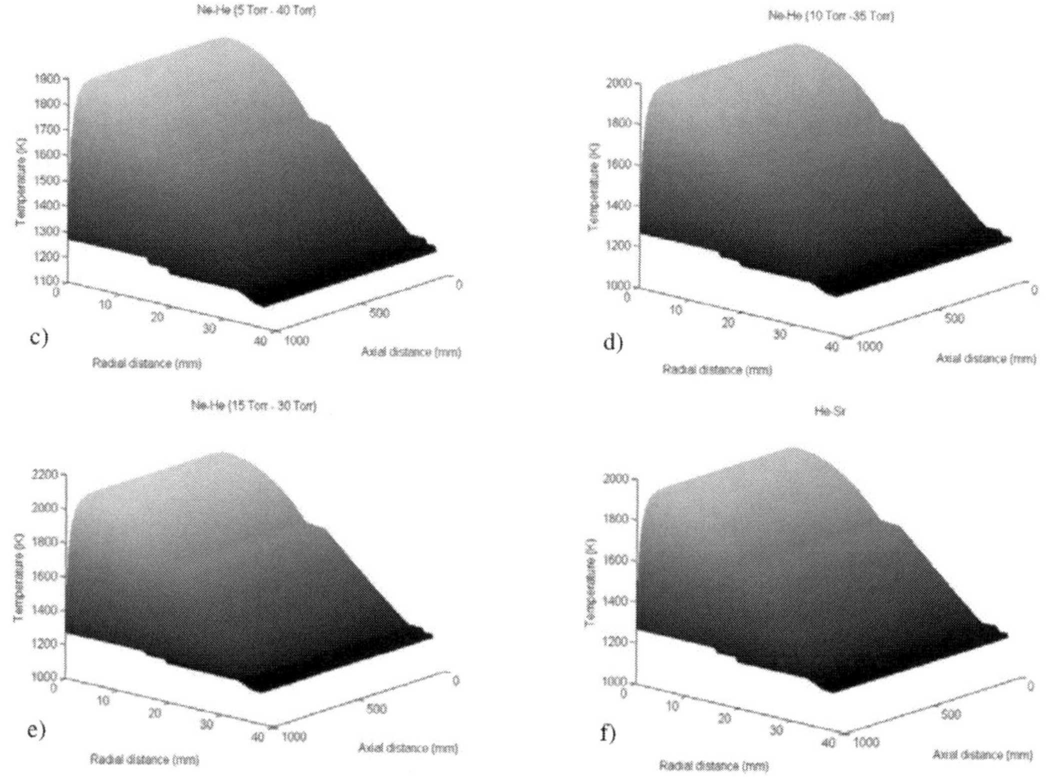

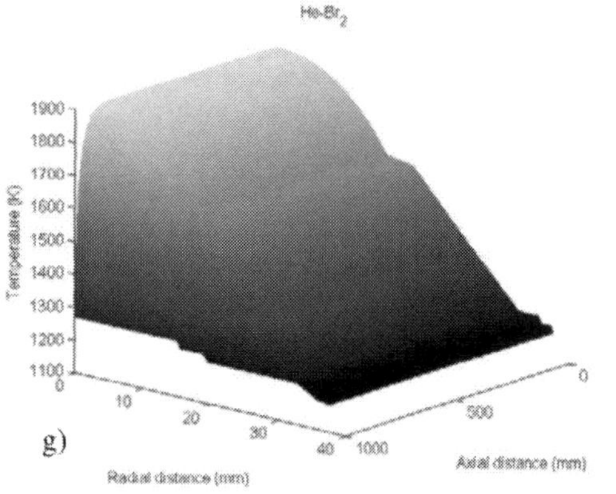

Figure 2. Two-dimensional gas temperature profiles for a nanosecond pulsed longitudinal discharge in different gas mixtures.

Conclusions

The results are presented of 2D numerical modeling (r, z) of the gas temperature in a nanosecond pulsed longitudinal discharge in various mixtures, considering axial symmetry and uniform power input. The development of a 2D numerical model for determining the gas temperature in this type of gas discharge is a further and more complicated step in approaching the real experimental conditions for the discharge studied. A new discharge tube design with incompact ZrO_2 active-volume insulation was developed in order to increase additionally the operation temperature; its investigation is in progress.

Acknowledgements

The first author is supported by the Sofia University Foundation under Grant No 154/2011.

Appendix A. Description of the 2D model

For the four zones considered the 2D heat-conduction equations have the following form:

$$\frac{1}{r}\frac{\partial}{\partial r}\left(rk_i(T^{(i)})\frac{\partial T^{(i)}}{\partial r}\right)+\frac{\partial}{\partial z}\left(k_i(T^{(i)})\frac{\partial T^{(i)}}{\partial z}\right)+q_v=0,\quad R_{i-1}<r<R_i,\quad 0<z<L,\ R_0=0,\qquad \text{(A.1)}$$

where $q_v = 1.9\ \mathrm{W.cm^{-3}}$ is the power deposited into the discharge zone per unit volume and i = 1, 2, 3, 4.

The equations are solved considering the following boundary conditions:

$$\left.\frac{\partial T^{(1)}}{\partial r}\right|_{r=0}=0,\quad \left.T^{(4)}\right|_{r=R_4}=T_w,\quad 0\le z\le L,$$

and for i = 1, 2, 3

$$T^{(i)}(R_i-0,z)=T^{(i+1)}(R_i+0,z),\quad 0\le z\le L,$$

$$\left.k_i(T^{(i)})\frac{\partial T^{(i)}}{\partial r}\right|_{R_i-0}=\left.k_{i+1}(T^{(i+1)})\frac{\partial T^{(i+1)}}{\partial r}\right|_{R_i+0},\quad 0\le z\le L,$$

$$\left.T^{(i)}\right|_{z=0}=\left.T^{(i)}\right|_{z=L}=T_{w,i},\quad R_{i-1}\le r<R_i,$$

$$\left.T^{(i)}\right|_{z=0}=\left.T^{(i)}\right|_{z=L}=0.5\left(T_{w,i}+T_{w,i+1}\right),\quad r=R_i,$$

$$\left.T^{(4)}\right|_{z=0}=\left.T^{(4)}\right|_{z=L}=T_{w,4},\quad R_3<r\le R_4,$$

References

[1] Temelkov K A, Vuchkov N K, Ekov R P and Sabotinov N V 2008 *J. Phys. D: Appl. Phys* **41** 105203

[2] Temelkov K A, Vuchkov N K, Freijo-Martin I and Ekov R P 2010 *J. Phys. D: Appl. Phys* **43** 075206

[3] Handbook 1976 *Tables of physical constants*, ed I K Kikoin (Moscow: Atomizdat) pp 257-61

Plasma deposited polymers as gas sensitive films

E Radeva[1,3], V Georgieva[1], J Lazarov[1], L Vergov[1] and N Donkov[2]

[1] Georgi Nadjakov Institute of Solid State Physics, Bulgarian Academy of Sciences, 72, Tzarigradsko Chaussee, 1784 Sofia, Bulgaria
[2] Emil Djakov Institute of Electronics, Bulgarian Academy of Sciences, 72, Tzarigradsko Chaussee, 1784 Sofia, Bulgaria

E-mail: eradeva@issp.bas.bg

Abstract. The possibility is presented of producing thin plasma polymers with desired properties by using nanofillers. Composite films are synthesized from a mixture of hexamethyldisiloxane (HMDSO) and detonation nanodiamond particles (DNDs). The chemical structure of the composite consists of DNDs distributed in the polymer matrix. The effect of DNDs on the humidity and ammonia sorptive properties of the polymers obtained is studied by measuring the mass changes as a result of gas sorption by using a quartz crystal microbalance (QCM). The results show that, in view of building a sensing element for measuring humidity, ammonia or other gases, it is possible to maximize the sensor sensitivity to a certain gas by using an appropriate concentration of DNDs in HMDSO. Thus, a high degree of sensor sensitivity, together with short response time and minimum hysteresis, can be achieved. Composites of plasma-polymerized HMDSO with DNDs can be used as gas sensitive layers for the development of quartz resonator sensors.

1. Introduction

Since the chemical and physical properties of plasma polymers (PPs), and especially of silicon containing PPs, may be tailored for particular needs, PPs have found a permanent place in electronic sensors [1-5]. Nanotechnology has made it possible to replace classical sensors materials by PPs and fabricate composite films with organic and inorganic materials [6,7]. Nanodiamond structures are of interest due to the combination of the unique properties inherent to diamond with the particles' specific surface structure. Detonation nanodiamonds (DNDs) with predominant particles' size of 4 nm can be used as fillers in PPs [6-8]. On the other hand, PPs appear to be the most appropriate technical solution for realization of gas sensors using a quartz crystal microbalance (QCM). These sensor elements combine the specific sensing properties of PPs with the high mass-sensitivity of the acoustic devices [2-5].

This study is an attempt to modulate the structure and properties of plasma polymers by synthesis of DND-based composites from a mixture of hexamethyldisiloxane (HMDSO) and DNDs. The chemical structure of the polymers obtained was established. The effect of DNDs on the humidity and ammonia sorptive properties of the polymers obtained was studied by measuring the mass changes as a result of gas sorption by applying a QCM-technique.

[3] To whom any correspondence should be addressed.

2. Experimental

The plasma polymerization technology used for obtaining plasma polymerized hexamethyldisiloxane (PPHMDSO) was described previously [4]. The polymerization of the HMDSO (Merck, 99.99%) was performed at 0.08 mA/cm^2 current density (27.12 MHz) and vacuum of about 133 Pa for 15 min. The composite deposition was carried out in the same equipment as the polymer films. The following procedure was applied: 0.10 g DND powder was added to 100 ml HMDSO and shaken for 15 min in an ultrasonic apparatus; further, the mixture was stirred (275 r.p.m) continuously at room temperature. The plasma conditions were the same – 0.08 mA/cm^2 current density and vacuum of about 133 Pa for 15 minutes. The films were deposited on KBr as well as on the quartz resonators.

The FTIR absorbance spectra of the plasma polymer films were recorded on a Bruker FTIR spectrometer in the 4000-400 cm^{-1} spectral region with a resolution of 2 cm^{-1}. The humidity-frequency measurements of the sensor system and the methodology were described in [2], while those of the ammonia-frequency measurements were presented in [9]. All studies were carried out at room temperature (26 °C).

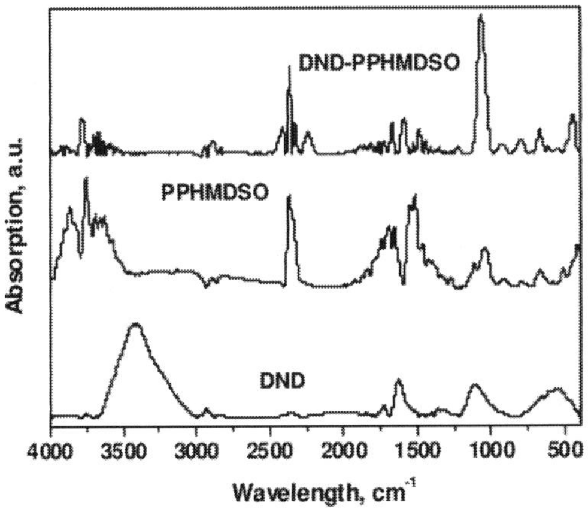

Figure 1. FTIR spectra of DND, PPHMDSO and DND-PPHMDSO.

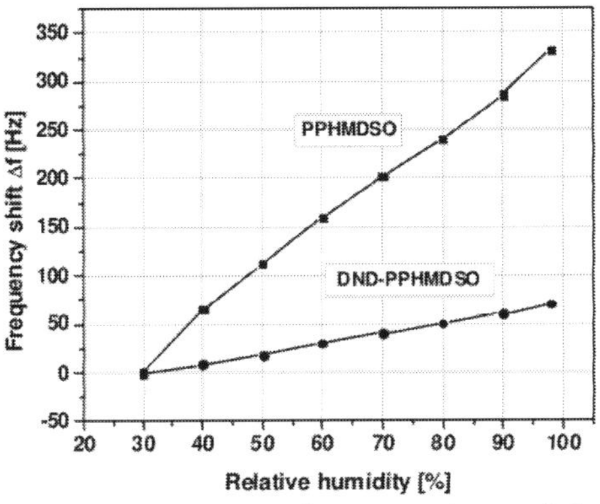

Figure 2. Humidity-frequency characteristics (HFCs) of the coated resonators.

3. Results and discussion

3.1. FTIR study of plasma polymers

The modification of the PPHMDSO structure achieved by adding DNDs filler is seen in the FTIR spectra (figure 1).

The characteristic peaks of sp^3 bonded carbon at 1550 cm^{-1} and sp^2 bonded carbon at 1160 cm^{-1} of the DND are easily observed. Thus, the DND-PPHMDSO polymer is characterized by functional groups of the type –OH, >C=O, Si–O–Si, Si–O–C and the formation of new (C-Si-C) bonds and different tetrahedral CH_3 (the bands in the region 2800 – 3000 cm^{-1}). The change observed of the characteristic polymer bands can be explained by penetration of DND particles in the polymer matrix.

3.2. Humidity sorptive properties of the plasma polymer films

Humidity-frequency characteristics (HFCs) of quartz resonators with PPHMDSO and DND-PPHMDSO are shown in figure 2. The two curves were obtained by measuring the frequency shift when increasing the relative humidity (RH) at a constant temperature of 25^0C. The sensitivity to humidity of a polymer coated quartz resonator is the frequency change as a function of the percentage change in RH and depends on the chemical composition and structure of the film. This is the slope of the HFCs in figure 2.

The average sensitivities to humidity for the resonators with PPHMDSO and DND-PPHMDSO are 5 Hz/%RH and 1 Hz/%RH, respectively. The results show that the sensitivity to humidity of the QCM with pure polymer is larger than that with DND-PPHMDSO, but a decrease in the response time and a hysteresis effect are observed. Therefore, by varying the DNDs concentration in the mixture one could find a compromise between the sensing element's sensitivity, response time and hysteresis.

3.3. Ammonia sorptive properties of the plasma polymer films

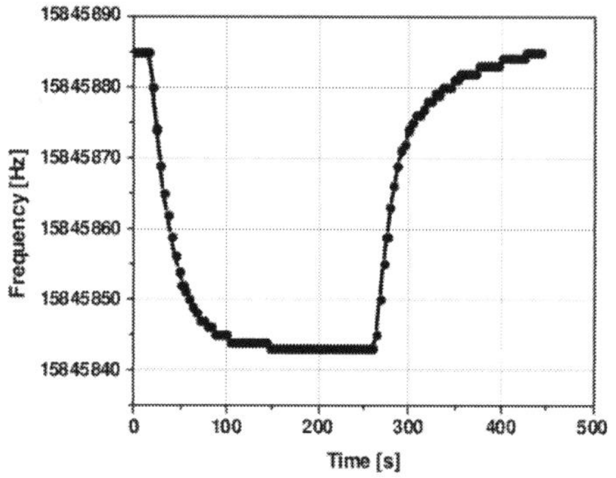

Figure 3. Time-frequency dependence of a resonator with PPHMDSO at 500 ppm NH_3.

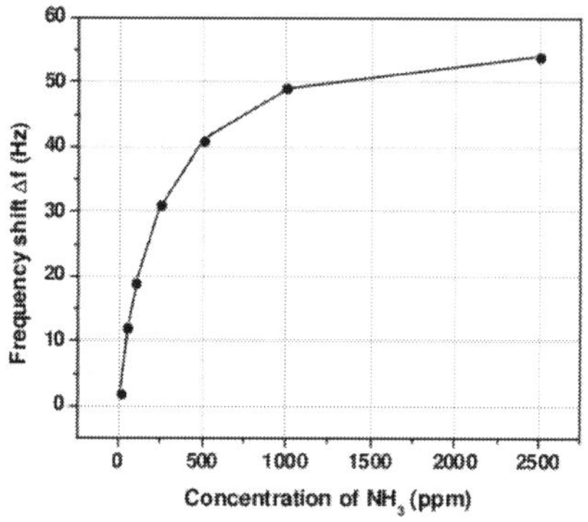

Figure 4. Frequency dependence versus NH_3 for a PPHMDSO coated resonator.

The time-frequency characteristics of the sensing elements at a certain ammonia concentration were also studied. Figure 3 shows the dependence for a QCM with PPHMDSO (film thickness of 640 nm) at 500 ppm NH_3 concentration.

The ammonia sensitivity of the resonator with PPHMDSO at different concentrations ranging from 10 to 2500 ppm is presented in figure 4. The mass adsorbed was calculated on the basis of the frequency shift measured according to Sauerbrey equation for AT-cut quartz resonators [10] (table 1).

The resonators covered with DND-PPHMDSO have lower sensitivities, compared with those with pure PPHMDSO (table 2), but the response time is shorter.

The results imply that, by choosing an optimum concentration of DNDs in HMDSO for plasma polymerization, it is possible to increase the sensor sensitivity to ammonia vapor or particular gases.

Table 1. Dependence of the adsorbed mass versus NH_3 concentration.

Ammonia cfoncentration [ppm]	Adsorbed mass [ng]
10	0.44
50	2.66
100	4.21
250	6.86
500	9.07
1000	10.85
2500	11.95

Table 2. Ammonia sensitivities of resonators with PPHMDSO and DND-PPHMDSO.

Ammonia concentration [ppm]	Δf [Hz]	
	PPHMDSO	DND-PPHMDSO
10	2	-
50	12	1.8
100	19	3
250	31	4.9

Conclusions

A study is presented of composite films produced by plasma polymerization from mixture of DNDs and HMDSO. The chemical structure of the composite consists of DND nanoparticles distributed in the polymer matrix thus creating new bonds. In view of fabricating a sensing element for measuring humidity, ammonia vapor or other gases, it is possible to maximize the sensor sensitivity to a particular gas by using the appropriate concentration of DNDs in HMDSO. Thus, a high degree of sensor sensitivity together with short response time and minimum hysteresis could be achieved. Composites of plasma polymerized HMDSO with DND particles could, therefore, be used as gas sensitive layers for the development of quartz resonator sensors.

References

[1] Saloum S and Alkhaled B 2010 *Acta Physica Polonica A* **117**/3 484
[2] Radeva E and Spassov L 1998 *Vacuum* **51** 217
[3] Avramov I, Rapp M, Krawczak P and Radeva E 2002 *IEEE Sensors J.* **2**/3 150
[4] Tsankov D, Radeva E, Hinrichs K, Roseler A and Korte E.-H. 2005 *Thin Solid Films* **476** 174
[5] Mugumura H, Kase Y, Murata N and Matsumura K 2006 *J. Phys. Chem. B* **110** 26033
[6] Shenderova O A, Zhirnov V V and Brenner D W 2002 *Solid State Mater.Sci.* **27**(3/4) 227
[7] Dolmatov V Yu 2007 *Russian Chemical Reviews* **70**/7 607
[8] Bogatyreva G P, Voloshin M N and Padalko V I 2008 *Diamond & Related Materials* **17** 213
[9] Georgieva V, Stefanov P, Raicheva Z, Atanassov M, Tincheva T, Manolov E and Vergov L 2009 *J. Optoelec. Adv. Mater.* **11** 1363
[10] Sauerbrey G 1959 *J. Physik* **155** 206

Journal of Physics: Conference Series **356** (2012) 012014 doi:10.1088/1742-6596/356/1/012014

Optical properties of chalcogenide Ge-Te-In thin films

A Zaidan[1,2,3], V Ivanova[1] and P Petkov[1]

[1] Thin Film Technology Laboratory, Department of Physics, University of Chemical Technology and Metallurgy, 8, Kl. Ohridski Blvd., 1756 Sofia, Bulgaria
[2] Department of Physics, Faculty of Science and Technology, Airlangga University, 60115 Surabaya, Indonesia

E-mail: zaidan@unair.ac.id

Abstract. Thin films of the chalcogenide $(GeTe_4)_{1-x}In_x$ with various compositions (x = 0, 5, 10, 15, 20 at %) were deposited under vacuum on glass substrates by thermal evaporation. The optical transmission and reflection spectra of the films at normal incidence were investigated in the spectral range from 800 to 2600 nm. Using the transmission spectra, the optical constants (refractive index (n) and extinction coefficient (κ)) were calculated based on Swanepoel's method. The optical band gap (E_g^{opt}) was also estimated using Tauc's extrapolation procedure.

1. Introduction

Chalcogenide glasses have received much consideration because of their interesting applications in optics, optoelectronics and electronics due to their unique properties (optical transparency in the infrared region, high refraction index, low phonon energies). Chalcogenide glasses applications include phase-change materials [1, 2, 3, 4], sensors [5, 6], optical circuits, gratings, waveguides [7, 8, 9], and many others.

Chalcogenides based on Ge-Te have been extensively studied by many authors because of their potential for use as phase-change materials for optical recording. The Ge-Te system has the ability to form glasses at timescales and temperatures that are appropriate for certain data-transfer rates and laser powers available [10].

In the present study, thin films of the system $(GeTe_4)_{1-x}In_x$ with x = 0, 5, 10, 15, and 20 at % were deposited by thermal evaporation onto floatglass substrates. The transmission and reflection spectra of the thin films were measured. Using the transmission spectra, the optical constants and film thickness were determined using Swanepoel's method. The absorption coefficient and the optical band gap were also estimated and discussed.

2. Experimental details

2.1. Sample preparation and characterization

Various compositions of bulk samples of the system $(GeTe_4)_{1-x}In_x$ with x = 0, 5, 10, 15, and 20 at % were prepared by melt quenching using 5N purity elements of Ge, Te and 4N purity of In. All

[3] To whom any correspondence should be addressed.

Journal of Physics: Conference Series **356** (2012) 012014 doi:10.1088/1742-6596/356/1/012014

elements were heated in evacuated ampoules at a constant heating ratio of 3 K/min up to 1200 K. Finally, the ampoules were quenched into ice-cooled water to avoid crystallization.

All thin film in this study were deposited by conventional thermal evaporation at a residual gas pressure of $1{,}33\times10^{-4}$ Pa and temperature 800-900 K onto floatglass substrates using the respective bulk composition as a source material. The thickness of the films obtained was about 400-600 nm. The transmission and reflection spectra of the thin films were measured using a Jasco UV-VIS-NIR spectrophotometer (Model V-670) at room temperature in the wavelength range from 800 nm to 2600 nm.

2.2. Calculation of the thickness, optical constants, absorption, and optical band gap

The optical constants (refractive index, n, and extinction coeffcient, κ) of the thin films were determined based on a method proposed by Swanepoel [11], which use the interference fringes from the optical transmission data. In this method, the thin film considered is a homogeneous and uniform film with thickness d, refractive index n and absorption α deposited on a thick transparent substrate. The transparent substrate has thickness several orders of magnitude larger than d, refraction index n_s and absorption coefficient $\alpha_s = 0$.

Swanepoel [11] has shown that the transmittance T of a uniform thin film of thickness d, refractive index n, and absorption coeffcient α, deposited on a substrate with a refractive index n_s is given by:

$$T = \frac{Ax}{B - Cx\cos\phi + Dx^2}, \tag{1}$$

where $A = 16n^2 n_s$, $B = (n+1)^3(n+n_s^2)$, $C = 2(n^2-1)(n^2-n_s^2)$, $D = (n^2-1)^3(n-n_s^2)$, $\phi = 4\pi nd/\lambda$ and $x = \exp(-\alpha d)$. From equation (1), the transmission values at maxima and minima of the interference fringes can be obtained by setting the interference condition $\cos\phi = 1$ for maxima (T_M) and $\cos\phi = -1$ for minima (T_m). Using T_M and T_m, approximate value of the optical constants, film thickness and absorption coeffcient can be obtained [11].

3. Results and discussion

3.1. Transmission spectra and refractive index of the thin films

Transmission spectra of the system $(GeTe_4)_{1-x}In_x$ with x = 0, 5, 10, 15, and 20 at % in the wavelength range from 800 nm to 2600 nm are shown in figure 1. It is obvious that the system is transparent in the near infrared and infrared ranges and has transmission spectra edge from 900 nm to 1200 nm. This infrared transmittance behavior can be explained by the high percentage of tellurium in the system. It is known that most glasses containing tellurium are opaque in the visible region and have a wide transparency range in the infrared region [12]. The wide infrared region is the reason why the transmission of the system $(GeTe_4)_{1-x}In_x$ in the wavelength range from 800 nm to 2600 nm is only up to 85 %.

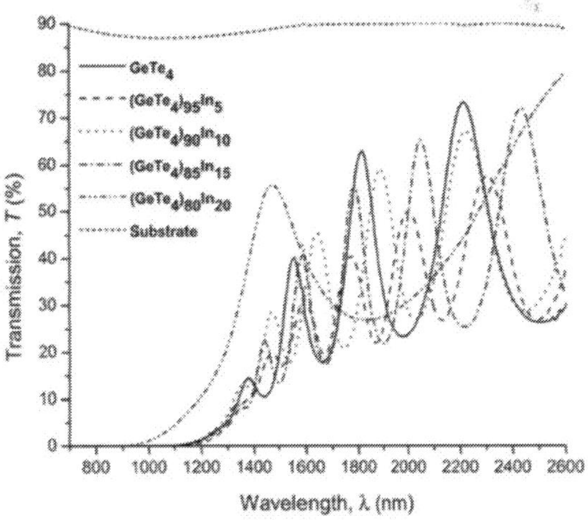

Figure 1. Transmission spectra of chalcogenides $(GeTe_4)_{1-x}In_x$ with x = 0, 5, 10, 15, 20 at % in wavelength range from 800 nm to 2600 nm.

Using the transmission spectra, the optical constants and the thickness of the films were calculated numerically using the PUMA (Pointwise Unconstrained Minimization Approach) code [13]. The

Journal of Physics: Conference Series **356** (2012) 012014 doi:10.1088/1742-6596/356/1/012014

numerical calculation of the refractive index shows that for the chalcogenide system $(GeTe_4)_{1-x}In_x$ with $x = 0, 5, 10, 15$, and 20 at %, the refractive index in the wavelength range from 1000 nm to 2600 nm is following an exponential relation. At 2600 nm, our thin films have high refractive index ($2{,}41 < n < 3{,}36$). This makes Ge-Te-In films suitable for infrared optics applications, such as waveguides in the infrared region. The thin Ge-Te-In film has a high refractive index because it contains germanium which has a very high refractive index in the infrared region and is one of the most common infrared materials.

3.2. Absorption coefficient and optical band gap energy

The absorption coefficient of thin films can be obtained using the extinction coefficient, *k*, and the expression $\alpha = 4\lambda k/\lambda$.

The absorption spectra in figure 2 show that absorption is lower at higher wavelengths and almost zero in the infrared region. This perfectly makes sense due to the film transparency in the infrared region. Using the absorption coefficient obtained, the optical band gap can be calculated by means of Tauc's equation [14]:

$$\alpha h\nu = B\left(h\nu - E_g^{opt}\right)^n,$$

where *B* is a parameter that depends on the transition probability, E_g^{opt} is the optical band gap energy and *n* is equal to 1/2 for allowed direct optical transitions and 2, for indirect transitions. The results of the optical band gap calculation are summarized in table 1.

Generally, for the system $(GeTe_4)_{1-x}In_x$, the optical band gap value decreases upon adding indium from 0% to 20%. This decrease suggests a structural change in the films. Another reason for the decrease may be the generation of an impurity band because of indium doping in the band tails region. The impurity band merges with the conduction or valence band and causes the formation of a new band tail. This new formation of a band tail narrows down the band gap.

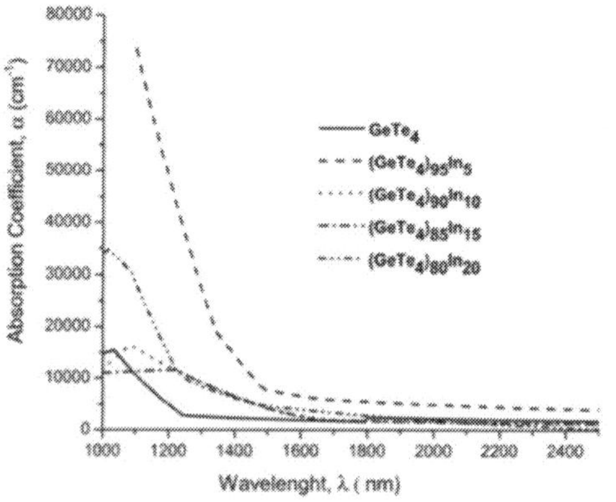

Figure 2. Absorption coefficient of chalcogenide $(GeTe_4)_{1-x}In_x$ with $x = 0, 5, 10, 15, 20$ at %.

Table 1. Optical band gap, E_g^{opt} and refractive index of chalcogenide $(GeTe_4)_{1-x}In_x$ with $x = 0, 5, 10, 15, 20$ at %.

Thin film	E_g^{opt}(eV)
$GeTe_4$	0.84
$(GeTe_4)_{95}In_5$	0.73
$(GeTe_4)_{90}In_{10}$	0.55
$(GeTe_4)_{85}In_{15}$	0.61
$(GeTe_4)_{80}In_{20}$	0.41`

Conclusions

The study of the optical properties of chalcogenide systems $(GeTe_4)_{1-x}In_x$ with $x = 0, 5, 10, 15, 20$ at % shows that the thin films are opaque in the visible region and highly transparent in the near infrared and infrared region. The refractive index of the films is high ($2{,}41 < n < 3{,}36$) and follows an exponential relation in the wavelength range from 1000 nm to 2600 nm. The films' optical band gap depends on the composition; its value decreases from 0.84 eV to 0.41 eV upon the addition of indium from 0% to 20%.

References

[1] Boniardi M et all 2011 *Solid-State Electr.* **58** 11-6
[2] Sousa V 2011 *Solid-State Electr.* **88** 807-13

[3] Kozyukhin S A, Popov A I and Voronkov E N 2010 *Thin Solid Films* **518** 5656-8
[4] Abrutis A, Plausinaitiene V, Skapas M, Wiemer C, Salicio O, Longo M, Pirovano A, Siegel J, Gawelda W, Rushworth S and Giesen C 2008 *Microelectr. Eng.* **85** 2338-41
[5] Kolev K, Popov C, Petkova T, Petkov P, Mihailescu I N and Reithmaier J P 2009 *Sensors and Actuators B: Chemical* **143** 395-9
[6] Vassilev V S and Boycheva S V 2005 *Talanta* **67** 20-7
[7] Rode A V, Zakery A, Samoc M, Charters R B, Gamaly E G and Luther-Davies B 2002 *Appl. Surf. Sci.* **197**/8 481-5
[8] Hu J, Tarasov V, Carlie N, Petit L, Agarwal A, Richardson K and Kimerling L 2008 *Optical Mater.* **30** 1560-6
[9] Florea C, Sanghera J S and Aggarwal I D 2008 *Optical Mater.* **30** 1603-6
[10] Meinders E R, Mijiritskii A V, van Pieterson L and Wuttig M 2006 *Optical Data Storage Phase-Change Media and Recording* (Dordrecht Springer)
[11] Swanepoel R 1983 *J. Phys. E: Sci. Instrum.* **16** 1214
[12] Ray Hilton A 2010 *Chalcogenide Glasses for Infrared Optics* (New York The McGraw-Hill)
[13] Birgin E G, Chambouleyron I and Martnez J M 1999 *J. Computational Phys.* **151** 862
[14] Morigaki K 2010 *Physics of Amorphous Semiconductors* (London Imperial College Press)

Influence of burst pulses on the film topography in picosecond pulsed laser deposition of $LaAlO_3$

Erik Thelander[1] and Bernd Rauschenbach

Institute of Surface Modification, 15, Permoserstraße, 04318 Leipzig, Germany

E-mail: erik.thelander@iom-leipzig.de

Abstract. Using a commercial picosecond laser in a pulsed laser deposition setup for the deposition of $LaAlO_3$ thin films, we showed that it is possible to obtain a wide range of different surface topography just by altering the temporal output of the laser pulses. In single-pulse mode, a highly structured surface is obtained independent of the laser fluence. As the number of burst pulses is increased from 2 to 10, the surface roughness gradually decreases to almost atomically smooth as investigated with scanning electron and atomic force microscopy. Studies by X-ray photoelectron spectroscopy revealed no significant variation in the film composition indicating that the burst mode only tunes the topography without affecting other parameters. The surface roughness modification is independent of the background pressure in a wide range, which opens up the possibility of using different reactive atmospheres, which could be used to tune the properties of the material even further.

1. Introduction

The use of femtosecond (fs) laser pulses in pulsed laser deposition (PLD) has been thoroughly investigated in the last couple of years [1,2]. Since fs-pulses are of shorter duration than the electron-phonon coupling time, this is a cold ablation process, i.e. the ablated material is not melted. This was thought to be beneficial in the reduction of ejects of molten micrometer droplets, a typical contaminant in PLD processes. Instead, a new interesting feature was discovered; the deposited films showed a characteristic surface topography consisting of coalesced nanoparticles. This results in coatings with a high surface roughness, which is either desired or detrimental depending on the application in mind. The nanoparticles arise from the extreme material states that are induced when light of ultrashort duration interacts with matter at energy densities over the ablation threshold [3-5]. They therefore need to be considered as an inherent property of fs-PLD. Nevertheless, it has been shown that by adjusting the temporal output of the fs-pulses, in the so-called burst mode fs-PLD [6], it is possible to tune the surface topography and even to obtain epitaxial thin films. The reduction of nanoparticles in burst mode fs-PLD is attributed to the interaction of subsequent laser pulses with the slow moving part of the plasma plume which could further ionize or evaporate the particles in the plasma. Interestingly, the use of laser pulses with picosecond (ps) duration has not been extensively investigated even though it is known from micromachining that it is also a cold ablation process. From the few investigations available, one can distinguish two different regimes: when using a laser repetition rate < 4MHz the deposited films consists of nanoparticles and are identical to those deposited with

[1] To whom any correspondence should be addressed.

fs-pulses [7,8], but when using a repetition rate > 4MHz a new ablation mechanism is initiated, the so-called thermal ablation regime [9,10]. Here the laser pulses are delivered so fast that the heat cannot dissipate between the pulses and the material is heated to a critical temperature where it melts and, therefore, this is no more a cold ablation process. This, however, implies that there exists a possibility to alter the surface topography just by changing the repetition rate of the laser, something that is easier said than done since most lasers are optimized for a relatively narrow repetition rate range. A possible solution would be the use of burst pulses which take advantage of a very high instant repetition rate in combination with a lower pulse sequence repetition rate. In this paper, we report the use of ps-pulses in PLD of $LaAlO_3$. Specifically, we discuss the effect of the number of burst pulses on the surface roughness and film composition.

2. Experimental conditions

In the experiments, a Nd: YAG laser operating at 355 nm was used for pulsed laser ablation of $LaAlO_3$. The pulse duration was 9 ps with a pulse sequence repetition rate of 80 kHz. Pulse sequences with up to 10 burst pulses can be achieved by an internal Pockels cell. The burst pulses are delivered at 50 MHz and the burst sequences can be repeated with the repetition rate mentioned above. In the experiments, the number of burst pulses was varied between 1 and 10. The laser power was 1.80 W, yielding fluences in the range between 0.73 J/cm² and 0.073 J/cm² with losses in the beam path included. This is illustrated in figure 1 where 1, 5 and 10 burst pulses are shown schematically. An increase in the number of burst pulses within the pulse sequence group from 1 to10 reduces the energy for each individual pulse by a factor 10, and, therefore, the fluence as well. The burst pulses have a 20 ns gap between each pulse, while the pulse sequence is repeated every 12.5 µs. The thin film deposition was carried out in a UHV-chamber with a base pressure of 10^{-8} mbar. During deposition, 1 sccm of O_2 was let into the chamber resulting in a pressure of ~10^{-5} mbar. A 1×1 mm^2 silicon wafer was used as a substrate, placed directly over the laser spot with a target to substrate distance of about 4 cm. The substrate was kept at room temperature and rotated to even out the inhomogeneities in the plasma plume. A $LaAlO_3$-target (99.99%) was used and scanned in a pre-programmed way to allow for a homogeneous ablation of the target. All samples were deposited for 20 minutes. The film topography was investigated by scanning electron microscopy (SEM) at 2.5 keV acceleration voltage and scanning force microscopy (AFM) in tapping mode. The composition of the films was investigated by XPS.

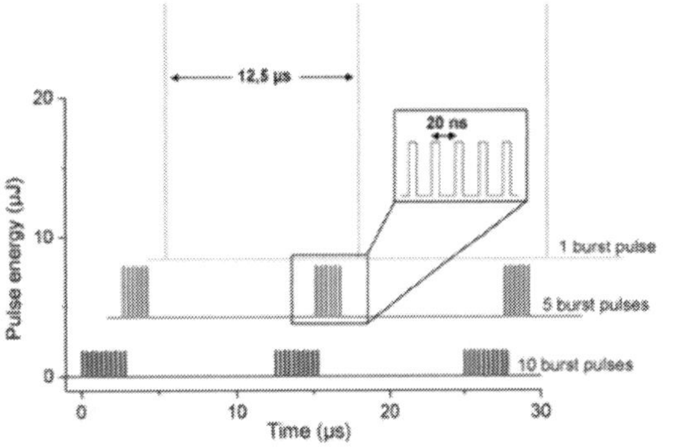

Figure 1. Illustration of the burst mode setup for 1, 5 and 10 burst pulses. Note that the pulse energy is reduced as the burst number is increased.

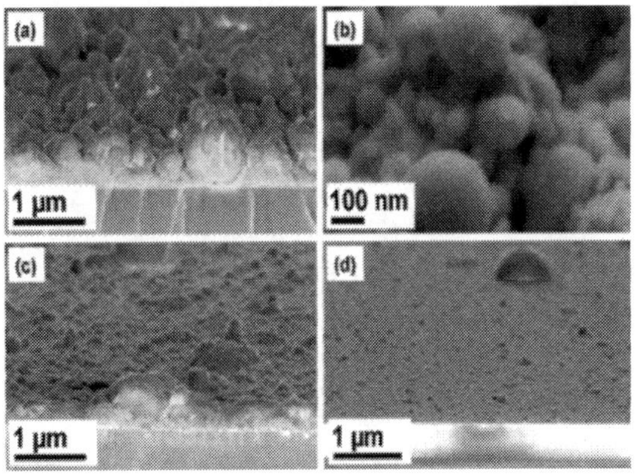

Figure 2. SEM images of the surface topography of thin films fabricated with 1 burst pulse (a), 5 burst pulses (c) and 10 burst pulsse (d). High magnification of sample in (a) reveals the coalescence of nanoparticles (b).

3. Results and discission

In figure 2, SEM images are shown of $LaAlO_3$ films prepared by ps-pulsed laser ablation with burst pulse number 1, 5 and 10. The topography prepared with only 1 burst pulse, figure 2a, shows a very rough surface with particles ranging from 0.1 µm to 1 µm. At higher magnification, in figure 2b, it is seen that the film is built through coalescence of particles. The topography of the films highly resembles what is normally reported for fs-PLD of dielectrics [11]. For single-pulse experiments at lower fluencies (not shown here), the topography looks similar all the way down to the ablation threshold for single-pulse mode, showing that a reduction of the laser fluence in the single-pulse mode is not a possible route to obtain smooth thin films of $LaAlO_3$.

Although the laser fluence reduction has no beneficial effect on the film smoothness in single-pulse mode, a clear effect can be seen when applying burst pulses, as evidenced in figures 2c and d, where samples fabricated by 5 and 10 burst pulses are presented, respectively. As can be seen, as the number of burst pulses is increasing and, hence, the laser fluence is reduced, the surfaces get smoother. The samples are not free of nanoparticles, but the films look much more homogenous than in the sample with only one pulse. It should be noted that films prepared with 2, 3 and 8 pulses agree well with this general trend but are not shown here. This is in good agreement with PLD experiments using fs-burst pulses [6], where reduction of the nanoparticles is reported as the number of burst pulses is raised, and even epitaxial growth of TiO_2 is achieved using 19 burst pulses. However, a direct comparison is hard to do since a number of important parameters differ from the experiment presented in this paper (pulse energy and number, wavelength).

To quantify the surface roughness and hence the ability to tune the topography, the samples were investigated with AFM. Figure 3 shows an AFM image of a film deposited by 10 burst pulses. Individual nanoparticles can be clearly seen with particles sizes of 100 nm or less, yielding a root-mean-square (RMS) roughness of 8.3 nm. Identical measurements were carried out for all samples; the roughness obtained is plotted against the number of burst pulses in figure 3b. As can be seen, the surface roughness decreases by more than a factor 20 as the number of burst pulses increases from 1 to 10. Since the surfaces are rough and the AFM tip has a finite geometry, the measurements are probably underestimating the surface roughness in general, particularly so for the roughest surfaces. Therefore, the trend should hold even if the absolute values are underestimated.

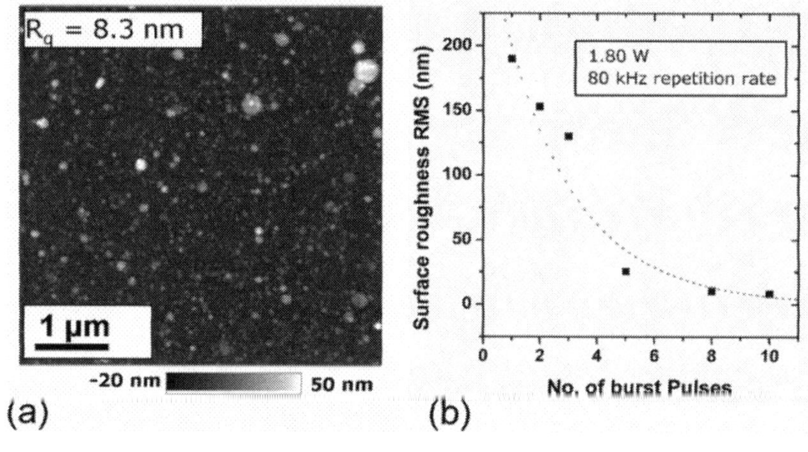

Figure 3. Typical AFM-measurement of a sample at 10 burst pulses (a). Surface roughness obtained by AFM displayed against the number of burst pulses (b). The dashed line is only intended as a guide for the eye.

The surface roughness reduction observed can be explained considering two different possibilities: heat accumulation in the target between subsequent burst pulses [9] leading to a change in the ablation mechanism, i.e. thermal ablation and plasma absorption of subsequent burst pulses [6]. The heat accumulation model is based on a step-wise temperature rise in the target when irradiated by laser pulses with high repetition rate. When the thermal conductivity is low, i.e. oxides, chalcogenide glasses, the heat absorbed cannot be dissipated and is accumulated. A critical temperature T_c is reached after a certain number of laser pulses, so that the target enters a new ablation regime which could have effect on the deposited samples [10]. The number of laser pulses needed depends on the repetition rate and material, but is in general many tens of pulses for a bad thermal conductor. Therefore, it is plausible to say that heat accumulation in the target is not the dominant mechanism since a) the maximum number of burst pulses is only 10, i.e. the thermal ablation threshold is not fully reached and b) there is a relatively long pause (12.5 μs) between each burst sequence to allow for any accumulated heat to dissipate.

However, the effect of plasma absorption by subsequent pulses needs to be considered. A calculation reveals that the nanoparticles in the plasma plume travel a distance of around 20 μm from the target surface in the time between each burst pulse assuming a propagation velocity for the nanoparticles of around 10 000 m/s. Considering the dimensions of the laser beam (20 μm Gaussian beam radius and 60° incident angle) a large part of the nanoparticles can actually be in the area where absorption processes occur. Therefore, it is likely that a part of the plasma plume body is re-evaporated/ionized by subsequent burst pulses. Optical observation of the plasma plume also indicates plasma absorption due to a broadening of the plasma plume when the number of burst pulses increases (not shown here).

XPS measurements were carried out to determine the La/Al ratio in the deposited oxide films. A $LaAlO_3$ single crystal was used as a reference sample with a calibrated La/Al-ratio of 1; all other samples was normalized against it. The La/Al-ratio ranged between 0.92 and 0.81 for all samples with no apparent systematic dependency on the number of burst pulses. It is worth noting that all samples show a small La deficit, but in a first approximation the target transfer is near congruent. For picosecond pulses, it is known that a change of the ablation mechanism from a regime where the single pulse interaction dominates to a regime where thermal evaporation is dominant can alter the composition of the films produced [10]. Considering this, it is clear that the regime where thermal evaporation dominates was not reached even for the samples where a high number of burst pulses were used, since there is no compositional trend that supports this.

Conclusions

Pulsed laser deposition of $LaAlO_3$ using a picosecond laser oscillating at 355 nm was studied. Based on the experimental findings, the following conclusions can be drawn:

- Deposition with single pulses results in thin films covered with nanoparticles of size between 100 and 1000 nm and a surface roughness of more than 160 nm. The surface morphology is independent of the laser pulse energy and repletion rate.
- Deposition with bundled laser pulses in the so-called burst mode setup results in a reduction of the surface roughness. When the number of burst pulses is increased from 1 to 10, the surface roughness is reduced by more than a factor 20, demonstrating that the method could be used to modify the surface roughness of thin oxide films.

Acknowledgement

This work has been supported by Leipzig School of Natural Sciences BuildMoNa (Grant No. GS 185/1) and the European Social Fund (ESF) within Nachwuchsforschergruppe "Multiscale functional structures". Dr. F. Frost is gratefully acknowledged for the AFM measurements.

References

[1] Amoruso S, Ausanio G, Bruzzese R, Vitiello M and Wang X 2005 *Phys. Rev. B: Condens. Matter Mater. Phys.* **71** 033406

[2] Iannotti V, Amoruso S, Ausanio G, Barone a C, Campana C, Wang X and Lanotte L 2009

Appl. Surf. Sci. **255** 5224-7
[3] Miotello A and Kelly R 1999 *Appl. Phys. A: Mater. Sci. Process.* **69** S67
[4] Sokolowski-Tinten K, Bialkowski J, Cavalleri A, Von Der Linde D, Oparin A, Meyer-Ter-Vehn J and Anisimov S I 1998 *Phys. Rev. Lett.* **81** 224-7
[5] Zhigilei L V 2003 *Appl. Phys. A: Mater. Sci. Process.* **76** 339-50
[6] Murakami M, Liu B, Hu Z, Liu Z, Uehara Yand Che Y 2009 *Appl. Phys. Expr.* **2** 04251
[7] Teghil R, D'alessio L, Santagata A, Zaccagnino M, Ferro D and Sordelet D J 2003 *Appl. Surf. Sci.* **210** 307-17
[8] Wu M H, Mu R, Ueda A, Henderson D O and Vlahovic B 2005 *Mater. Sci. Eng. B* **116** 273-7
[9] Luther-Davies B, Rode a V, Madsen N R and Gamaly E G 2005 *Opt. Eng.* **44** 051102
[10] Salminen T, Hahtala M, Seppälä I, Laukkanen P and Niemi T 2010 *Appl. Phys. A: Mater. Sci. Process.* **101** 735-8
[11] Teghil R, D´Alessio L, De Bonis A, Galasso A, Ibris N, Salvi a M, Santagata A and Villani P 2009 *J. Phys. Chem. A* **113** 14969-74

Optical characterization of WO_3-VO_x thin films for application in electrochromic devices - "smart windows"

G Bodurov[1], T Ivanova[1], M Aleksandrova[2] and K A Gesheva[1, 3]

[1]Central Laboratory of Solar Energy and New Energy Sources,
Bulgarian Academy of Sciences, 72 Tsarigradsko Chaussee, 1784 Sofia, Bulgaria
[2]Technical University of Sofia, Faculty of Microelectronics,
8 Kliment Ohridski Blvd., 1000 Sofia, Bulgaria

E-mail: kagesh@phys.bas.bg

Abstract. WO_3-VO_x thin films were deposited by atmospheric pressure chemical vapor deposition (APCVD). A typical electrochromic device is a sandwichlike structure with two conductive glasses and an electrolyte layer. An electrochromic transition metal oxide film is deposited over one of the glass substrates thus forming the working electrode. The counter electrodes in the devices are bare conductive glasses, and the polymeric electrolyte is used to laminate the two glass substrates and to supply the devices with Li ions. The working electrode material is investigated related to its initial transmittance, vibrational properties, and structure.

1. Introduction

An electrochromic device (ECD) is an optical system consisting of two conductive glasses over one of which the functional layer, the electrochromic (EC) material, is deposited. The second conductive glass could be either bare, or an "ion-storage" film may be deposited to form the counter electrode. The two glasses are laminated by a polymeric electrolyte, in which in advance small alkali ions are introduced. The device elements are three: a working electrode (conductive glass covered by EC film); a counter electrode, and a polymeric layer laminating the two device electrodes. If a small voltage pulse is applied to the device, electrons from the conductive glass are injected into the film, ions from the polymeric electrolyte intercalate as well in the film structure, and, as a result, the newly formed film structure becomes absorptive. This film absorbs part of the solar light so that what we see is the complementary color. Following the first publication for the effect discovered in WO_3 by Satyen Deb [1], a large number of studies have been published in the area of EC materials and devices. The intensive work worldwide has resulted in many books on the topic [2-5]. Among transition metal oxides, tungsten trioxide appears to be the best electrochromic compound [2]. Its advantages are the high coloration efficiency, good stability, and relatively low cost. However, mixing tungsten oxide with other transition metal oxide can lead to improvement of their properties especially, in optical and electrochromic aspects. Moreover, our previous experience with the W-Mo oxide films obtained by atmospheric pressure chemical vapor deposition (a carbonyl process) resulted in a higher growth-rate as compared to the single APCVD WO_3 films [5]. This aspect is important in view of a production process.

[3] To whom any correspondence should be addressed.

In this work, we present the preparation and investigation of mixed APCVD-WO_3-VOx films, deposited on glass and conductive glass substrates. The vibrational properties are studied by Raman and FTIR spectroscopy. Only FTIR results are presented here, together with UV-VIS transmittance spectra.

2. Experimental details

WO_3-VO_x thin films were obtained by atmospheric pressure CVD in oxygen ambient. The corresponding carbonyl mixture is $W(CO)_6$:$V(CO)_6$ = 10:1. The substrate temperature was kept at 200°C. A sublimator temperature of 60 °C for WO_3-VO_x films was used. The ratio of the flow-rates of the carrying carbonyl vapors Ar flow and the flow-rate of the reactive oxygen gas was $Ar_{carbonyl}/O_2$ 1/32. The deposition time was 40 min. The thickness of th MoO_3-VO_x thin films is 110 nm.

FTIR measurements were performed in the spectral range of 350 - 1200 cm^{-1} by a Shimadzu FTIR Spectrophotometer (IR Prestige – 21). The samples studied were deposited on Si substrates and a bare Si wafer (p-Si, with orientation <100>) was used as a reference sample. The UV-VIS measurements were carried out by a UV 3600 Shimadzu spectrophotometer in the spectral range of 300 - 1000 nm.

3. Results and discussions

The WO_3-VO_x films were investigated by FTIR spectroscopy in as-deposited state and after annealing at 300, 400 and 500°C in air for 1 hour (figure 1). Single WO_3 films are also given as a comparison in figure 2. The WO_3 films were obtained at similar CVD process conditions. Table 1 presents the results.

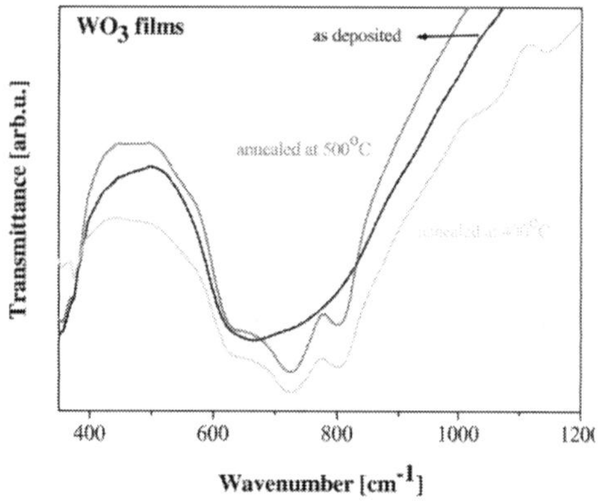

Figure 1. FTIR spectra of APCVD WO_3, films as deposited and annealed at 400 and 500°C.

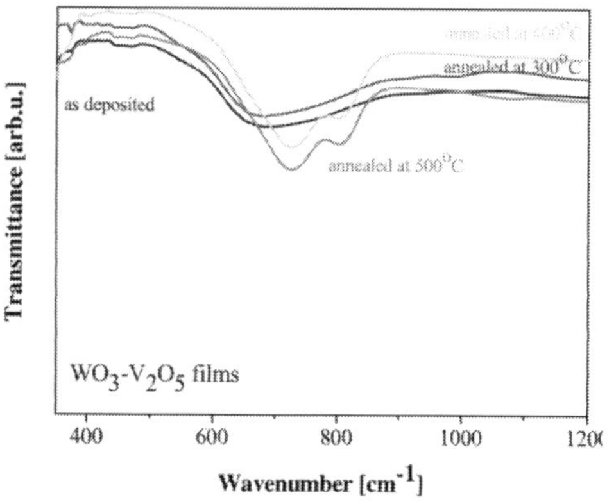

Figure 2. FTIR spectra of APCVD WO_3-V_2O_5 films, as deposited and annealed at 300, 400 and 500°C.

The FTIR study shows that the spectra of mixed oxide films are very similar in shapes to the FTIR spectra of pure tungsten oxide films, but have lower intensity. On the other hand, several weak absorption bands appeared which are clearly assigned to the vanadium oxide component.

The FTIR spectra of pure tungsten oxide films (figure 1) reveal broad absorption bands. In as-deposited state, there is band spreading in the spectral range 600-731cm^{-1}, where contributions from several deformation and stretching modes of WO_3 can be present [6]. After annealing at 400 and 500°C, the broad band splits into two strong lines at 724 and 802 cm^{-1}; this splitting can be observed in the spectra of the annealed mixed oxide samples. These two IR maxima are the most characteristic for the presence of monoclinic crystalline WO_3 [7]. As can be seen, the weak bands due to the stretching and bending vibrations of V-O bonds appear in the spectral range 400 – 470 cm^{-1} [8] in all spectra of WO_3-V_2O_5 films. In the range above 750 cm^{-1}, IR bands assigned to asymetric stretching mode of V-O-V vibrations can be found [9]. The FTIR study clearly indicates that in spite of the small amount of vanadium carbonyl in the precursor mixture, vanadium oxide is present into the films structure and influences their properties.

Table 1. FTIR data for APCVD WO_3-V_2O_5 and pure WO_3 films, deposited at similar technological parameters and additionally annealed at different temperatures.

Annealing temperature	WO_3 films	WO_3-V_2O_5 films	Assigned to
As deposited		370.0 very weak	W-O bonds
		439.6 and 468.8	bending V-O-V vibrations
	600.6-731.5 broad center 658.8	623-800 broad center 682.1	- 717, 803 cm^{-1} crystalline monoclinic WO_3 - out of plane W-O-W deformation mode~ 700 cm^{-1} - ~ 796 cm^{-1} stretching V-O-V bonds
	809.2 sh		stretching vibrations of W-O bonds in the crystalline monoclinic WO_3
300°C		371.0 clear	W-O bonds
		401.8 and 421.7	stretching V-O bonds
		458.3	bending V-O-V vibrations
		623-800 broad center 682.1	- 717, 803 cm^{-1} crystalline monoclinic WO_3 - out of plane W-O-W deformation mode ~ 700 cm^{-1} - ~ 796 cm^{-1} stretching V-O-V bonds
400°C	372.7 weak		W-O bonds
		400.3 and 420	stretching V-O bonds
		458.3 weak	bending V-O-V vibrations
	630.8 clear		stretching W-O in disordered W-O framework or W---O---W bridging stretching band
	724.0 strong	725.5 strong	W-O stretching vibrations in the crystalline monoclinic WO_3
	801.9 strong	804.9 strong	W-O stretching vibrations in the crystalline monoclinic WO_3
	1049.3 sh		Si-O bonds due to Si substrate
500°C	372.7 v. weak	371.2 weak	W-O bonds
		390.0 weak	stretching V-O bonds
		441.5 and 470.5	bending V-O-V vibrations
	627.8 clear		stretching W-O bonds
	724.0 strong	728.6 strong	W-O stretching mode, monoclinic WO_3
	801.9 strong	804.9 strong	W-O stretching mode, monoclinic WO_3
		1070.6 clear	Si-O bonds due to Si substrate

The optical behavior of the APCVD WO_3-V_2O_5 films as a function of the annealing temperature is shown in figure 3 and figure 4. Figure 3 presents the optical transmittance of films deposited on ordinary glass, while in figure 4, the substrate is a conductive glass (glass covered by SnO_2:Sb, a very thin conducting layer). The initial transmittance of WO_3-V_2O_5 film on glass is 78%; it then increases with the annealing temperature up to 88% at 400°C, and drops significantly to 67 % after annealing at 500°C (at 550 nm). The decrease of the optical transmittance after high temperature treatment was observed for MoO_3-WO_3 films [5].

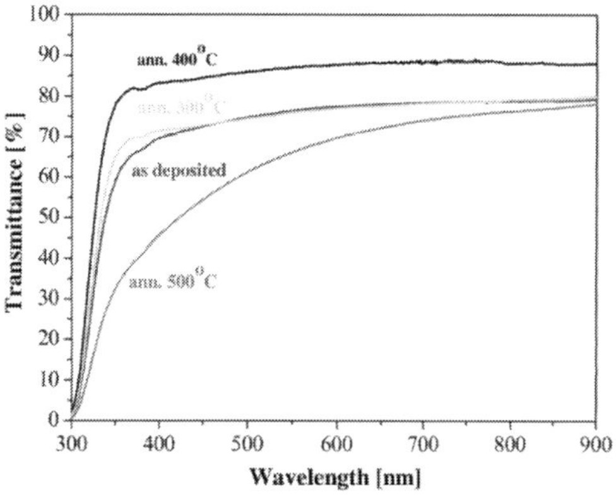

Figure 3. UV-VIS spectra of APCVD WO_3-V_2O_5 films annealed at different temperatures. The substrate used is ordinary glass.

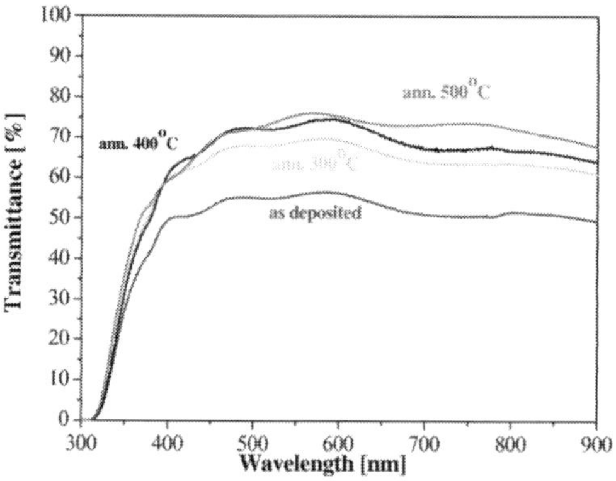

Figure 4. UV-VIS spectra of APCVD WO_3-V_2O_5 films annealed at different temperatures. The substrate used is conductive glass.

The films deposited on conductive glass exhibit lower transparency which is due to the substrate. The annealing at 400°C and 500°C resulted in spectra with transmittance of 74 and 76%, respectively, at 550 nm. The transmittance for as deposited film was 56%, and for the ones annealed 300 °C, 69 %.

Further experiments will use the W-V oxide films as working electrode in a three electrode configuration cell to perform voltametric measurements (current density vs voltage dependence). From the voltamograms obtained, the color efficiency and optical modulation will be determined.

Conclusions

The proposed APCVD technology is found to be suitable for deposition of mixed W-V oxide films. The WO_3-VO_x thin films are highly transparent. The films are expected to serve as successful working electrodes in electrochromic devices.

References

[1] Deb S K 1969 *Appl. Opt. Suppl.* **3** 192
[2] Granqvist C G 1995 *Handbook of Inorganic Electrochromic Materils* (Elsevier Science, Amsterdam)
[3] Monk P M S, Mortimer R J and Rosseinsky D R 1995 *Electrochromism: fundamentals and applications* (VCH)
[4] Monk P M S, Mortimer R J and Rosseinsky D R 2007 *Electrochromism and Electrochromic devices* (Cambridge University Press)
[5] Gesheva K A 2008 *Thin Film Optical Coatings for Effective Solar Energy Utilization: Spectrally Selective Surfaces and Energy Control Coatings* (Nova Publishers NY)
[6] Dickz-Reyes J, Delgado-Macuili R J, Dorantes – Garcia V, Balderas-Lopes J A and Arita-Ortega J A 2010 *Mater. Sci. Eng. B* **174** 183
[7] Gesheva K, Szekeres A and Ivanova T 2003 *Solar Enegy Mater. Solar Cells* **76** 563
[8] Surca A and Orel B 1999 *Electroch. Acta* **44** 3051
[9] Ban C and Whittingham M S 2008 *Solid State Ionics* **179** 1721

Optical characterization of Zn coatings deposited on low carbon steel substrates

G Bodurov[1], N Boshkov[2], L Lutov[3], T Ivanova[1] and K A Gesheva[1,4]

[1] Central Laboratory of Solar Energy and New Energy Sources,
Bulgarian Academy of Sciences, 72, Tsarigradsko Chaussee, 1784 Sofia, Bulgaria
[2] Institute of Physical Chemistry, Bulgarian Academy of Sciences,
Acad. G. Bonchev Str., Bl. 11, 1113 Sofia, Bulgaria
[3] Faculty of Chemistry, St. Kliment Ohridski University of Sofia,
1, J. Bourchier Blvd., 1164 Sofia, Bulgaria

E-mail: kagesh@phys.bas.bg

Abstract. We studied the optical reflectance profile of zinc layers with thickness of 2-10 μm electrodeposited on low carbon steel in the spectral range of solar radiation (0.3 - 2.5μm) and in the thermal infrared (2.5 - 15μm). Spectral selectivity is an important parameter of the solar absorbers used in solar thermal collectors. It is defined by low reflectance in the range 0.3 - 2.5 μ, and high reflectance in the thermal infrared, 2.5 - 15 μm. The Zn coatings are thick and their surface is rough. If the surface features size is in the order of the visible wavelengths, then absorption of light at the surface is possible. The investigations revealed the formation of ZnO on the Zn coating surface. The Raman study showed a weak peak around 512 cm^{-1}, which can be attributed to the ZnO oxide phase. The longer wavelengths penetrate the oxide surface and reach the bulk part of the Zn coating where they are reflected. As a result, a step-wise profile is expected in the reflectance spectrum within the 0.3 -15 μm range, with a change from a comparatively low reflectance in the visible to a high reflectance in the infrared.

1. Introduction

Electrodeposited zinc coatings on steel find wide practical utilization in various industrial applications [1–3]. Under atmospheric conditions, zinc corrodes by a factor of up to 100 less than steel [4]. This metal is also less noble compared to iron at ambient conditions and will predominantly corrode to protect the substrate [5]. Various techniques have been applied for improving the zinc coatings performance - alloying with other metals, conversion coatings prepared from environmentally friendly solutions [6], composite coatings, incorporating organic or non-organic micro- or nano-sized particles etc. [7-9]. Steel is a material used very often in solar absorbers. Many studies have been conducted on spectrally selective solar absorbing coatings deposited on stainless steel. The cost-effectiveness of the overall technology would be raised if a technology with ordinary steel is developed ensuring non-corrosive properties of the solar absorbers system by means of a coating (in the present study low carbon steel is used). At the same time, the absorber efficiency will increase if this coating exhibits a certain degree of spectral selectivity of the reflectance. APCVD deposited black Mo [10] has been

[4] To whom any correspondence should be addressed.

subjected to long-term testing at 500°C for 2000 hours. Other, less expensive, technologies have been developed and reported. In [11], the authors considered chemical-solution derived spectrally selective absorbers and employed an antireflection (AR) approach to develop optically selective coatings by deposition of silica, alumina and mixture of silica-titanium. In order to make the coatings more flexible, an organic compound is often added [12] to the film structure, thus forming a hybrid film structure. In the technology considered here, an organic substance was added to fill the large pores of the coating with anticorrosion purposes and, in some cases, to smooth the shiny surface.

The technological process used is usually employed for implementing anti-corrosive coatings. The process is simple, fast, environmentally friendly and suitable for large-scale applications. We introduce below some basic information for solar absorber systems. The spectrally selective optical coatings are characterized by a coefficient of absorptance of the solar radiation and by acoefficient of thermal re-radiation, called in short solar absorptance and thermal emittance. The purpose is to make the solar absorber system (a substrate with a coating) absorptive in the range where sun radiates (0,3 - 2,5 μm). Since the solar absorber systems are not transparent (the substrate is metallic), the solar radiation is either reflected by the coating, or absorbed in the optical system (a substrate covered by a coating):

$$1 = R(\lambda, T, \theta, \phi) + \alpha(\lambda, T, \theta, \phi). \qquad (1)$$

Solar radiation falls under angles θ, ϕ on the surface of a coating and is absorbed or scattered at different angles in the hemisphere above the coating surface. According to Kirchhoff's law, at thermodynamic equilibrium with the environment, the spectral coefficient of absorption and the thermal emittance at a certain wavelength are equal, $e(\lambda, T, \theta, \phi) = \alpha(\lambda, T, \theta, \phi)$; equation (1) can then be written as:

$$1 = R(\lambda, T, \theta, \phi) + e(\lambda, T, \theta, \phi). \qquad (2)$$

It is obvious that if a coating reflects highly in the infrared, it will not emit in this spectral range.

2. Experimental details

The coatings were electrodeposited at cathodic current density of 2 A/dm^2 on low-carbon steel substrates (20×10×1 mm) from electrolyte containing $ZnSO_4.7H_2O$ (150 g/l), NH_4Cl (30 g/l), H_3BO_3 (30 g/l) and two additives: AZ1 (wetting agent – 50 ml/l) and AZ2 (brightener – 10 ml/l).

Normal reflectance and full reflectance spectra were measured by a Shimadzu UV-VIS-NIR spectrophotometer. FTIR spectra were obtained by a FTIR IRPrestige-21 spectrometer at a high signal to noise ratio 40000:1, accumulation 200 scans, high-energy ceramic light source, resolution 4 cm^{-1}. The FTIR system covers the spectral range 350 – 7300 cm^{-1}.An SRM 8000 reflectance attachment was used. The Raman measurements were performed on a Raman spectrophotometer equipped with a He-Ne laser (633 nm, laser power 7 mW) and an objective ×50, in the range 100 - 4000 cm^{-1} and with accumulation time 30 s × 3 (for N1 and N2) and 50 s × 3 (for steel plates).

3. Results and discussions

The samples were optically characterized by measurements of the reflectance spectra in the range 0.3 – 2.5 μm and in the infrared range of 1.3 – 25 μm. Two sets of samples were investigated denoted as 82 (Zn:ZnO coatings with a thickness of about 10 μm), and 83 (Zn:ZnO coatings with a thickness of about 2 μm), each containing 10 samples. Figure 1 illustrates the very high (73-80%) infrared reflectance of almost all samples. Sample 9 with the highest (95%) IR reflectance (the yellow line in figure 1) exhibited the lowest R in the visible (the blue line in figure 2). Selected samples, with the highest reflectance in the infrared range and the lowest one in the visible, were annealed in view of improving the reflectance profile in the range 0,3 – 25 μm.

Journal of Physics: Conference Series **356** (2012) 012017 doi:10.1088/1742-6596/356/1/012017

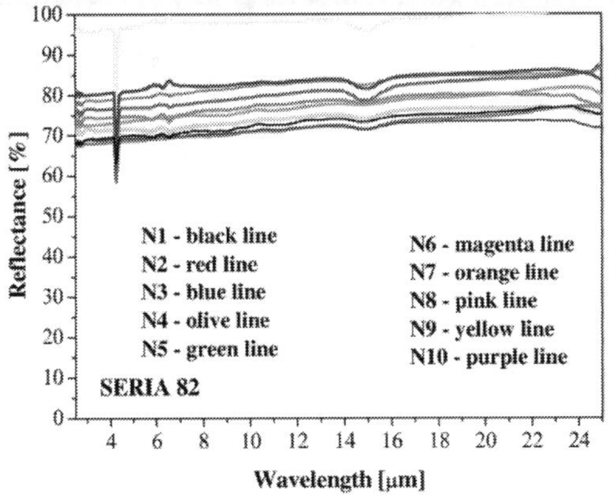

Figure 1. Infrared spectral reflectance for samples of set 82.

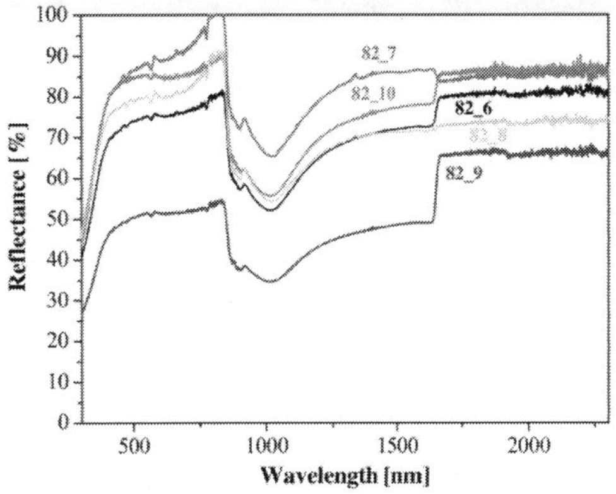

Figure 2. Visible reflectance of selected samples (showing the best IR reflectance) of set 82.

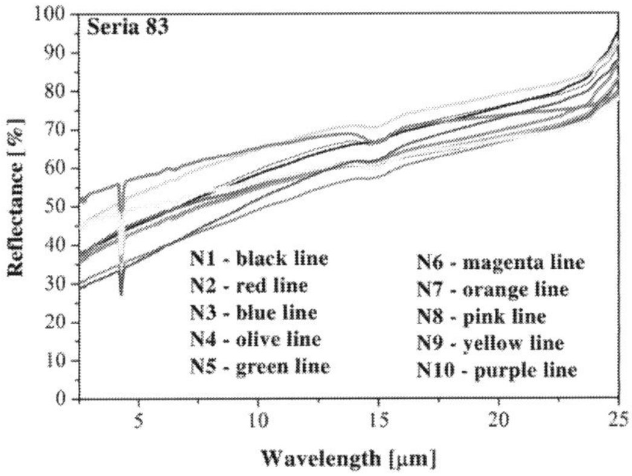

Figure 3. Infrared spectral reflectance for samples of set 83.

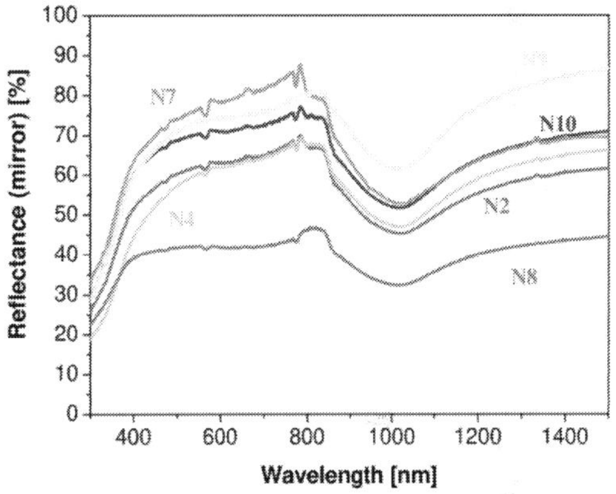

Figure 4. Reflectance spectra in the VIS for samples of set 83.

Figure 3 presents the resulting infrared reflectance spectra of the samples of set 83. As is seen, the samples show a decreasing reflectance towards the shorter wavelengths. The technological difference between sets 82 and 83 is in the deposition time and, respectively, in thickness of the zinc coatings. Samples 8 and 5 exhibited the highest IR reflectance.

Figure 4 shows that the lowest visible reflectance was observed for sample 8. This sample has the highest reflectance in the infrared (figure 3, pink line).

The process causing a lowering in the visible reflectance could be light trapping at the rough surface of the coatings. If the degree of roughness is in the order of the wavelength, then absorption is possible due to the multiple reflectance of light

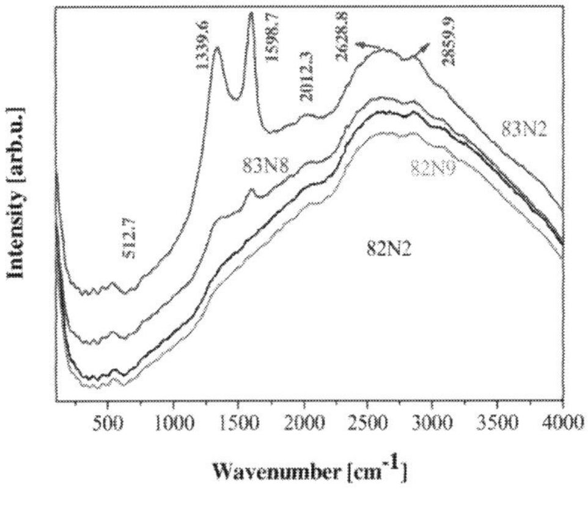

Figure 5. Raman spectra of selected samples (83N2, 83N8, and 82N2, 82N9), selected on the basis of the reflectance spectra presented in figures 2-5.

between the crystallites walls. Another contribution might arise from the graded index profile of the structure air/ZnO/Zn – the light enters the optical system meeting media with gradually increasing density (air, ZnO and finally Zn surface). It is interesting to note that better selectivity is observed for samples with higher numbers, which were prepared last in order, probably because the amount of some ingredients of the solution has decreased. We present here only the first experimental results. Further studies are necessary on the dependence of the films structure, morphology, and the chemical composition on the process parameters.

The Raman spectra of selected samples with optimal properties are given in figure 5. The Raman peaks observed centered at 1339, 1598 and 2012 cm^{-1} are related to structural water in the film. The Raman peak in all samples at 512 cm^{-1} is attributed to the oxide phase. We envisage to conduct studies on the effect of annealing at temperatures of a few hundred degrees in order to increase the oxide fraction.

Conclusions

A very simple technological process is developed for deposition of coatings on low-carbon steel. The films possess spectral selectivity of their reflectance in the range of 0.3 – 25 μm. The investigations reveal that some of the films show a decreased visible reflectance and an increased infrared one. The selected technology is promising for the development of solar absorbing coatings. Optimization of films structure and chemical composition, as related to the process parameters optimization, will result in improved spectral selectivity.

References

[1] Brenner A 1963 *Electrodeposition of Alloys. Principles and Practice* (Academic Press, New York)
[2] Zhang J, Yang Z, An M, Li W and Tu Z 1995 *Plat. Surf. Finish.* **82** 135
[3] Vourlias G, Pistofidis N, Chaliampalias D, Pavlidou E, Patsalas E, Stergioudis G, Tsipas D and Polichroniadis E K 2006 *Surf. & Coat. Techn.* **200** 6594
[4] Zhang X G 1996 *J. Electrochem. Soc.* **143** 1472
[5] Marder A R 2000 *Prog. Mater. Sci.* **45** 191
[6] Rafaey S A M, Abd El-Rehim S S, Taha F, Saleh M B and Ahmed R A 2000 *Appl. Surf. Sci.* **158** 190
[7] Hovestad A, Heesen R J C H L and Janssen L J J 1999 *J. Appl. Electrochem.* **29** 331
[8] Koleva D, Boshkov N, Raichevski G and Veleva L 2005 *Trans. Instit. Metal. Finish.* **83** 188
[9] Boshkov N, Tsvetkova N, Petrov P, Koleva D, Petrov K, Avdeev G, Tsvetanov Ch, Raichevsky G and Raicheff R 2008 *Appl. Surf. Sci.* **254** 561
[10] Gesheva K A and Chain E E 2008 *Black Molybdenum Spectrally Selective Surfaces* (Novapublishers N.Y.)
[11] Bostrom T K, Wackelgard E and Westin G 2005 *Solar Energy Mater. Solar Cells* **89** 197
[12] Tadanaga K, Iwashita K, Minami T and Tohge N 1996 *J. Sol-Gel Sci. Techn.* **6** 107

Hybrid sol-gel silica films with (TiO_2-CeO_2) binary nanopowders

M Zaharescu[1], M Nicolescu[1], M Gartner[1,3], A Barau[1], L Predoana[1], M Anastasescu[1], M Stoica[1] and A Szekeres[2]

[1] Ilie Murgulescu Institute of Physical Chemistry, Romanian Academy, 202 Splaiul Independentei, 060021 Bucharest, Romania
[2] Institute of Solid State Physics, Bulgarian Academy of Sciences, 72, Tzarigradsko Chaussee, 1784 Sofia, Bulgaria

E-mail: mfgartner@yahoo.com

Abstract. In the present work the preparation of hybrid sol-gel silica coatings doped with binary TiO_2-CeO_2 nanopowders was studied. The oxide powder was embedded in the hybrid matrix either by in-situ generation or by previously prepared powder dispersion. The main objective of the work was to establish a correlation between the method of generation of the dopant particles in the system and the properties of the films. The films were deposited on silicon wafer and glass substrates by the 'dip-coating' method and characterized in the as-prepared stage and after annealing at 120 °C. The optical and morphological properties of the films deposited on glass and silicon wafer were determined by spectroscopic ellipsometry (SE) and atomic force microscopy (AFM).

1. Introduction

It is well known that hybrid coatings with complex compositions are usually prepared using several substituted siloxane precursors. The most common precursors for the hybrid films preparation include: phenyl-trimethoxysilane (PTMOS), methyltriethisxysilane (MTEOS), 3-glycidoxypropyl-trimethoxysilane (GPTMS), trimethoxysilyl-propyl-methacrylate (TSPM), methylmethacrylate (MMA) and methacylic acid (MA) [1-10], as well as mixtures of organically substituted and un-substituted Si-alkoxides.

In this work, we used the sol-gel method approach to prepare hybrid coatings in the SiO_2-TiO_2-CeO_2 (abbreviated as SiTiCe) system.

The binary TiO_2-CeO_2 powder was selected for investigation taking into consideration its very good anticorrosive properties. The main objective was to establish a correlation between the method of embedment of the dopant particles in the system and the optical and morphological properties of the resulting films.

[3] To whom any correspondence should be addressed.

2. Experimental

2.1. Film preparation

The system chosen consists of methyltriethoxysilane (MTEOS) as a SiO_2 matrix and TiO_2-CeO_2 powder in a TiO_2:CeO_2 = 4:1 ratio used as a dopant. The oxide powder was embedded in the hybrid material either by "in-situ generation" or by dispersion of powder previously prepared ("powder dispersion"). The thermal treatment of the coatings was realized at the low temperatures of 120 °C. In the case of "in situ" particles generation, the solution for the matrix was previously prepared and mixed with the solution containing the TiO_2 and CeO_2 precursors.

The preparation of the TiO_2-CeO_2 powder was reported earlier [10]. $Ti(iOPr)_4$ (Aldrich) and $Ce(NO_3)_3 \cdot 6H_2O$ (Aldrich) as starting precursors in EtOH medium were used. The metal oxide powder preparation was achieved using an NH_4OH NH_3 25% aqueous solution. The resulting powder was filtered, washed with distilled water, dried in the oven for 48 h at 100^0C and then annealed for 1 h at 400°C to eliminate the water and organic residues. In both cases, the solutions were prepared corresponding to the 90 mol% SiO_2 – 10 mol% (TiO_2-CeO_2) molar composition. From both type of solution prepared as mentioned above, thin films on glass and silicon substrates were deposited by "dip-coating" with a withdrawal rate of 5 cm/min. For all types of samples, a thermal treatment of 30 min at 120 °C with a heating rate of 1°C/min was applied for better film consolidation.

2.2. Characterization methods

The TiO_2-CeO_2 powders were characterized by XRD and FTIR in the conditions mentioned before [10]. The optical and morphological properties of the films deposited on glass and silicon wafers were determined by spectroscopic ellipsometry (SE) and atomic force microscopy (AFM). The SE measurements were performed in a wide spectral range using Woollam equipment consisting of a VASE ellipsometer for the UV-VIS-NIR range. The AFM measurements were performed in order to examine the surface morphology by model EasyScan2 of Nanosurf® AG Switzerland.

3. Results and discussion

Under the experimental conditions mentioned above, quasi-crystalline powder with anatase structure was obtained, as illustrated by the XRD and FT-IR spectra presented in figure 1.

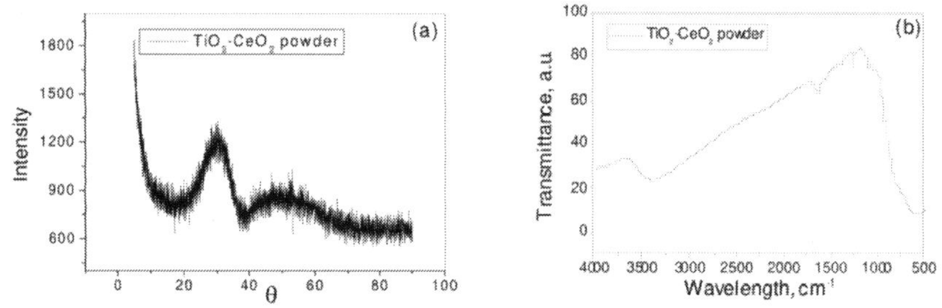

Figure 1. XRD (a) and FT-IR (b) spectra of TiO_2-CeO_2 powder.

The AFM images of coatings deposited on glass and silicon substrates before and after thermal treatment show adherent, homogeneous, continuous and smooth films, as presented in figure 2. It could be observed that SiTiCe films deposited on glass (figure 2, a-d) generally exhibit higher roughness values as compared to those on silicon (figure 2, e-h). In the case of glass substrates, higher roughness is obtained for the coatings prepared by "powder dispersion" in comparison with the films prepared by "in-situ generation". This could be due to the nanopowder agglomeration in the hydrophobic hybrid SiO_2 matrix. A more uniform morphology is observed for the films deposited on silicon by the "in situ" method (figure 2, e and f) as compared with those deposited by "powder dispersion" (figure 2, g and h). For the films deposited on silicon substrates, the roughness remains rather low, less than 8 nm.

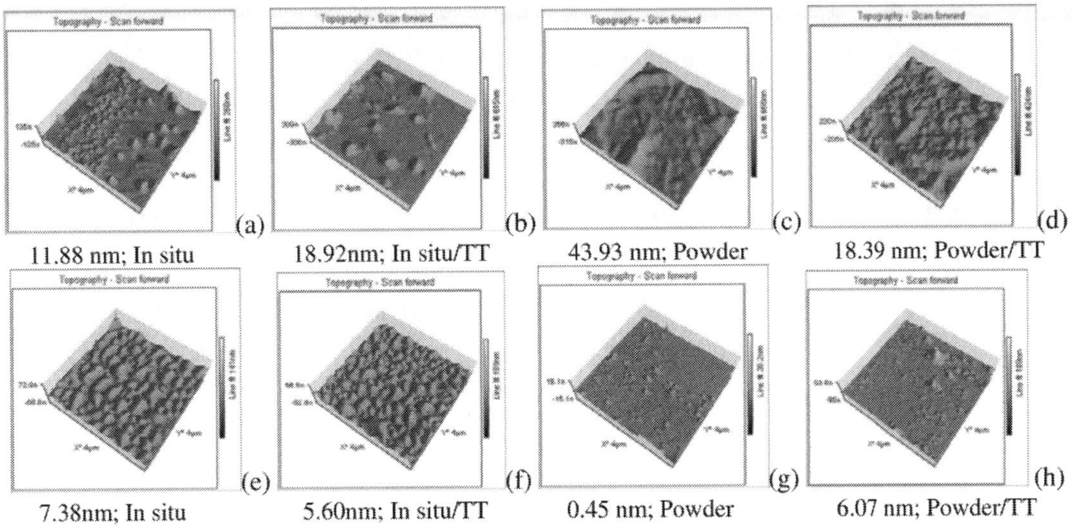

Figure 2. AFM images showing 3D topographic view of the SiTiCe films deposited on glass (a-d) and silicon (e-h); below each image the RMS roughness values and method of added powders are presented. TT means thermally treated films. Images from the left columns (a, b, e and f) were obtained using "in situ" procedure, while those in the right columns (c, d, g and h), by the "powder dispersion" method.

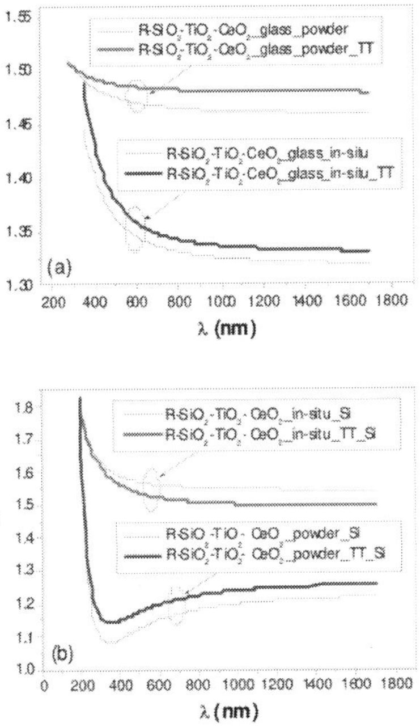

Figure 3. Refractive indices of the coatings deposited on: (a) glass and (b) silicon.

Table 1. Thickness (d_{film}) and band gap (E_g) of the SiO_2-TiO_2- CeO_2 films.

Substrate	Sample	d_{SiTiCe} (nm)	E_g (eV)
Si	SiTiCe_in-situ	50.35	3.71
	SiTiCe_in-situ_TT	48.90	3.62
	SiTiCe_powder	2.26	3.80
	SiTiCe_powder_TT	3.12	4.23
Glass	SiTiCe_in-situ	57.62	2.98
	SiTiCe_in-situ_TT	51.54	2.97
	SiTiCe_powder	87.02	3.38
	SiTiCe_powder_TT	89.24	3.53

The analysis of the experimental SE data used the following model: substrate (glass or silicon) / Si-Ti-Ce film / roughness. In the case of films deposited on silicon, an interfacial layer composed by native SiO_2 (2-3 nm thick) was added. The SiTiCe film thickness and optical band gap of the films deposited on glass and silicon substrates are presented in the table 1, while the refractive index dispersion is presented in figure 3. It can be seen that: the thickness and the refractive index of the films studied depend on the preparation method and on the type of substrate used; the films deposited by "powder dispersion" exhibit the highest thickness differences between the coatings deposited on silicon and those on glass, due to their low adherence to the Si substrate; no significant differences in the optical and microstructural properties of the films were noticed after thermal treatment, except for the coatings deposited on Si by "powder dispersion"; the optical band gap energy (E_g) increases after thermal treatment for the samples prepared by "powder dispersion".

Conclusions

Sol-gel coatings of the SiO_2-TiO_2-CeO_2 (SiTiCe) system deposited on silicon and glass substrates were successfully prepared by both "in situ" and "powder dispersion" methods. The embedment of the pre-synthesized oxides is a better way of determining the composition of the particles introduced in the hybrid matrix. The "in situ" generation allows coatings to be obtained with a more homogeneous distribution of particles the coating. When pre-synthesized particles were used, their tendency to aggregate in the hybrid hydrophobic silica matrix was observed. The spectroscopic ellipsometry results showed a strong influence of the method of dopant generation and substrate on the optical properties of the resulting films. The preparation of these films could be expanded on other substrates, such as aluminum or magnesium due to their potential applications as corrosion protective layers.

Acknowledgement

The financial support of Romania-Bulgaria Interacademic exchange program is gratefully acknowledged.

References

[1] Phani A R, Gammel F J, Hack T and Haefke H 2005 *Materials and Corrosion* **56** 1105
[2] Zandi-Zand R, Ershad-langroudi A and Rahimi A 2005 *Progress in Organic Coatings* **53** 286
[3] Zandi-Zand R, Ershad-langroudi A and Rahimi A 2005 *J. Non-Cryst. Solids* **351** 1307
[4] Shen G X, Chen Y C, Lin L, Lin C J and Scantlebury D 2005 *Electrochimica Acta* **50** 5083
[5] Nguyen V, Perrin F X and Vernet J L 2004 *Materials and Corrosion* **55** 659
[6] Gallardo J, Duran A and De Damborenea J J 2004 *Corrosion Sci.* **46** 795
[7] Sheffer M, Groysman A, Starosvetsky D, Savchenko N and Mandler D 2004 *Corrosion Sci.* **46** 2975
[8] Pepe A, Aparicio M, Cere S and Duran A 2004 *J. Non-Cryst. Solids* **348** 162
[9] Fedrizzi L, Rodriguez F J, Rossi S, Deflorian F and Di Maggio R 2001 *Electrochimica Acta* **46** 3751
[10] Zaharescu M, Wittmar A, Teodorescu V, Andronescu C, Wittmar M and Veith M 2009 *Zeitschrift fur Anorganische und Allgemeine Chemie* **1915-1924** 635

Surface topography and optical properties of Ge-Sb(As)-S-Te thin films by atomic-force microscopy and variable angle spectroscopic ellipsometry

N Dulgheru (Nedelcu)[1], M Anastasescu[1], M Nicolescu[1], M.Stoica[1], M Gartner[1,3], V Pamukchieva[2], A Szekeres[2] and K Todorova[2]

[1] Institute of Physical Chemistry, Romanian Academy,
Spl. Independentei 202, 060021 Bucharest, Romania
[2] Institute of Solid State Physics, Bulgarian Academy of Sciences,
72 Tsarigradsko Chaussee, 1784 Sofia, Bulgaria

E-mail: mfgartner@yahoo.com

Abstract. Thin films were evaporated from powdered $Ge_xSb(As)_{40-x}S_{50}Te_{10}$ (x=10, 20 and 27) glassy materials and their optical properties and surface morphology were studied by spectroscopic ellipsometry, performed in the UV-VIS-NIR range, and AFM imaging. For both kinds of quaternary systems, the optical constants (n and k) values decreased with increasing the Ge content in the films, their values being smaller for the $Ge_xAs_{40-x}S_{50}Te_{10}$ compositions. All films were transparent (k = 0) above 800 nm. The optical band gap energy increased with the Ge content; its value was larger for the $Ge_xAs_{40-x}S_{50}Te_{10}$ compositions. The AMF images revealed surfaces which are cracks-free and fully covered with uniformly distributed grains having particular structure depending on the film composition. For all the evaporated films, the average mean square (RMS) roughness did not exceed 5 nm, giving evidence for sufficiently high smoothness.

1. Introduction

The ternary Ge-Sb-S and Ge-As-S glasses with nonstoichiometric compositions have been the subject of intensive studies and their properties are well established. The introduction of a small amount of Te into these glasses alters the glassy structure and optical properties. An important advantage of tellurium-containing glasses is that due to the heavier mass of Te the transmission in the infrared region is shifted toward the longer wavelengths. Therefore, telluride glasses and thin films from quaternary systems are potential candidates for integrated optics due to their high refractive index and photosensitivity, as well as their transparency in the infrared spectral region. For optical device application, the homogeneous surface morphology and small roughness are very important in order to reduce to minimum the optical loss by light scattering and the optical transmission deterioration. While a large number of papers have dealt with thin films from ternary glasses [1-3], there have been few reports in the literature concerning the structure and properties of thin films prepared from quaternary glasses Ge-Sb(As)-S-Te [4-6].

[3] To whom any correspondence should be addressed.

In this paper we present results concerning the effect of the glass composition on the optical properties and surface morphology of $Ge_xSb(As)_{40-x}S_{50}Te_{10}$ chalcogenide thin films evaporated from the powdered parent glassy material. As tools, spectroscopic ellipsometry (SE) and atomic-force microscopy (AFM) were applied [7-10].

2. Experimental details

The bulk glasses with composition of $Ge_xSb(As)_{40-x}S_{50}Te_{10}$ (x=10, 20, 27 atomic percent) were prepared by the conventional melt-quenching method from 5N purity elements. The syntheses were performed in evacuated quartz ampoules (residual pressure of 10^{-3} Pa) using a rotary furnace. The ampoules were heated up to 950°C and were kept at this temperature for 24 h while rotating the furnace so that the melt be homogenized. After 24 hours, the ampoules were pulled out from the furnace and were quenched in air. The materials thus obtained were powdered.

The films were deposited on glass substrates by vacuum evaporation of powdered parent glassy materials. The thickness of the films was measured by a quartz microbalance during evaporation and determined more precisely from the SE data analysis.

The ellipsometric measurements were performed on a Variable Angle Spectroscopic Ellipsometer (VASE) (J.A.WoollamCo.) within the wavelength region of 250–1700 nm and at different incidence angles varying from 50° to 70°. The optical parameters, refractive index n, extinction coefficient k, absorption coefficient α, optical band gap energy E_{og}, and the thicknesses of the films were evaluated from the SE data analysis. The film thickness and the optical constants were obtained with an accuracy of ±0.2 nm and ±0.005, respectively. In the SE data simulation, a two-layer optical model was used as the actual surface roughness was included in the second top-layer with a composition of 50 % film and 50 % voids. In the spectral range of 400-1700 nm, where the films were expected to be transparent, the optical constants were determined using Cauchy's equation. Below 400 nm, general oscillator approximation was applied to obtain the optical parameters. An iterative least-squares method was used for minimizing the difference (mean square error) between the experimental data and the theoretical one.

The surface morphology of the films was studied by recording AFM images in intermittent contact mode (Easy Scan2 from Nanosurf) using sharp tips (NCLR type) with a radius of curvature of less than 10 nm (typically 7 nm).

3. Results and discussion

The dispersion curves of the refractive index n and extinction coefficient k for the compositions studied are presented in figure 1. The common feature is that the n and k values of the $Ge_xSb_{40-x}S_{50}Te_{10}$

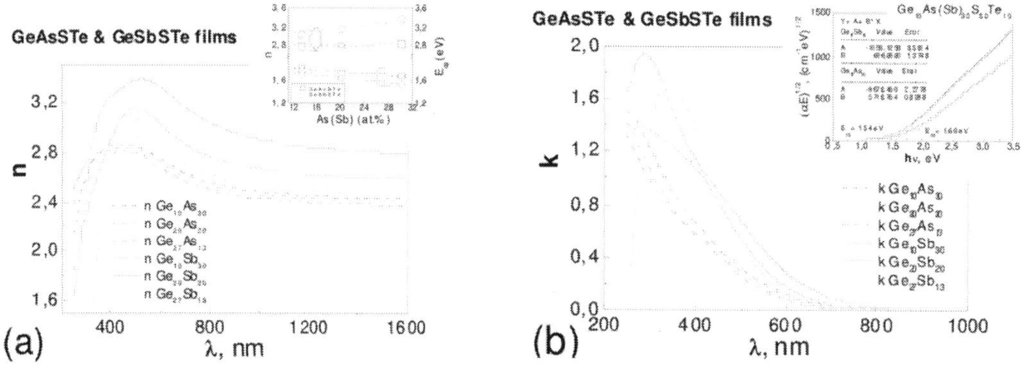

Figure 1. Dispersion curves of the refractive index n (a) and extinction coefficient k (b) of films with $Ge_xSb_{40-x}S_{50}Te_{10}$ and $Ge_xAs_{40-x}S_{50}Te_{10}$(x=10, 20, 27) compositions. Inset in (a) presents the n(λ=500 nm) and E_{og} values vs. As(Sb) at. %.

Journal of Physics: Conference Series **356** (2012) 012019 doi:10.1088/1742-6596/356/1/012019

compositions are larger than those for $Ge_xAs_{40-x}S_{50}Te_{10}$ ones. For both compositions, the *n* and *k* values tend to decrease as the Ge content increases. For all films, the *k* values approach zero above 800 nm indicating good transparency in the spectral range of 800-1700 nm.

The results from the SE data simulations, namely, the thickness of each sub-layer and the refractive index *n* taken at λ= 500 nm of the chalcogenide films studied are presented in table 1. As is seen, a surface layer of considerable thickness representing the surface roughness is obtained, it is thicker for the $Ge_xSb_{40-x}S_{50}Te_{10}$ compositions. The thickness of this top-layer is in good correlation with the peak-to-valley values obtained from the corresponding AFM image analysis, for the Sb-series of films, but exhibits some discrepancies for the As-based series due to the presence of superficial larger grains.

Table 1. Sub-layers thickness, refractive index *n* and optical band gap E_{og} of the chalcogenide films studied obtained from the ellipsometric modelling.

Glassy composition	Thickness of 1st layer [nm]	Thickness of 2nd layer (roughness) [nm]	*n* [λ = 500 nm]	E_{og} [eV]
$Ge_{10}As_{30}S_{50}Te_{10}$	1557.94	6.62	2.84	1.69
$Ge_{20}As_{20}S_{50}Te_{10}$	1887.67	10.38	2.82	1.72
$Ge_{27}As_{13}S_{50}Te_{10}$	1546.82	8.91	2.79	1.80
$Ge_{10}Sb_{30}S_{50}Te_{10}$	1433.60	14.70	3.38	1.54
$Ge_{20}Sb_{20}S_{50}Te_{10}$	1385.67	19.90	3.16	1.65
$Ge_{27}Sb_{13}S_{50}Te_{10}$	1341.78	27.33	3.08	1.72

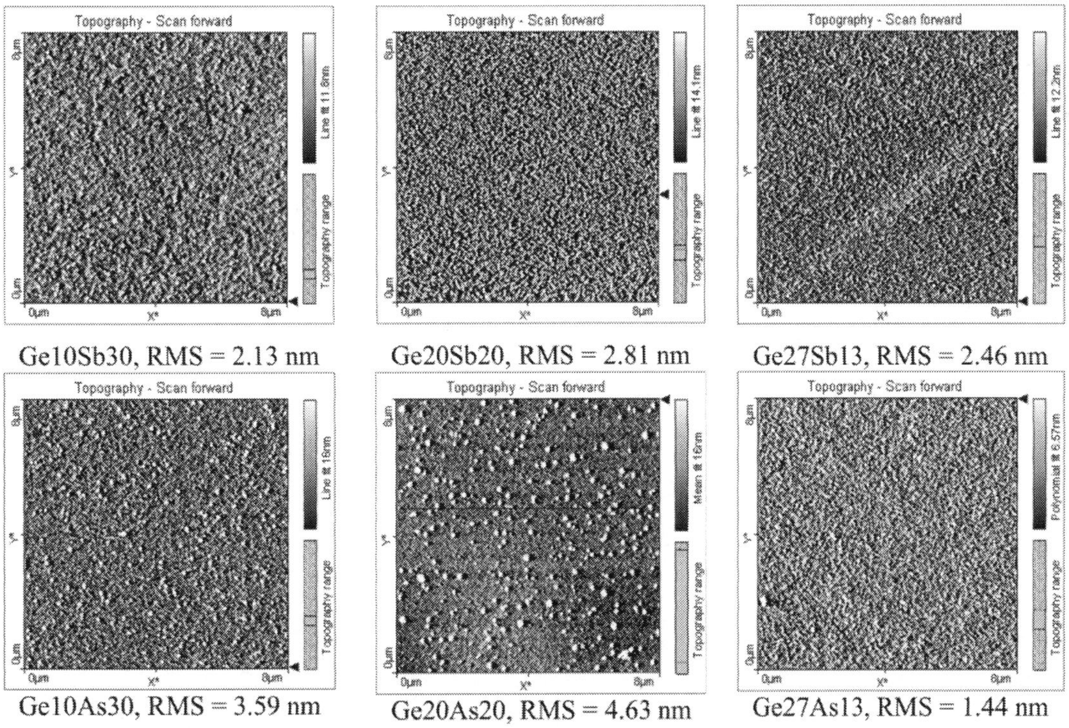

Figure 2. 2D AFM images (shaded map) at the scale of (8μm×8μm) showing the topography of the $Ge_xSb_{40-x}S_{50}Te_{10}$ and $Ge_xAs_{40-x}S_{50}Te_{10}$ (x=10, 20, 27) films; the corresponding composition and root mean square (RMS) roughness values are presented below each image.

The optical band gap energy E_{og} was derived from the absorption coefficient α values (α=4$\pi k/\lambda$) by constructing $(\alpha h\nu)^{1/2}$ versus photon energy ($h\nu$) Tauc plots for indirect transitions (inset in figure 1b). The linear part of the Tauc plot was extrapolated to α = 0; the intersection with the energy axis gave the optical bandgaps E_{og} of the films. The E_{og} values are also presented in table 1. One can see that $Ge_xSb_{40-x}S_{50}Te_{10}$ compositions possess band gap energies lower that those for $Ge_xAs_{40-x}S_{50}Te_{10}$ compositions (inset in figure 1a). For both kinds of compositions, the E_{og} values increase with increasing the Ge content in the film.

Figure 2 presents the 2-dimensional topographic AFM images taken from a 8μm×8μm scanning area for the quaternary chalcogenide films, for both series with $Ge_xSb_{40-x}S_{50}Te_{10}$ (first row) and, respectively with $Ge_xAs_{40-x}S_{50}Te_{10}$ (second row), together with their corresponding root-mean-square (RMS) roughness values. As can be seen, no major defects on the surface are visible at the investigated scale. The films with Sb-containing compositions exhibit smoother surfaces as compared to those with As-containing compositions. Larger grains are formed on the surface of films with $Ge_xAs_{40-x}S_{50}Te_{10}$ compositions, the biggest size grains being on the surface of the film with 20 at. % As content.

Conclusions

Quaternary chalcogenides films with continuous, homogeneous, smooth and cracks-free surface hwere deposited by evaporation on glass substrate. The optical constants values (n and k) were determined by analyzing the SE data. They show a tendency to decrease as the Ge content increases for both chalcogenide systems. The films are transparent in the infrared spectral range starting from ~700 nm. The transmittance values are within 80-85 %. The optical band gap energy values of the Ge-Sb-S-Te system are lower in comparison with those for the Ge-As-S-Te system.

The AMF images revealed films surface fully covered with uniformly distributed grains and having particular structure depending on the film composition. For all films evaporated, the average roughness does not exceed 5 nm, giving evidence for a sufficiently low roughness.

Acknowledgements

The financial support of Romania-Bulgaria Inter-academic exchange program is gratefully acknowledged.

References

[1] Pamukchieva V, Levi Z and Savova E 1998 *Semicond. Sci. Technol.* **13** 1309-12
[2] Gonbeau D, Guimon M-F, Skordeva E, Dedryvere R and Arsova D 2006 *Physica B* **371** 302–8
[3] Pamukchieva V and Szekeres A 2008 *Opt. Mater.* **30** 1088–92
[4] Main C and Owen A E 1998 *J. Non-Cryst. Solids* **227–230** 810–14
[5] Pamukchieva V, Szekeres A, Svab E, Fabian M, Revay Z and Szentmiklosi L 2008 *J. Phys.: Conf. Series* **113** 012054
[6] Vigreux-Bercovici C, Bonhomme E and Pradel A 2007 *J. Non-Cryst. Solids* **353** 1388–91
[7] Aljihmani L, Vassilev V and Petkov P 2003 *J. Optoelectr.& Adv. Mater.* 1187–92
[8] Vengel T and Kolomiets B 1957 *Sov. Phys. - Tech. Phys.* **2** 2314-23
[9] Ovshinsky S 1968 *Phys. Rev. Lett.* **21** 1450-3
[10] Fritzsche H 1972 *Annual Rev. Mater. Sci.* **2** 697-744

Planar homogeneity of the electrical properties of $YBa_2Cu_3O_7/La_{0.7}Sr_{0.3}MnO_3$ bi-layers

T Nurgaliev[1], B Blagoev[1, 3], E Mateev[1], V Štrbík[2], Š Beňačka[2], V Šmatko[2], Š Gaži[2] and Š Chromik[2]

[1] Emil Djakov Institute of Electronics, Bulgarian Academy of Sciences, 72, Tsarigradsko Chausse, 1784 Sofia, Bulgaria
[2] Institute of Electrical Engineering, SAS, Dubravska cesta 9, 84104 Bratislava, Slovakia

E-mail: blago_sb@yahoo.com

Abstract. The $YBa_2Cu_3O_7/La_{0.7}Sr_{0.3}MnO_3$ (YBCO/LSMO) microstructures represent model systems for investigating the interplay between two fundamental condensed-matter phenomena – superconductivity and ferromagnetism. Such microstructures make it possible to realize new types of devices for superconductor spintronics. In this work, the planar homogeneity of the electrical properties of YBCO/LSMO microstructures was investigated because of its importance for application in complex devices.

1. Introduction

Combining superconducting (SC) and ferromagnetic (FM) materials in joint heterostructures leads to a variety of effects due to various coupling mechanisms across the interface [1-4]. The ferromagnet/superconductor heterostructures are model systems for studying the interplay between two fundamental condensed-matter phenomena – superconductivity and ferromagnetism. The unique physical characteristics of such structures open possibilities for a great number of applications, such as magnetic sensors, magnetic recording and spintronic devices.

The high-temperature superconducting (HTS) material $YBa_2Cu_3O_7$ (YBCO, $T_c \sim 90$ K) and the ferromagnetic (FM) material $La_{0.7}Sr_{0.3}MnO_3$ ($T_{Curie} \sim 350$ K) are relatively well studied in view of creating HTS/FM $YBa_2Cu_3O_7/La_{0.7}Sr_{0.3}MnO_3$ (YBCO/LSMO) heterostructures. However, growing high-quality heterostructures of multicomponent materials, such as YBCO and LSMO, is not a trivial problem, as a minor difference in the technological conditions along the substrate can lead to a significant worsening of the quality and the planar homogeneity of the structure.

In this paper we report studies on the planar homogeneity of the electrical properties of YBCO/LSMO microstructures bearing in mind their importance for applications more complex devices.

2. Experimental

RF and DC magnetron sputtering techniques were used for growing bi-layers of ferromagnetic (FM) manganite $La_{0.7}Sr_{0.3}MnO_3$ and high temperature superconducting (HTS) $YBa_2Cu_3O_7$ thin films on

[3] To whom any correspondence should be addressed.

$SrTiO_3$ (100) single crystal substrates. First, the FM LSMO thin film (with thickness $t_F \sim 30$ nm) was deposited by RF magnetron sputtering on a 10×10 mm^2 STO substrate. Thee, the HTS YBCO thin film with the thickness $t_{SC} \sim 70$ nm was grown on top of the LSMO film using DC magnetron sputtering equipment. The growing procedure and growing conditions of the HTS YBCO and FM LSMO films and sandwiches were close to those used in the case of $LaAlO_3$ substrate [5, 6].

Afterwards, 28 microstrip motives of width 10 μm and length 5 μm were formed by photolithography and Ar ion beam etching on the 10×10 mm^2 $SrTiO_3$ substrate. The superconducting properties of the microstrips located in different areas of the substrate were investigated using four probe DC measurements.

3. Results and discussion

The single LSMO films deposited on the STO substrates were in FM state at room temperature. This was confirmed by the response of suspendent samples to the strongly inhomogeneous magnetic field of a permanent magnet. The temperature dependence of the resistance of an as-prepared LSMO/STO sample is shown in figure 1. It is typical for ferromagnetic manganite films of good quality. The insulator–metal transition temperature of the film is $T_{IM} \sim 320$ K, which means that the Curie temperature T_{Curie} of the sample is greater than 320 K.

The critical temperature T_c of the YBCO film in as-prepared YBCO/LSMO/STO bi-layers, measured using the four-probe method, was about 85 K (figure 2). The temperature dependence of the resistance R of these bi-layers was nearly linear at $T > T_c$ due to the electrical shunting of the LSMO film by the YBCO one (which has a significantly lower R and a linear temperature dependence of the resistance).

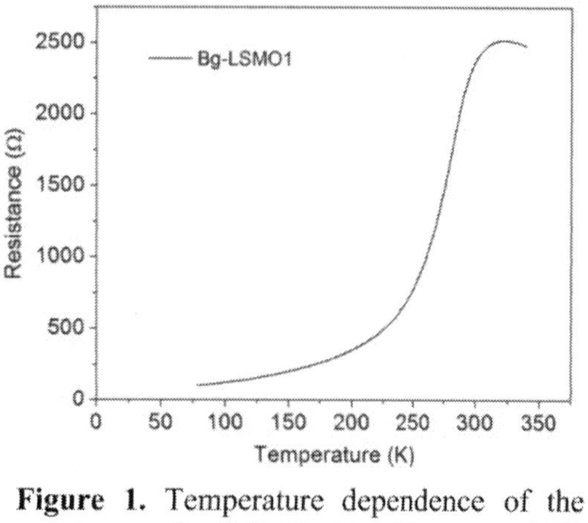

Figure 1. Temperature dependence of the resistance of a LSMO thin film on a STO substrate before patterning.

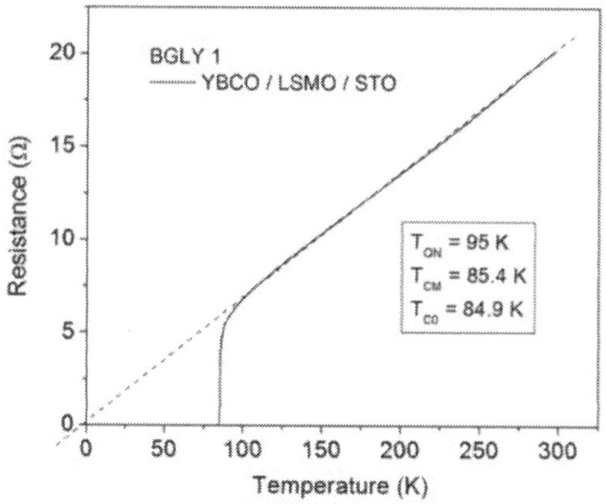

Figure 2. Temperature dependence of the resistance of a YBCO/LSMO/STO bi-layer on a STO substrate before patterning.

The YBCO/LSMO/STO bi-layers prepared were patterned using photolithography and Ar ion beam etching. The configuration (chip – figures 3 and 4) contained the current carrying microstrips, the narrow microstrip lines and the square shaped contact areas being used for measurement of the voltage drop in four probe measurements. Visual examination of the optical microscopy image of the chip showed its good quality and absence of large scale defects.

The temperature dependences of the resistance of the microstrips at different planar loctions in the substrate are shown in figures 5-7. The selection of microstrips at the corner, at the edge and in the middle of chip was made randomly. The electrical characteristics of the microstrips (figure 5) located in the central area of substrate were close to those of the as prepared YBCO/LSMO/STO bi-layer (figure 2). They demonstrated a nearly linear temperature dependence of the resistance R and a superconducting transition temperature T_c ~ 85 K. Such microstrips are suitable for further focused ion beam machining in order to use them in more complex devices.

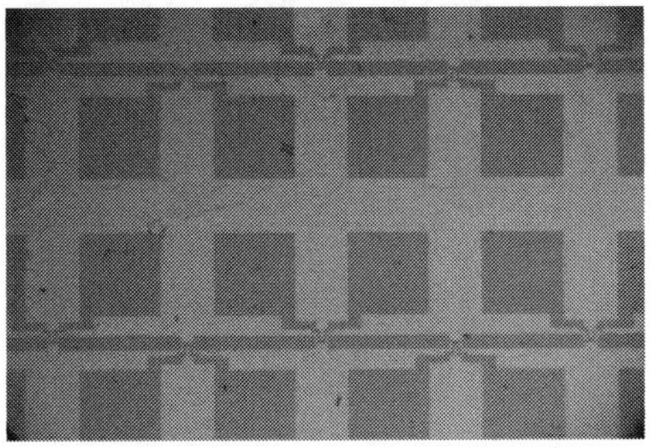

Figure 3. Fragment of the structures implemented.

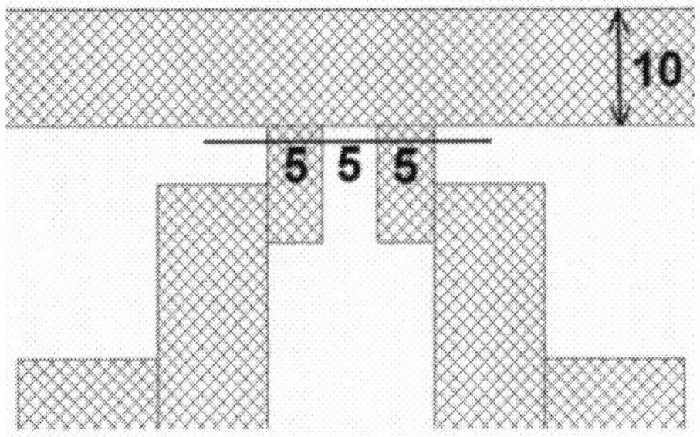

Figure 4. Details of a microstrip in the structures implemented.

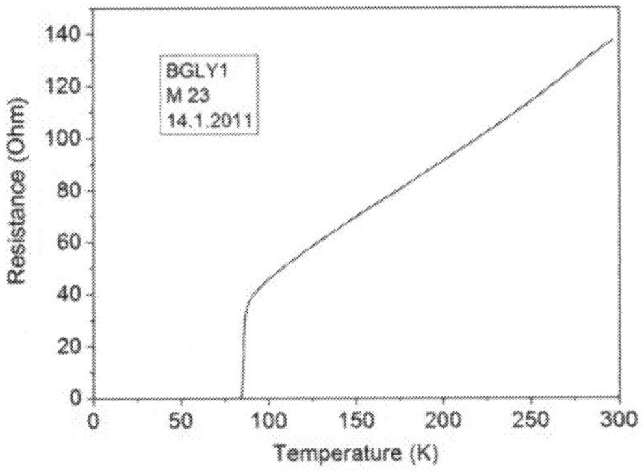

Figure 5. Temperature dependence of the resistance of a microstrip located in the central part of the substrate.

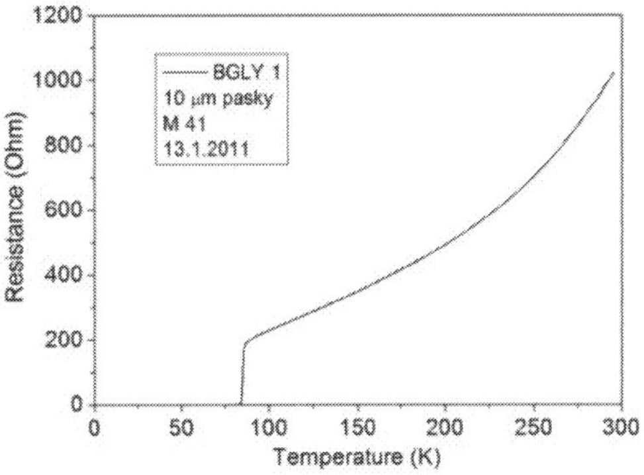

Figure 6. Temperature dependence of the resistance of a microstrip located at the edge of the substrate.

The resistance of the microstrip at the edge of the substrate was higher possibly due to the higher resistance of the YBCO component (figure 6). The electrical shunting capability of such YBCO layer is not sufficient to "transform" the temperature dependence of the resistance to a linear one, as is the case of as prepared bi-layers or of microstrips in the central area of the substrate. In spite of this, the microstrips demonstrate a sharp superconducting transition.

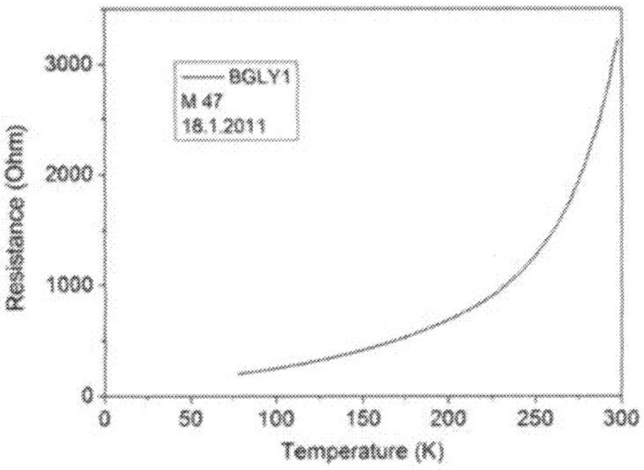

Figure 7. Temperature dependence of the resistance of a microstrip, placed at the corner of the substrate.

Microstrips at the corners of the chip were less or more degraded – some of them even fully, demonstrating a lower value of the critical temperature or absence of a superconducting transition (figure 7). Differences in the technological conditions (especially in what concerns the temperature regime) at the corner of the substrate in comparison with those in the middle of the substrate during the processes of growing and etching the bi-layer may cause such a degradation of the microstrips in the corner area.

Conclusions

In this paper, the planar homogeneity was investigated of the electrical properties of YBCO/LSMO microstructures. The analysis of the temperature dependence of the microstrip resistances in different areas of the sample revealed that the sample is practically homogeneous, although a certain inhomogeneity of the electrical parameters is only observed close to the corners of the structures. Thus, the planar homogeneity of the structures prepared is sufficient for the study of more complex superconductor/ferromagnet/superconductor microstructures.

Acknowledgments

This work was supported by the European Social Fund under the project BG051PO001.3.3.04/54/2009 and by the Slovak Grant Agency VEGA under projects No. 2/0144/10 and 2/0164/11.

References

[1] Bergeret F S, Volkov A F and Efetov K B 2004 *Phys. Rev. B* **69** 174504
[2] Przyslupski P, Tsarou A, Dluzewski P, Paszkowicz W, Minikayev R, Dybko K, Sawicki M, Dabrowski B and Kimaball C 2006 *Supercond. Sci. Technol.* **19** S38
[3] Habermeier H U, Soltan S and Albrecht J 2007 *Physica C* **460-462** 32
[4] Soltan S 2010 *Physica C* **470** 2006
[5] Nurgaliev T, Strbik V, Miteva S, Blagoev B, Mateev E, Neshkov L, Benacka S and Chromik S 2007 *Cent. Eur. J. Phys.* **5** 637
[6] Nurgaliev T, Mateev E, Blagoev B, Miteva S, Neshkov L, Strbik V, Uspenskaya L S, Benacka S, Chromik S and Nedkov I 2010 *J. Phys.: Conf. Ser.* **234** 01202

Properties of SFS heterostructures prepared by a focused-ion-beam technique

V Štrbík[1,3], Š Beňačka[1], V Šmatko[1], Š Gaži[1], Š Chromik[1], E Mateev[2], B Blagoev[2] and T Nurgaliev[2]

[1] Institute of Electrical Engineering, Slovak Academy of Sciences, Dúbravská cesta 9, 841 04 Bratislava, Slovakia
[2] Emil Djakov Institute of Electronics, Bulgarian Academy of Sciences, 72, Tsarigradsko Chaussee, 1784 Sofia, Bulgaria

E-mail: elekstrb@savba.sk

Abstract. We present a study of superconductor-ferromagnet (SF) bilayers and superconductor-ferromagnet-superconductor (SFS) heterostructures of nanometer dimensions prepared by a gallium focused-ion-beam (FIB) technique. The SFS heterostructures were implemented on the basis of high-T_c superconducting $YBa_2Cu_3O_x$ (YBCO) and ferromagnetic $La_{0.67}Sr_{0.33}MnO_3$ (LSMO) thin films deposited by magnetron sputtering on $SrTiO_3$ (100) single crystal substrates. SFS weak link junctions require weak link dimensions in the range of nanometer size realizable by FIB patterning. On the other hand, the focused gallium ion beam might cause unacceptable degradation of the superconducting and ferromagnetic thin film properties. Our results confirm such influence of the FIB technology. However, protection of the structures by a gold thin film may effectively solve this problem, as is presented in the paper.

1. Introduction

Superconductor-ferromagnet (SF) bilayer and SFS heterostructures, in a form of weak links or Josephson junctions, are very attractive objects for the study of the interplay between superconductity and ferromagnetism [1]. In addition, their potential applications are very promising in cryoelectronic or superconducting spintronic circuits [2] (e.g. qubits, 0-, pi-junctions, spin valves, etc.). However, utilization of high-T_c superconductors (HTS) has until now been only partially successful due to the very small coherence length of HTS. In the case of SFS junctions with high-T_c cuprate superconductor $YBa_2Cu_3O_x$ (YBCO), manganites appear as convenient ferromagnetic materials (e.g. $La_{0.67}Sr_{0.33}MnO_3$ (LSMO) - ferromagnetic perovskite half metal, which can be totally spin polarized) enabling one to create high-quality thin films and SF interfaces necessary for the study of their physical properties. Nevertheless, there are two essential differences in the transport properties of SF and SN (N-normal metal) heterostructures in close proximity. In SN structures, the penetration depth of Cooper pairs (CP) into the N metal is determined by the N metal coherence length $\xi_N = (\hbar D_N/2\pi k_B T)^{1/2}$, where $D_N = v_F l/3$ is the diffusion coefficient of N, v_F is the Fermi velocity of carriers, l the mean free length, T is the temperature, and $\hbar$ and k_B are the Planck and Boltzmann constants. In case of SF bilayers, the

[3] To whom any correspondence should be addressed.

Journal of Physics: Conference Series **356** (2012) 012021 doi:10.1088/1742-6596/356/1/012021

penetration depth of CP into a ferromagnet is much shorter $\xi_F = (\hbar D_F/2\pi E_{ex})^{1/2}$ since the magnetic exchange energy $E_{ex} \gg k_B T$ in a ferromagnet. Fortunately, in addition to the short coherence length ξ_F (~1-2 nm) « ξ_N, it was found [3] that in case of presence of magnetic inhomogeneity F, there may exist a triplet component (with spins of CP oriented in the same direction) with a penetration depth close to ξ_N (the so-called long-range proximity effect (LRPE) in FS structures).

Nevertheless, the realization of high-quality high-T_c SF structures or SFS Josephson weak links manifesting LRPE is a complicated task. The magnetic inhomogeneity, in the so-called series geometry, must be localized immediately at the SF interface, which is experimentally extremely difficult; otherwise the triplet current amplitude is negligibly small. It was found recently [4] that, in comparison with the serial geometry, the amplitude of the triplet current component may be considerably enhanced in the so-called *lateral geometry* (figure 1), whereby the magnetic inhomogeneity may not be necessarily localized near the SF interface. In this paper, we report on the first attempt to prepare and study the properties of YBCO/LSMO/YBCO nanometer heterostructures using gallium focused ion beam (FIB) patterning in view of establishing whether this technology is a suitable tool for implementing such superconductor weak link structures.

2. Experimental

RF and DC magnetron sputtering were used for growing the ferromagnetic $La_{0.7}Sr_{0.3}MnO_3$ (LSMO) and superconducting $YBa_2Cu_3O_x$ (YBCO) bilayer heterostructures [5]. The polycrystalline thin films were directly deposited by RF off-axis single magnetron sputtering on $SrTiO_3$ (STO) single-crystal substrates. During the sputtering in Ar:O_2 (1:1) atmosphere with total pressure of 5.3 Pa, magnetron RF power of 30 W and substrate temperature of 780°C were used. The LSMO layer thickness was about 30 nm. DC off-axis double magnetron sputtering was used for deposition of YBCO films on top of the LSMO layer. The film deposition was carried out at substrate temperature T_{dep} = 780 °C followed by annealing at temperature T_A = 530°°C. The YBCO layer tickness was 70 nm.

The patterning of the basic structure was carried out by photolithography and argon ion beam etching with cooling of the substrate. The cross section of the *lateral geometry* is shown in figure 1. Figure 2 is a SEM image of the realized YBCO/LSMO nanostrip with YBCO laterally removed within a length L < 100 nm [6]. Before patterning, the FS bilayer heterostructure was investigated by the dc four-point method, as well as by a contactless inductive method. After patterning, only the dc method was used. X-ray diffraction analysis (XRD) was used to characterize the growth of both films.

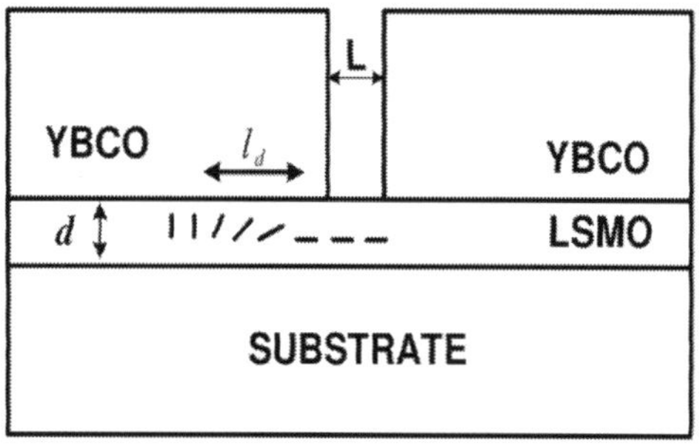

Figure 1. Cross section of the lateral geometry for long range proximity effect study. The domain wall width l_d and the YBCO separation L, and LSMO layer of thickness d are shown.

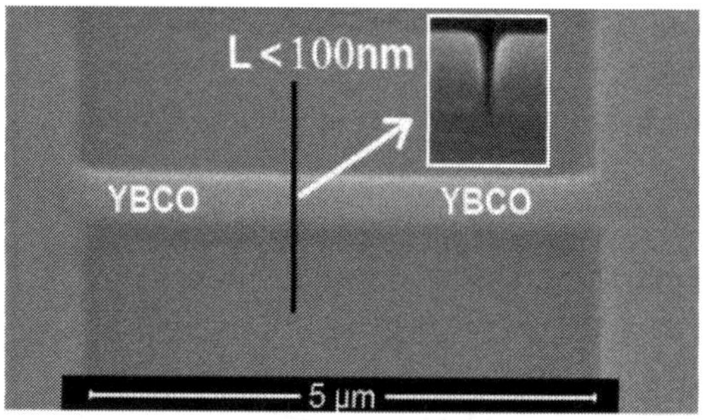

Figure 2. SEM picture of a SFS junction consisting of a YBCO/LSMO nanostrip (5×0.3 µm^2) with laterally removed YBCO within length L < 100 nm (inset).

3. Results and discussions

The *R-T* measurement of the LSMO and YBCO single layers indicate very good properties of both thin films. The resistance of a 70-nm thick single YBCO film linearly decreases between 300-100 K with the ratio $R_{300K}/R_{100K} = 3$, and the resistance, extrapolated to temperature $T = 0$, was $R_{ex}(0) \approx 0$ as a rule. The zero resistance critical temperature T_{c0} of the as-prepared single layer YBCO films reached 90 K. The LSMO single films exhibited a metal-to-insulator transition at temperature $T = 320$ K, indicating ferromagnetic properties at room temperature. The resistivity of the LSMO single layers at 77 K was $\rho_{LSMO} \approx 300$ μΩ cm, while the resistivity of the YBCO single layers just above the transition temperature was $\rho_{YBCO} \approx 40$ μΩcm (figure 3).

The YBCO/LSMO bilayers, patterned by argon ion beam etching into 5-μm wide strips, had a zero resistance critical temperature of about 84.5 K; the width of the phase transition was $\Delta T_c < 1$ K. To realize nanostrips of length 5 μm and width below 0.5 μm (figure 2, without the cross line), we applied the FIB technique (using 30 keV focused Ga ions and current in the range of 10 pA). Strong deterioration of the YBCO film properties after FIB application was observed if the bilayer was not covered by a protecting film (figure 4). Covering the bilayer by a 50-nm thick gold layer was sufficient to suppress the impact of the focused Ga ion beam (figure 5).

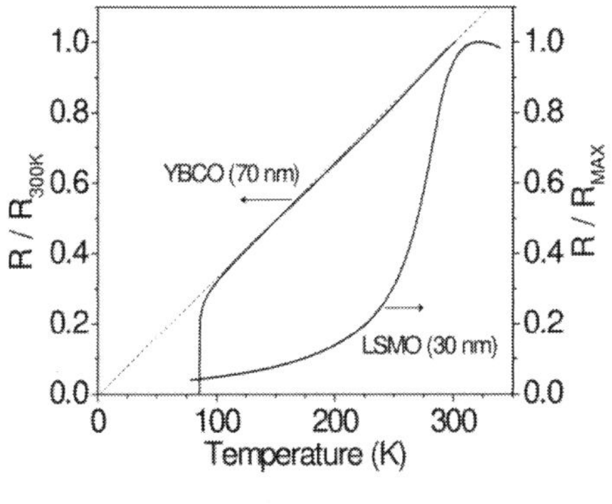

Figure 3. Normalized *R-T* dependences of YBCO and LSMO single layers, both with good electrical parameters.

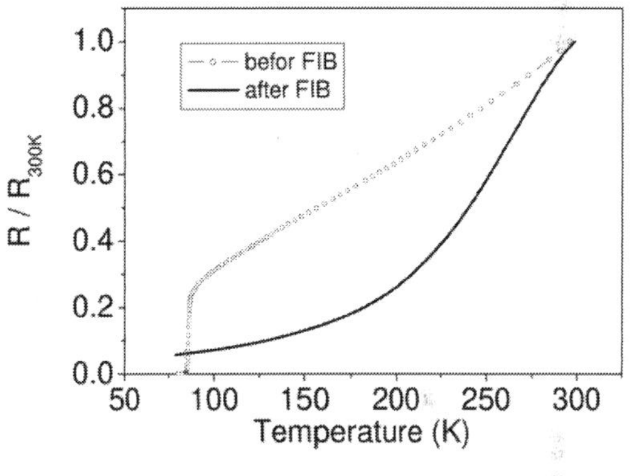

Figure 4. Normalized *R-T* dependences of the YBCO/LSMO bilayer before and after etching by FIB when the microstrip was not protected by Au.

To complete the SFS structure of lateral geometry (figure 1), cross-lines of length $L \leq 100$ nm were realized. The final aim of these attempts wass to remove completely the YBCO film in the cross-line (figure 2, inset) to observe the LRPE in the SFS structure. This step is crucial as the local removal of YBCO is extremely difficult without removing the very thin LSMO film beneath the YBCO). This is the most critical step of the SFS sample preparation, because during the sample irradiation by 30 keV Ga^{+3} ions, the properties of superconductor and manganite films can be strongly affected. We applied this procedure on the 28 microstrip (5×5 μm^2) motives on 1 cm^2 of STO substrate. After calibration of the FIB etching time (from the microstrip resistance measurement during the FIB application), it was possible to adjust the time for weak link preparation. Figure 6 shows the *R-T* dependence of the SFS junction with critical temperature of the nanostrip $T_{c0n} > 76$ K, resistance of the weak link plateau $R \approx 4$ Ω and critical temperature of the weak link $T_{cWL} \approx 57$ K. In addition, the *R-T* dependence of the structure above the microstrip critical temperature is similar to the LSMO *R-T* dependence, in agreement with the result in [7]. Bearing in mind the estimated flux-flow type *I-V* characteristics and the weak dependence of the junction critical current on the magnetic field, it is not clear whether the coupling of the YBCO through LSMO of length $L < 100$ nm is not the result of a residual YBCO film in the cross-line (the critical current density of the SFS structure is in the range of MA/cm^2) or indicates the presence of LRPE. Clarifying this will be the aim of following studies. The important output of these preliminary results is that applying FIB to a chosen SFS motive did not affect the properties of the other 27 motives located on the STO substrate when the sample was covered by a 50 nm thick gold layer.

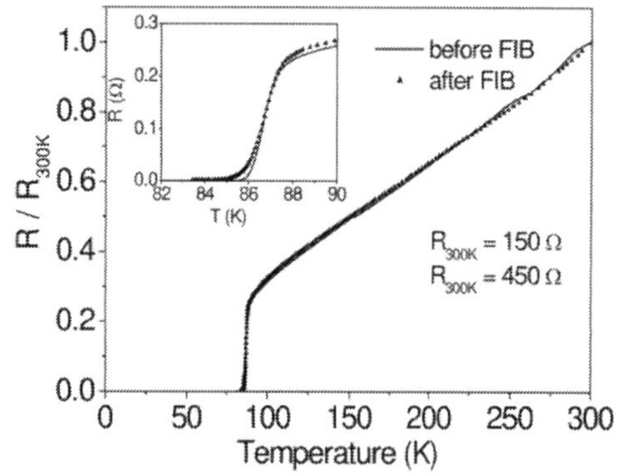

Figure 5. *R-T* dependences of microstrip before and after FIB application when the microstrip was covered by a 50-nm thick Au film.

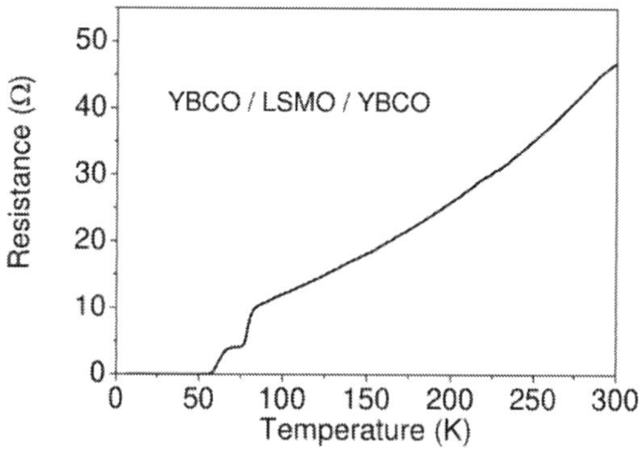

Figure 6. *R-T* dependence of the SFS junction with resistance plateau ($R \approx 4\ \Omega$) of the junction weak link part and $T_{cWL} \approx 57$ K.

Conclusions

The realization of high quality superconducting weak links or Josephson junctions based on high-T_c superconductors is at present a very difficult technical task. Progress is expected using advanced microcircuit technologies (e.g. FIB) for preparation of nanometer-sizeSNS or SFS structures. The SFS heterostructures, in addition, offer the opportunity of new physical effects, LRPE, as well as new modes of operation in cryoelectronic or superconducting spintronic circuits. We presented preliminary results on the high-T_c superconducting SFS structure in the so-called lateral geometry. We showed that FIB is a suitable procedure for implementing nanometer size devices, while the problems of films degradation due to FIB irradiation may be solved by protecting the structure by a gold layer with thickness about 50 nm.

Acknowledgement

This work was supported by the Slovak Grant Agency VEGA under projects No. 2/0144/10 and 2/0164/11 and by the European Social Fund under the project BG051PO001.3.3.04/54/2009.

References

[1] Bergeret F S, Volkov A F and Efetov K B 2005 *Rev. Mod. Phys.* **77** 1321
[2] Tao Y C and Hu J G 2010 *J. Appl. Phys.* **107** 041101
[3] Buzdin A I 2005 *Rev. Mod. Phys.* **77** 935
[4] Kupferschmidt J N and Brouwer P W 2009 *Phys. Rev. B* **80** 214537
[5] Blagoev B, Mateev E, Strbik V, Nurgaliev T and Uspenskaya L S 2010 *J. Phys.: Conf. Series* **223** 012015
[6] Štrbík V, Beňačka Š, Gaži Š, Šmatko V, Chromik Š, Dujavová A and Vávra I 2011 *J. Eletrical Eng.* **62** 109
[7] Lin J G, Hsu D, Song M Y, Chiang C H and Chang W C 2010 *JAP* **107** 09E130

Impedance measurements of epitaxial and polycrystalline LSMO thin films

E Mateev, B Blagoev[1] and T Nurgaliev

Emil Djakov Institute of Electronics, Bulgarian Academy of Sciences,
72, Tsarigradsko Chaussee, 1784 Sofia, Bulgaria

E-mail: blago_sb@yahoo.com

Abstract. The results are presented of measurements of the complex surface impedance dependence on the frequency in the 1 – 100 kHz range. $La_{2/3}Sr_{1/3}MnO_3$ (LSMO) single- and polycrystalline films were investigated at temperatures bellow the Curie temperature. The two types of layers show opposite behaviour – inductive and capacitive, correspondingly. Independently, information on the bulk resistance of the grains in the polycrystalline material and the intergrain complex resistance was obtained.

1. Introduction

The perovskite manganites have been the object of experimental and theoretical studies due to several reasons [1]: colossal-magnetoresitance applications [2, 3], bolometric purposes and applications as a cathode material in solid oxide fuel cells [4].

The typical usage of the perovskite magnetoresistive materials is based on DC measurements. However, the same measurements can be done in AC (above 1 kHz) regime. AC measurements can be more efficient because the $1/f$ noise is strongly reduced, which makes the study of the AC complex resistance interesting in view of practical applications. At the same time, such a study my yield valuable knowledge on the individual contribution of grains and grain boundaries to the low-field magnetoresistance [5, 6].

We present the results of a study of the complex impedance frequency dependence in the range 1 – 100 kHz.

2. Experimental

The $La_{2/3}Sr_{1/3}MnO_3$ (LSMO) system was chosen as the film material because its Curie temperature is well above room temperature. Two types of substrates were used: monocrystalline – $SrTiO_3$ (STO) and sapphire Al_2O_3 (ALO) substrates. RF magnetron sputtering of a solid LSMO target at 4 Pa 50% Ar and 50% O_2 gas mixture was used for film deposition on 5×10 mm substrates. The substrate temperature was 780 °C and the deposition was followed by an annealing step in O_2 at 80 kPa and 500 °C. The LSMO layer grown on the STO substrate was epitaxial and 30 nm thick. The pure sapphire substrate (without a buffer layer) does not stimulate the epitaxial growth of LSMO layers. Thus, the LSMO films (with a thickness of 40 nm) grown on the ALO substrate were polycrystalline and consisted of grains of submicron size.

[1] To whom any correspondence should be addressed.

The four probe-method was used to measure the complex surface impedance of the LSMO films. The complex value of the voltage between the test probes were obtained by a SR530 lock-in amplifier in differential mode and was divided by the complex value of the voltage drop on a referent resistor measured in single-ended mode. The results are the modulus and phase angle of the complex surface impedance, as shown in figures 1 and 2. The measurements were conducted at 50 mV input voltage: the measurements at 500 mV input voltage yielded the same result.

The variation of the modulus of the complex surface impedance on STO is negligible, while its mean value is 450 Ω/sq.

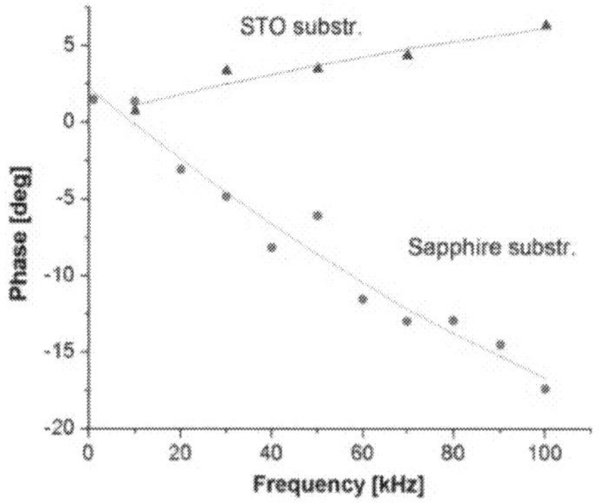

Figure 1. Frequency dependence of the complex surface impedance phase of LSMO films.

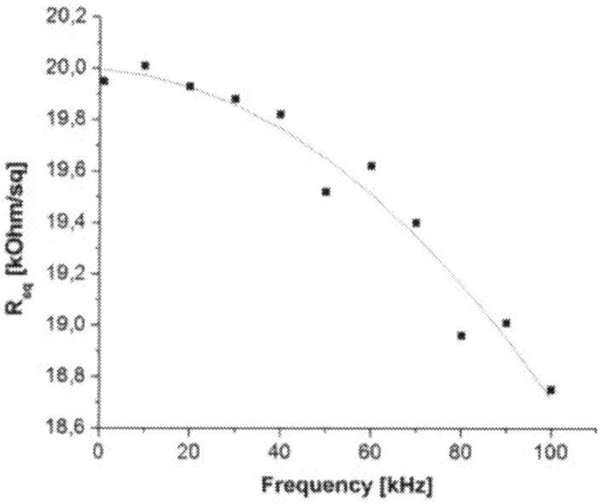

Figure 2. Complex surface impedance modulus of the LSMO film on a sapphire substrate versus frequency.

3. Results and discussion

The complex surface impedance of the LSMO film grown on a STO substrate shows slightly inductive behavior (the phase of the complex impedance is positive). The impedance of the film grown on the sapphire substrate, on the contrary, shows a well-expressed capacitive behavior.

In framework of the classical electrodynamics, the skin depth δ of a conducting plate, where the main part of an AC current is flowing, is expressed as $\delta = (2\rho^{-1}\mu_r\mu_0\omega)^{-1/2}$, where μ_0 is magnetic permeability of vacuum, μ_r is the relative magnetic permeability of the material, ρ is the resistivity of the material, ω is the angular frequency of AC current. The estimations show that the skin depth of the LSMO material ($\rho \geq 1$ mΩ cm) at low frequencies (≤ 100 kHz) exceeds by far the thickness of the films used in our experiments. This implies a uniform distribution of AC current across our films.

Rahmouni et al [4] proposed an equivalent circuit for the case shown in figure 3A. Here, R_g represents the contribution to the impedance of the grains bulk of the polycrystalline material, R_{gb} is the contribution to the impedance of the grain boundaries and C_{gb} is the contribution of the capacitance between the grains (grain boundary capacitance). Using the data shown in figures 1 and 2, one can calculate the active and reactive parts of the impedance for different frequencies and fit them with the formulae for the active and reactive parts of the impedance corresponding to the equivalent circuit shown in figure 3A with fitting parameters R_g, R_{gb} and C_{gb}:

$$Z_{compl} = R_g + R_{gb}/(1 + i\omega R_{gb} \times C_{gb}).$$

The result is shown in figure 4. The optimum values of the fitting parameters are:

$$R_g = 3.435 \text{ k}\Omega,\ R_{gb} = 16.524 \text{ k}\Omega \text{ and } C_{gb} = 35.54 \text{ pF}.$$

The mean square error of the complex impedance modulus is about 1% of the mean value of the complex impedance modulus.

It should be pointed out that the circuit shown in figure 3A is a rather rough approximation. The real situation is better represented by the circuit shown in figure 3B. These two cases cannot be distinguished by complex impedance measurements. Nevertheless, two parameters obtained from the circuit 3A are valid for the circuit 3B, too: the ratio R_g/R_{gb} = 0.2079 shows that 17.2% of the DC surface resistance is contribution of the bulk of the grains and 82.8% is contribution of the grain boundaries. Thus, only the bulk part of the surface resistance alone is greater than the total surface resistance of the single crystal layer. The time constant $R_{gb}C_{gb}$ = 587.32 ps is the mean time constant for all grain boundaries. If the grain boundaries are treated as a homogeneous medium between two highly conductive electrodes, the time constant equals $\rho\varepsilon_0\varepsilon$, where ρ is the volume resistivity of the material and ε is its dielectric permittivity.

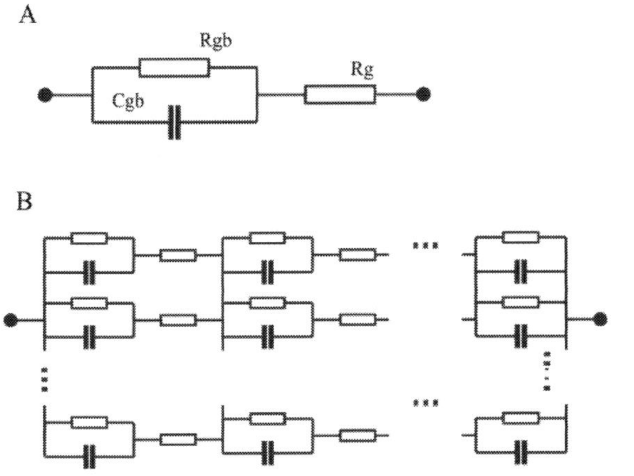

Figure 3. Simplified equivalent circuits for calculation of the polycrystalline film complex impedance.

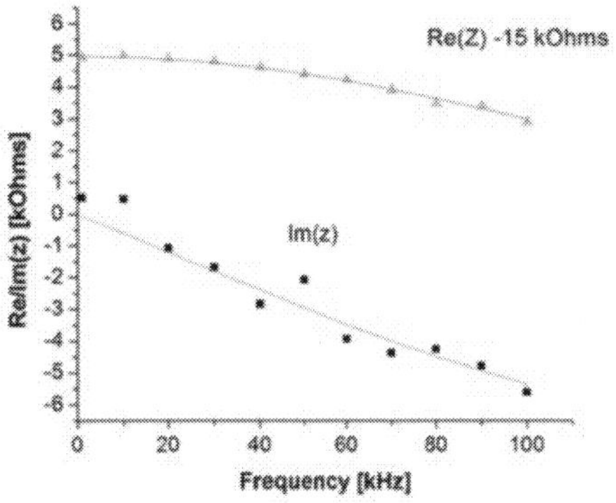

Figure 4. The active and reactive parts of the measured complex surface impedance and the corresponding fitting curves of polycrystalline LSMO film (LSMO on sapphire substrate). The active part is displaced for better observation.

Conclusions

The complex surface impedance of the single crystal layer exhibits a slightly inductive behavior (the phase of the complex impedance is positive), while the impedance of the polycrystalline layer shows a capacitive one. The time constant of the intergrain transition is τ = 587.32 ps. The contribution of the grain volume to the DC resistance of the polycrystalline layer (3.435 kΩ) is substantially higher than the DC resistance of the single crystal layer. The contribution of the grain boundaries the DC resistance of the polycrystalline layer is predominant. It can be reduced (and as a result, the total resistance) using a film with larger grains.

Acknowledgments

This work was supported by the European Social Fund under project BG051PO001.3.3.04/54/2009.

References

[1] Strbik V, Spankova M, Reifers M, Kovac J and Benacka S 2010 *J. Phys.: Conf. Series* **223** 012027
[2] Nagaev E L 1996 *Phys. Usp.* **39** 781
[3] Yang S Y, Kuang W L, Liou Y, Tse W S, Lee S F and Yao Y D 2004 *J. Magn. Magn. Mater.* **268** 326
[4] Rahmouni H, Nouiri M, Jemai R, Kallel N and Rzigua F 2007 *J. Magn. Magn. Mater.* **316** 23
[5] Shim In-Bo, Park Key-Taeck, Kim Chul-Sung and Oh Young-Jei 2001 *J. Magn. Magn. Mater.* **226-230** 733
[6] Lee H M, Shim In-Bo and Kim C S 2004 *J. Magn. Magn. Mater.* **282** 291

Vacuum deposited GeSbSe thin films for photonic applications

J Dikova[1], R Todorov and Tz Babeva

Acad. J. Malinowski Institute for Optical Materials and Technologies,
Bulgarian Academy of Sciences, Bl. 109, Acad. G. Bonchev Str., 1113 Sofia, Bulgaria

E-mail: julita@clf.bas.bg

Abstract. The potentialities are studied of $Ge_{28}Sb_{12}Se_{60}$ thin films deposited by thermal evaporation in vacuum for preparing 1D photonic crystal arranged as a multilayered stack and operating in the NIR. Thin films of PMMA (poly (methyl methacrylate)) deposited by spin-coating are chosen as a second counterpart of the stacks. The morphological and optical characterization results show that both kinds of single layers have a smooth surface and exhibit a large difference in their refractive indices, which presupposes a high optical contrast of the system. Further, a 13-layer stack built by alternating deposition of the two materials is prepared. The formation of an omnidirectional reflectance band (ODR) centered at $\lambda = 1.5$ μm with a maximum value of R = 96.3% is demonstrated. The ODR's width at half maximum is 320 nm. On the basis of the results obtained, the possible ways of optimizing further the system studied are outlined. Finally, the prospects of practical applications of the multilayered reflector developed are discussed.

1. Introduction

The 1D photonic crystals arranged as multilayered stacks of thin films of two materials with different dielectric constants have lately found ever widening applications in optical technologies. This has mainly been due to the recently achieved progress in designing the well-known Bragg's stacks, which allow one to obtain a photonic band gap (PBG) in different spectral ranges, similarly to the 3D photonic structures [1]. As is known, in the frequency ranges of the PBG, the media are characterized by an "omnidirectional reflection" (ODR), i.e., reflectance of nearly unity for every angle of incidence and light polarization state [2]. The key prerequisite for this is the presence of a sufficiently high optical contrast of the system, which is determined by the difference in the refractive indices of the two materials building the structure of the 1D photonic crystal.

In searching for suitable combinations of materials for preparing multilayered reflectors operating in the NIR spectral range, it has been demonstrated that the chalcogenide glasses and organic polymers offer unique possibilities for creating and processing jointly structures with high optical contrast, which explains the continuously growing interest toward these media [3]. In the most successive devices so far, thin films of As_2Se_3 coupled with thermo-plastic polymers have been used [4]. This glass possesses numerous valuable properties, namely, high refractive index and transmittance in a wide spectral range in the IR; however, it has some disadvantages, such as low glass transition

[1] To whom any correspondence should be addressed.

temperature T_g and possibility for photodegradation at 1550 nm, as already shown for As_2S_3 [5]. To overcome these deficiencies, the efforts of scientists have been focused on synthesizing and using of high-temperature multicomponent glasses with high refractive indices and transparency in the IR. In this respect, the GeSbSe glassy system seems very promising, especially the composition $Ge_{28}Sb_{12}Se_{60}$, which has T_g of about 285 °C, high transparency in the spectral range 1-16 µm and a refractive index of 2.61 at λ = 5 µm. This excellent optical glass has been widely studied recently and some potential applications have already been developed, for example as optochemical and biosensors operating in the near and mid-IR [6, 7], etc.

It is the aim of the present paper to study the potentialities of vacuum deposited $Ge_{28}Sb_{12}Se_{60}$ thin films coupled with organic polymer layers for preparing 1D photonic crystal with a photonic band gap in the NIR spectral region, centered at about λ = 1550 nm.

2. Experimental

The high-refractive-index component of the multilayered structures was obtained by thermal evaporation of preliminarily synthesized bulk material with composition $Ge_{28}Sb_{12}Se_{60}$. The synthesis of the glass from the pure constituent elements (99.999%) was performed in a quartz ampoule by a standard melt quenching method at 1000°C and continuous homogenization during 10 hours. The films were deposited at a rate of about 0.5 nm/s on BK-7 optical glass plates and Si wafers in a high-vacuum unit at a residual pressure of 5×10^{-3} Pa. The thickness of the samples was calculated according to the criteria for fabrication of a quarter-wave stack, known to be the best structure for producing high-reflectivity optical coatings [8]. Thus, for the wavelength of interest λ = 1550 nm, a thickness value of 140 nm was obtained.

The low-refractive-index counterpart of the multilayered system studied was prepared by spin coating of PMMA dissolved in dichlorethane (Aldrich). In preliminary experiments, the concentration of the solution and the spin rate and duration were varied in order to obtain films with thickness of 260 nm, as calculated by using the above mentioned criteria [8]. This was achieved by a solution containing 3.5 wt. % PMMA coated at 3000 rpm for 30 s. To remove the extra solvent, the as-prepared films were annealed at 60°C for 30 min.

After optical and morphological characterization of the two kinds of single layers, a multilayered stack comprising 13 layers (7 of chalcogenide and 6 fof polymer) was prepared by alternating vacuum evaporation of $Ge_{28}Sb_{12}Se_{60}$ and spin coating of PMMA with target thicknesses of 140 nm and 260 nm, respectively. During this step-by-step fabrication of the stack, transmittance and reflectance spectra at normal and oblique light incidence were taken periodically by a Cary 5E high-precision UV-VIS-NIR spectrophotometer (ΔT = 0.3%, ΔR = 0.5%). For the oblique measurements, a high-quality Glan-Taylor polarizer was used to obtaine linear polarization and a self made Al-mirror [9] was used as a reference one (error in R for oblique incidence of less than 1%).

The optical constants (refractive index, n, and extinction coefficient, k) and the thickness d of the single layers were calculated based on the spectrophotometric measurements at normal incidence in the interval 800-2400 nm using an already-developed calculating procedure [10]. The optical constants were used for characterization of both kinds of single layers, as well as for modeling and optimizing the optical behavior of the stacks.

Morphological investigations were performed by a Philips 515 scanning electron microscope.

3. Results and discussion

Figure 1 shows the spectral dependences of the refractive indices of 140 nm $Ge_{28}Sb_{12}Se_{60}$ and 260 nm PMMA films deposited on BK-7 glass plates at the experimental conditions chosen. The large difference in the refractive index values in the whole spectral range studied is clearly seen. For example, at the wavelength of interest λ =1550 nm, n is 2.72 for GeSbSe and 1.49 for PMMA, which determines a high optical contrast of 1.23. This value gives one reasons to expect that a multilayered reflector could be built by a relatively small number of alternating single layers. Simultaneously, the SEM micrograph presented in figure 2 reveals the smooth surface morphology of a chalcogenide thin film, which is an important prerequisite for the good quality of high-reflectivity optical coatings. A similar result was obtained for the polymer sample surface (the picture was not presented for simplicity). Therefore, the experimental conditions used allowed us to produce GeSbSe and PMMA films with properties appropriate for preparing a 1D photonic crystal in the form of a multilayered system operating in the NIR, in accordance with the aim of the study.

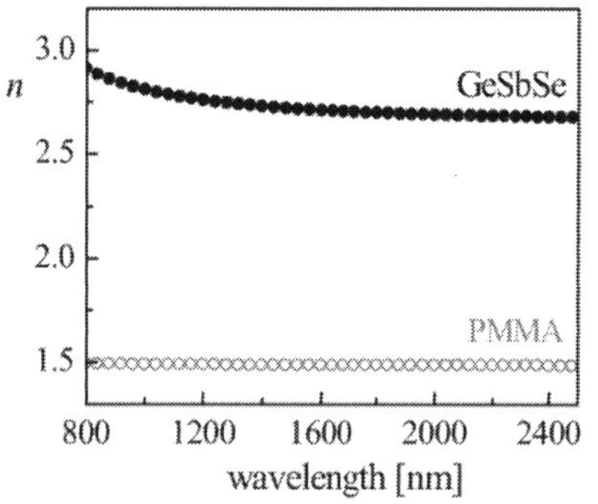

Figure 1. Dispersion curves of refractive indices of 140 nm GeSbSe and 260 nm PMMA films.

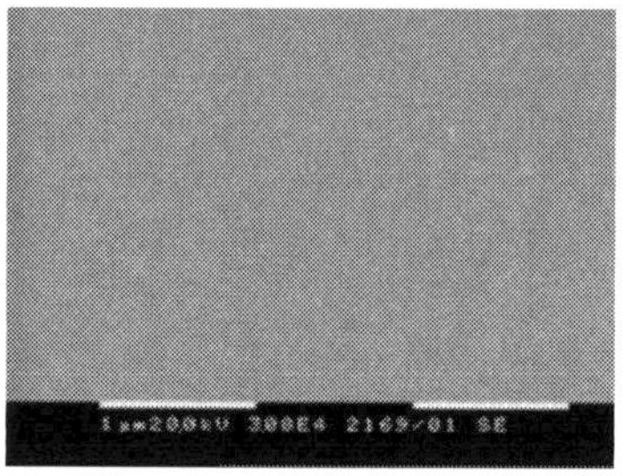

Figure 2. SEM micrograph of 140 nm GeSbSe film deposited on a BK-7 glass plate.

Figure 3 shows the transmittance (a) and reflectance (b) spectra measured at normal incidence during the step-by-step fabrication of a multilayered stack by alternating deposition of $Ge_{28}Sb_{12}Se_{60}$ and PMMA films with thicknesses 140 nm and 260 nm, respectively. The gradual formation of a high-reflectance band centered at 1650 nm is clearly seen. As it was expected due to the high optical contrast, a broad band with a maximal value of R = 98.7 % is achieved by a relatively small number of layers (11 in this stack – 6 of GeSbSe and 5 of PMMA), in contrast with the data in the literature [11]. The transmittance value in the stop band is 0.35%. The small deviation from 100% is within the experimental error. A further increase in the number of layers with one more pair of constituents practically did not lead to any significant increase of reflectance (R changed by 0.5 %).

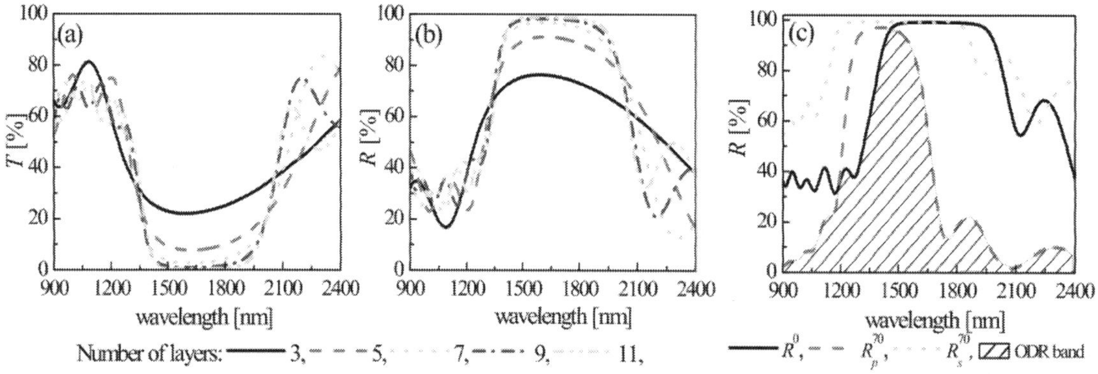

Figure 3. Transmittance (a) and reflectance (b) at normal incidence of the stack measured during preparation; (c) ODR band in a stack comprising 13 alternating GeSbSe and PMMA layers.

In fact, the stop bands in Figure 3(a) and (b) observed are to be expected since the constructed stack is a typical Bragg's reflector, in which photons are forbidden to proceed in certain spectral ranges for normal incidence of light [1]. To check the existence of an omnidirectional band, T and R were measured with s- and p-linearly polarized light varying the angle of incidence from 0 to 70°. It was established that by increasing the incident angle the band position is shifted to the shorter wavelengths, while the band is wider for s-polarization and narrower for p-polarization, as compared to the band for normal incidence. This is demonstrated in Figure 3(c), which presents the reflectance spectra of a 13-layers stack as measured at 0° and 70° angles of incidence. As seen, there is a wavelength region positioned around 1500 nm where the bands overlap and, therefore, can be regarded as an ODR band. The ODR's width at half maximum is about 320 nm and the *R* value in the maxima is 96.3%. Obviously, further optimization of the system studied is needed fin view of obtaining a value of *R* in the stop band close to 100%. Since the optical contrast is very high, more attention must be paid to improving the thickness control of the single layers, their mutual adhesion, etc. We believe that, after a successful optimization, the 1D photonic crystals studied will be appropriate for practical applications, for example, as biosensors for various bacteria, which has already been demonstrated [7] in a multilayered waveguide based on the same GeSbSe and GeSe glassy systems.

Conclusions

The results of the present study demonstrate the possibility of preparing a 1D photonic crystal arranged as a multilayered stack built by alternating deposition of thin films of $Ge_{28}Sb_{12}Se_{60}$ and PMMA. It is shown that both kinds of single layers are characterized by smooth surface morphology, which is an important prerequisite for the good quality of the optical coatings. Furthermore, the high optical contrast established of the system allows the formation of a band with omnidirectional reflectance (ODR) by a relatively small number of layers (7 of GeSbSe and 6 of PMMA). The photonic band gap obtained is centered at about λ =1.5 µm, the width at half maximum of the ODR is 320 nm and the *R* value in the maxima is 96.3%. The results obtained are very encouraging; nevertheless, further optimization is needed for bringing *R* in the band closer to 100%. This would open up opportunities for practical applications of the developed multilayered reflector in the NIR spectral region.

References

[1] Fink Y, Winn J, Fan Sh, Chen Ch, Michel J, Joannoplous J and Thomas E 1998 *Science* **282** 1679
[2] John S 1987 *Phys. Rew. Lett.* **58** 2486
[3] Freeman D, Grillet Ch, Lee M, Smith L, Ruan Y, Rode A, Krolikowska M, Hanic S, de Sterke C, Steel M, Luther-Davis B, Madden S, Moss D, Lee Y and Eggleton B 2008 *Photonics and Nanostrucures – Funadamentals and Applications* **6** 3
[4] Klement T, Ponnampalam N, Nguen H, DeCorby R 2006 *Optics Express* **14** 1790 Temelkuran B, Hart Sh, Benoit G, Joannoplous J, Fink Y 2002 *Nature* **420** 650
[5] Dwivedi P. Allen T, Tsui Y, Tonchev D, Haugen C, DeCorby R, McMiullin J, Kasap S 2003 Int. Conf. on Amorphous Materials (ICAMS), Campos do Jordano, Brazil
[6] Balan V, Vigreux C, Pradel A, Llobera A, Dominguez C, Alonso M, Garriga M 2003 *J. Non-Cryst. Solids* **354** 2757
[7] Ganjoo A, Jain H, Yu C, Irudayaraj J, Pantano C 2008 *J. Non-Cryst. Solids* **354** 2757
[8] Joanopoulus J, Meade R, Winn J 1995 in: *Photonic crystals: Molding the Flow of Light* (Princeton)
[9] Babeva Tz, Kitova S, Mednikarov B, Konstantinov I 2002 *Appl. Opt.* **41** 3840
[10] Konstantinov I, Babeva Tz, Kitova S 1998 *Appl. Opt.* **37** 4260
[11] Kuriki K, Shapira O, Hart Sh, Benoit G, Kuriki Y, Viens J, Bayindir M, Joannoplous J, Fink Y 2004 *Optics Express* **12** 1510

Structural properties of ZnO layers deposited on glass substrates by PECVD

S Kitova[1], R Kazakov and G Danev

Acad. J. Malinowski Institute of Optical Materials and Technologies,
Bulgarian Academy of Sciences, Acad. G. Bonchev Str., Bl. 109, 1113 Sofia, Bulgaria

E-mail: skitova@clf.bas.bg

Abstract. Thin ZnO layers were grown by metal-organic plasma-enhanced CVD (RF - 13.56 MHz) on glass substrates without and with ZnO seed film. The properties of the ZnO layers obtained were studied in dependence of the partial oxygen pressure by X-ray diffraction (XRD) and scanning electron microscopy (SEM). The results obtained show that by controlling the partial oxygen pressure during the layer growth one can control the crystallinity and texture of the layers obtained on pure glass substrates. It was found that ZnO layers deposited at low O_2 pressure have well-developed grain structures with a predominant *c*-axis phase and better crystalline quality than that of the samples obtained at high partial oxygen pressure. In the presence of ZnO seed films, the formation of *c*-axis phase and its quality are less dependent on partial P_{O_2}. Nanorods with good alignment and orientated vertically with respect to the substrate surface can be observed in the layers deposited on glass substrates with ZnO seeds and substrate temperature of 400 °C at low content of O_2 in the plasma. This behavior is interpreted in the framework of the so-called preferential nucleation and preferential grain growth.

1. Introduction

Zinc oxide (ZnO) thin films find a wide range of technological applications as sensors, heat mirrors, transparent electrodes, solar cells, piezoelectric and other optoelectronic devices [1-4]. So far, various deposition techniques have been applied, such as thermal evaporation, pulsed laser deposition, sputtering, sol-gel deposition, template assisted and solution processes, molecular beam epitaxy, metal-organic chemical vapor deposition MOCVD, in order to grow high-quality crystalline ZnO layers on substrates of practical importance, such as sapphire, Si wafer and glass substrates (for review see [3, 4]). Among them, MOCVD can offer high growth efficiency, large area uniformity and suitability for growing components containing volatile elements [5]. The plasma enhanced MOCVD (PECVD) technology is one of the most attractive for the synthesis of ZnO layers at low and moderate substrate temperatures and a minimal impact on the substrate. Diethylzinc and dimethylzinc are usually applied as precursors in the PECVD process [1, 6]. However, the use these precursors is not desirable because of their high toxicity and flammability. Zinc acetylacetonate ($Zn(acac)_2$) should be a good candidate for use in th low-temperature fabrication of crystalline ZnO layers because of its

[1] To whom any correspondence should be addressed.

non-toxicity and inflammability, low vaporization temperatures (150- 250 °C), high vapor pressure [6]. However, ($Zn(acac)_2$) has rarely been used as a precursor.

It is well known that structures of synthesized ZnO are dependent on the processing conditions, such as working pressure, substrate temperature, deposition power and gas mixture [1, 3, 7-9]. Selective growth of high-quality ZnO nanorods on Si substrates at 800-900 °C has been demonstrated using a pattern thin film of ZnO as a seed layer [10].

In this work, the structural properties of ZnO layers deposited on glass substrates with and without ZnO seeds were studied in dependence of the partial oxygen pressure. The layers are grown by PECVD using $Zn(acac)_2$ and O_2 as reactants.

2. Experimental

ZnO layers were grown in a PECVD equipment (GENUS 8720). The gas plasma was excited by a 13.56 MHz RF generator at RF power of 2000 W [11]. Oxygen gas and $Zn(acac)_2$ (Zn(C5H7O2)2.xH2O) (Sigma-Aldrich) were used as precursors. The Zn source was sublimated by heating up to 140 °C and carried by nitrogen gas to the growth chamber at a constant partial nitrogen pressure, P_{N_2}, of 70 mTorr. The stable precursor temperature resulted in a stable precursor flow rate, even if we could not directly measure it. O_2 gas was introduced separately into the reaction chamber in order to prevent any pre-reaction of the two precursors. Additional Ar gas was injected through a separately connected gas line in order to keep a constant total pressure, P_{tot}, of 280 mTorr. The temperature of the substrates, T_s, was 400 °C. The thickness of the layers of about 300-500 nm was measured by a profilometer type Talystep. The layers were deposited on Corning glass substrates, preliminary carefully cleaned. The cleaning procedure first involved washing by a 1% detergent solution and then with deionized water in an ultrasonic cleaner. Very thin ZnO seed films of about 3-5 nm were formed by spin coating (500 rpm, 180 s) of 20 mM $Zn(CH_3COO)_2.2H_2O$ (Sigma Aldrich) solution in ethanol followed by thermal treatment for 1 h at 400 °C in air.

The crystal structure of the layers was characterized by XRD diffraction (XRD), using a Philips (PW 1710) apparatus with Cu-Kα radiation separated by a graphite focusing monochromator. The intensity and full width at half maximum (FWHM) of a (hkl) plane were determined by profile fitting procedure using a Voight or Gaussian peak shape. Scherrer's equation was used to estimate the average crystallite size. The surface morphology of the layers was examined by a scanning electron microscope (SEM Philips 515).

3. Results and discussion

The results obtained show that the deposition rate R_d of the layers depends on the partial oxygen pressure P_{O_2}. As seen in figure 1, the layers deposited at low P_{O_2} grow more slowly that those deposited at higher P_{O_2}. The presence of a seed film on the glass substrate facilitates the layer growth at low P_{O_2}. The XRD analysis indicates that all the ZnO layers deposited at different P_{O_2} and at substrate temperature of 400 °C are polycrystalline. Three peaks are observed which can be indexed to $(10\bar{1}0)$, (0002) and $(10\bar{1}1)$ planes of the hexagonal wurtzite structure of the ZnO crystal. The results show that the crystallites are preferentially oriented along the c-axis direction perpendicular to the substrate surface, as the (0002) reflection is enhanced relatively to the usual $(10\bar{1}0)$ maximum reflection of ZnO. In order to quantify the preferred orientation (PO), often called texture, in the crystalline films, the texture coefficient (T_c) was calculated [11]. The sample with randomly oriented crystallites exhibits T_c(hkl) = 1, while the higher value of T_c indicates the larger abundance of crystallites oriented at the (hkl) direction.

Figure 2 presents the influence of the partial oxygen pressure on T_c, the FWHM and the grain size calculated from the XRD patterns of the highly oriented (0002) peaks. The high value of T_c of about 4.5, the narrow full-width of 0.22 – 0.35 and the grain size of 33 – 23 nm are indications of a predominant *c*-axis phase with good crystalline quality formed on pure glass substrates at oxygen partial pressure of up to 150 mTorr. As it is seen in figure 2, the further increthease of P_{O_2} in the plasma gases leads to a steep decrease in T_c to 1.6 and to FWHM broadening 1.5. Correspondingly, the average crystallite size, which is inversely proportional to the FWHM, decreases to the smallest value of 6 nm. These results is indicative of a lesser amount of a *c*-axis preferred phase with smaller size and worse crystalline quality grown at P_{O_2} higher than 150 mTorr. As a rule, as the oxygen partial pressure is increased, the crystalline quality degrades from a single oriented crystal to a polycrystalline structure. Further raising the oxygen pressure resulted in preventing the film growth on the glass substrates. It should be noted that we observed a similar dependence for ZnO layers deposited on a Si wafer [11], but the partial oxygen pressure influence is expressed more strongly in the case of amorphous glass substrates. The data presented in figure 2 indicate that T_c, FWHM and grain size are less dependent on the partial pressure P_{O_2} in the presence of ZnO seeds on the glass substrate.

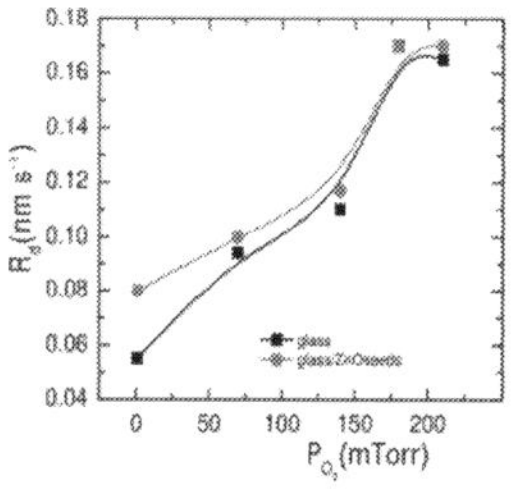

Figure 1. R_d vs the partial oxygen pressure P_{O_2}.

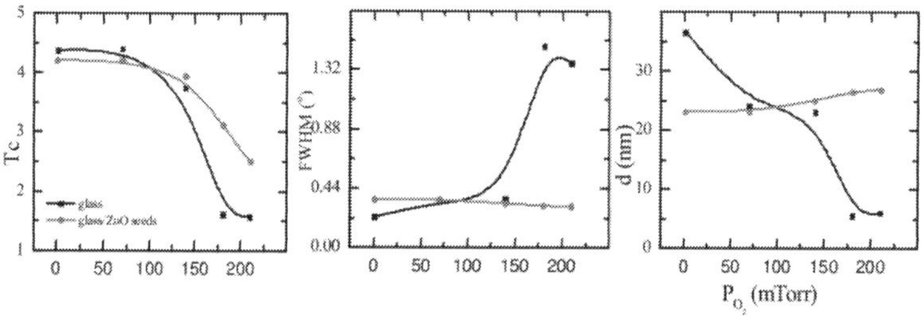

Figure 2. T_c, FWHM and grain size calculated from XRD patterns of the highly oriented (0002) peaks vs P_{O_2} for layers grown on glass substrates without and with ZnO seeds.

The results obtained are consistent with the data by SEM observations. Figure 3 shows top- and cross sectional- view SEM images of the layers grown on glass substrates with and without ZnO seeds at different partial oxygen pressures. The images show the well-developed grain structure of the layers deposited on pure glass substrates at low P_{O_2} in plasma and the denser and smoother surface of layers deposited at high P_{O_2} in plasma. Only the layers deposited at low P_{O_2} in plasma exhibit columnar structure. In accordance with the results by XRD stud,y all layers deposited on glass substrates with ZnO seeds exhibit a well developed grain structure. Nanorods with good alignment, vertically oriented to the substrate surface can be seen in the layers deposited at low content of O_2 in the plasma gases.

without ZnO seeds with ZnO seeds

Figure 3. Top- and cross sectional- view SEM images of the layers grown on glass substrates without and with ZnO seeds at the indicated P_{O_2}.

Most probably, the preferred orientation (PO) observed of the crystalline layers deposited and their related structure are connected with the so-called preferential nucleation and preferential grain growth. For the case of ZnO, *c*-axis PO nucleation has been considered to take place because of the lowest surface free energy of this plane among the hexagonal crystallographic planes [9]. At low P_{O_2}, when the layer deposition rate is low, once the condensation process begin, the initially condensed molecules form seed crystals serving as the nucleation sites with predominant *c*-axis PO; their PO determins the PO of the layers obtained. No preliminary created nucleation sites are necessary. On the contrary, initial nuclei with relatively more random orientation dominate in the substrates with ZnO seeds. Under this oxygen deficient condition, ZnO will grow preferentially on the pre-existing ZnO nuclei of the seed film or on the ZnO nuclei that have been formed during the beginning of the layer growth. In addition, due to the faster vertical growth rate along the *c*-axis compared to the lateral direction, rod shaped ZnO crystal are developed. In plasma with high O_2 content, ZnO has a high nucleation density and a rapid growth which leads to the formation of dense films with smaller grain size and more random orientation. A further excess of O_2 may induce defects in ZnO, which may influence the nucleation and the growth of the films. This results in a degradation of the crystal quality, or even to the inhibition of the grain growth.

Conclusions

Crystalline ZnO layers were obtained by PECVD at 400 °C on glass substrates with and without ZnO seeds. The SEM and XRD studies show that the crystallinity and texture of the ZnO layers depend strongly on the oxygen pressure in the case of layers grown on glass substrates without a seed film. It was found that ZnO layers deposited on pure glass substrates at low O_2 pressure have well developed grain structures with predominant *c*-axis phase and better crystalline quality than that of the samples obtained in plasma with high O_2 pressure. In the presence of ZnO seed films on the glass substrates, the formation of a *c*-axis phase and its quality are less dependent on partial pressure P_{O_2}. Nanorods with good alignment orientated vertically with respect to the substrate surface can be observed in the layers deposited on glass substrates with ZnO seeds at low content of O_2 in the plasma. This behavior can be explained by the so-called preferential nucleation and preferential grain growth. The above results can contribute to optimizing the growth condition of ZnO layers and achieving the performance required for the production of chemical- and bio-sensors or other electronic devices characterized by high efficiency and sensitivity and long lifetime.

Acknowledgements

The authors wish to thank Prof. D.Sc. I. Tomov for XRD measurements, and G. Maleshkova for the SEM observations. This study was supported by the Bulgarian National Science Fund Grant # DO02-207/2009.

References

[1] Ozgur U, Alivov Y, Liu C, Teke A, Reshchikov M, Dogan S, Avrutin V, Cho S, and Morkoc H 2005 *J. Appl. Phys.* **98** 041301
[2] Djurisic A and Chen A 2010 *Progress in Quantum Electronics* **34** 191-259
[3] Mende L and Driscoll L 2007 *Materialstoday* **10** 40
[4] Lu J, Chang P and Fan Zh 2006 *Mater. Sci. Eng.* **R52** 49
[5] Ma Y, Du G, Yin J, Yang T and Zhang Y 2005 *Semicond. Sci. Technol.* **20** 1198
[6] Karpina V, Lazorenko V, Lashkarev C, Dobrowolski V, Kopylova L, Baturin V, Pustovoytov S, Karpenko A, Erimin S and Lytvyn P 2004 *Cryst. Res. Technol.* **39** 980
[7] Iwanaga H, Shibata N, Nittono and Kasuga M 1978 *J. Cryst. Growth* **45** 228-32
[8] Liu X, Wu X, Cao H and Chang R 2004 *J. Appl. Phys.* **95** 3141-7
[9] Kajikawa Y 2006 *J. Cryst. Growth* **289** 387-94
[10] Conley J, Stecker L and Ono Y 2006 *J. Electr. Mater.* **35** 795
[11] Kitova S and Danev G 2010 *J. Phys.: Conf. Series* **253** 012031

Journal of Physics: Conference Series **356** (2012) 012025 doi:10.1088/1742-6596/356/1/012025

Correlation between the haze ratio and the surface roughness of electrochemically deposited nanostructured ZnO films

K Lovchinov[1,3], O Angelov[1], H Nichev[1], V Mikli[2] and D Dimova-Malinovska[1]

[1] Central Laboratory of Solar Energy and New Energy Sources,
Bulgarian Academy of Sciences, 72, Tsarigradzko Chaussee, 1784 Sofia, Bulgaria
[2] Centre for Materials Research, Tallinn University of Technology, Tallinn, Estonia

E-mail: lov4@abv.bg

Abstract. This work reports the synthesis of nanostructured ZnO by electrochemical deposition on glass substrates covered by a thin film of SnO_2 doped with F (SnO_2:F). Nanostructured ZnO films are deposited by an electrochemical process using a three-electrode potentiostatic system with saturated calomel electrodes as reference electrodes from an aqueous solution containing $ZnCl_2$, KCl, (pH 4.00) and an air flow or/and additional supplement of H_2O_2 or a ZnO powder suspension. The influence of the deposition conditions on the structural properties of the ZnO films prepared is studied by SEM, AFM and XRD. The SEM micrographs and the AFM images show that the ZnO films consist of nanograins with average size in the range of 52 – 83 nm depending on the deposition conditions. The XRD spectra demonstrate preferential (002) crystallographic orientation of the ZnO grains. The optical band gap values are in the range 3.26 – 3.46 eV. The spectra of total and diffused transmission are also measured. It is shown that the haze ratio depends on the surface roughness of the films.

1. Introduction

ZnO nanostructures have received increasing attention due to potential applications in solar cells, nanoscale electronics, optical and sensing devices, etc. [1-4]. In the fabrication of thin film solar cells of higher efficiency, ZnO films used as electrodes exhibit surface morphology suitable for light trapping and optical confinement.

This paper reports results from the study of the structural and optical properties of ZnO nanostructured films grown by electrochemical deposition on glass substrate coated by SnO_2:F.

2. Experimental

ZnO nanostructured films were deposited by an electrochemical process from an acid aqueous solution of $ZnCl_2$ (5×10^{-3} M) and KCl (0.1 M) with pH 4.0 at temperature of 60^0C in flowing air and a suspension containing ZnO powder as a precursor. A three-electrode electrochemical cell was used with saturated calomel electrodes (SCE) as reference electrodes, as described earlier in [5]. It is assumed that the ZnO powder suspension can serve as a source of seeding particles. For comparison, a layer was also prepared without adding the ZnO suspension (table 1, sample 2). SnO_2:F coated glass

[3] To whom any correspondence should be addressed.

Table 1. Structural and optical characteristics of the deposited ZnO films: optical band gap, E_g; position of the (002) peak in the XRD spectra, 2θ; FWHM of the (002) peak, Δ2θ; stress, σ; average grains size, *D*; thickness, *d*; average roughness, *R*; and haze ratio at different wavelengths. For comparison, the corresponding data for the SnO_2:F films are given as well.

Sample	*Eg*, (eV)	2θ, (deg.)	FWHM (deg.)	σ, (GPa)	*D*, (nm)	*d*, (nm)	*R*, (nm)	Haze, (%) λ=400nm	Haze, (%) λ=550nm	Haze, (%) λ=1000nm
1 –SnO_2:F	3.75	34.39	0.14	-0.25	59	910	29	30	13	3
2 –w/t drops	3.39	34.39	0.15	-0.24	56	550	315	82	65	23
3- 1 x drops	3.37	34.41	0.16	-0.12	52	580	89	85	82	35
3'-1x drops	3.38	34.38	0.14	-0.32	59	620	89	86	74	24
4- 2x drops	3.38	34.37	0.10	-0.38	83	700	147	88	86	34
5-2x drops+H_2O_2	3.46	34.39	0.12	-0.25	69	1570	82	90	73	29
6-3x drops	3.42	34.39	0.11	-0.25	76	750	299	93	82	50
6'-3x drops	3.41	34.34	0.13	-0.58	64	1000	225	91	82	50
7 -3x drops+ann.	3.26	34.48	0.13	0.36	64	1060	153	86	78	46
7'- 3x drops+ann.	3.27	34.50	0.13	0.47	64	850	78	91	77	31
8 -3x drops+H_2O_2	3.47	34.38	0.12	-0.32	69	900	75	89	65	22
9- 3x drops+Ni	3.51	34.38	0.12	-0.32	69	810	33	38	25	13

substrates were used as a working electrode (table 1, sample 1). In the case of sample 9, a thin Ni seeding layer (1 – 2 nm thick) was chemically deposited on the surface of the glass/SnO_2:F substrate before the electrochemical growth of ZnO. A spectrally pure graphite electrode was used as anode. The solution was stirred by air bubbling. The deposition was carried out by varying the redox potential of the system. Since the potential of Zn in the electrolyte is –1.05 V vs. SCE, the deposition process of ZnO was carried out at –0.9 V vs. SCE, thus preventing metal Zn deposition. ZnO films of good quality were obtained at a redox potential within the range +0.30 and +0.40 V vs. SCE. Oxygen saturation of the solution was maintained by air bubbling. Zinc peroxide (ZnO_2) was formed on the samples with bad adhesion to the substrate at a redox potential higher than +0.4 V. The duration of the ZnO deposition was 60 min. The ZnO films prepared were 0.55 – 1.57 μm thick. Different regimes were explored: sample 2 – without drops (dr.), sample 3 – 1×dr., sample 4 – 2×dr., sample 5 – 2×dr. + H_2O_2, samples 6 and 6' – 3×dr., samples 7 and 7'– 3×dr. + annealing at 300 °C, sample 8 – 3×dr. + H_2O_2 and sample 9 – 3×dr. + Ni seeding layer.

The film structure was studied by X-ray diffraction spectrometry (XRD) using a Brucker D8 Advance spectrometer with Cu K_α radiation: λ Cu $K_{\alpha 1}$ =1.540560 Å and λ Cu $K_{\alpha 2}$ = 1.544426 Å (intensity half of that of λ Cu $K_{\alpha 1}$). The instrumental broadening in 2θ geometry was 0.04^0. The surface morphology and the thickness of the deposited films were imaged by a Philips 515 scanning electron microscope (SEM). The average roughness was determined by AFM. The transmittance and the reflectance spectra, the diffuse transmittance and the diffuse reflectance were measured by a Shimadzu UV-3600 spectrophotometer in the range of 300 -2600 nm employing a 60 mm integrating sphere in the case of diffuse reflectance.

3. Results and discussion

XRD spectra of some of the deposited ZnO samples are shown in figure 1. For comparison, the XRD spectrum of the SnO_2:F film is displayed as well. The diffraction patterns demonstrate that the deposited ZnO films are polycrystalline with reflections corresponding to the (100), (002), (101), (102) and (103) planes of a wurtzite structure with the *c*-axis perpendicular to the substrate. The XRD peak of the (002) plane is the most intensive one. In the case of the as-deposited layers, its position is slightly shifted to the lower 2θ compared to the ZnO powder (2θ = 34.44°), which is evidence for the presence of stress, σ, in the layers, as confirmed by the calculated values of σ (table 1). The average size of the grains, *D*, is estimated using the FWHM of the (002) peak), Δ2θ. The value of the optical band, E_g, is calculated using the transmittance and reflectance spectra assuming a direct gap for ZnO [6]. The values of E_g are typical for ZnO – 3.39 – 3.46 eV in the case of the as-deposited samples.

Adding H_2O_2 results in an increase of E_g, (samples 5 and 8). After annealing (samples 7 and 7'), the value of E_g decreases to 3.26 – 3.27 eV compared to the values of the as-deposited samples. This is probably related to the removal of $Zn(OH)_2$ that could be present in the as-deposited layer in electrodeposition of ZnO from aqueous solution. The diffuse reflectance spectra, R_{diff}, are shown in figure 1 c. The values of the diffuse reflectance, R_{diff}, of the ZnO layers grown are higher than that of the SnO_2:F coated glass substrate. It is seen that the R_{diff} values increase when the quantity of the ZnO suspension added to the electrolyte is raised and when a Ni intermediate film is used. After annealing, the values of R_{diff} decrease slightly in the whole range of the spectrum.

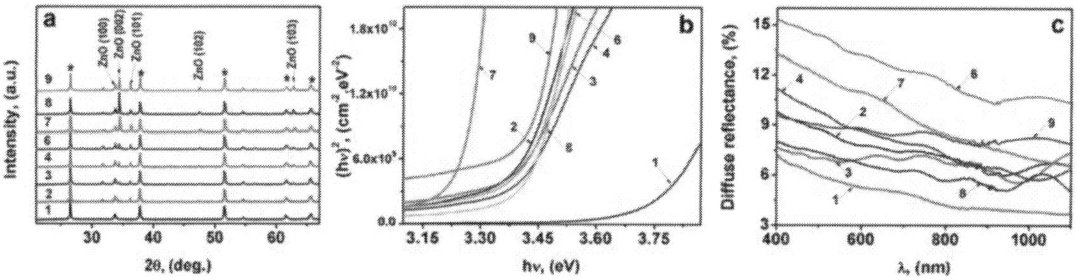

Figure 1. XRD spectra (a), $(\alpha E)^2$ versus energy, E, (b) and diffuse reflectance spectra of the ZnO layers deposited at different conditions (c). The corresponding curves (1) of SnO_2:F are given, too. The XRD peaks of SnO_2:F are marked by *.

Figures 2 and 3 display examples of the SEM and the AFM images of the SnO_2:F substrate and of some of the samples. The ZnO layer grown without adding a ZnO suspension (sample 2) consists of grains with size of about 56 nm (very inhomogeneous grains size, D) and has an average surface roughness of ~82 nm (figure 2b). It is seen that the grain size in the layers increases as different quantities of the suspension are added, which is in agreement with the data obtained by XRD (table 1). Annealing of the film at 350°C for 1 hour results in changes in the morphology – the individual grains grow into conglomerates (figure 2d) and the surface roughness of this sample decreases.

Figure 2. SEM plane views of SnO_2:F film (a) and ZnO layers – samples: 2- (b), 6 - (c), 7 - (d) and 9 - (e). The marker corresponds to 500 nm.

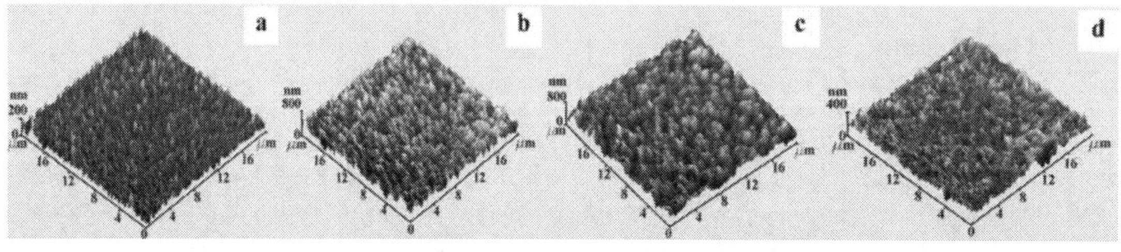

Figure 3. AFM images of a SnO_2:F film (a) and ZnO layers - samples: 6 - (b), 7 - (c) and 9 - (d).

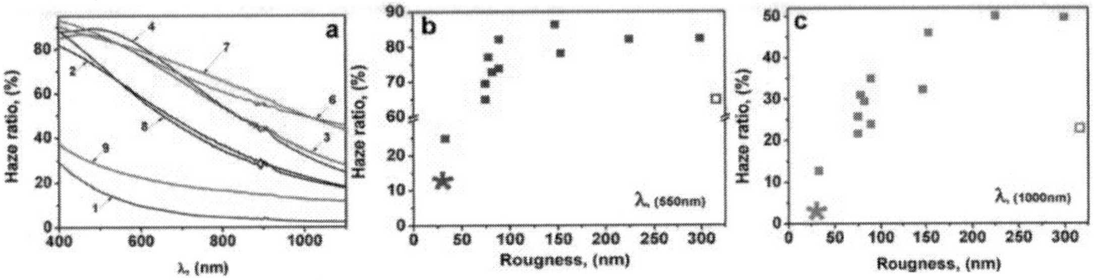

Figure 4. Spectral dependence of the haze ratio (a) and dependence of the haze ratio on the surface roughness at wavelengths 550 nm (b) and 1000 nm (c). The signs * and □ correspond to SnO_2:F (samples) and to the sample obtained without adding drops of ZnO suspention (sample 2), respectively.

The ZnO layer grown on the SnO_2:F coated by a thin Ni seeding film has a totally different shape of the grains – they look like spheres and the average roughness decreases significantly (figure 2e and figure 3d), however it is still higher than that of SnO_2:F film.

The spectral dependence of the haze ratio of the ZnO samples in the range of 400 - 1000 nm is presented in figure 4a. The values of the haze ratio of the electrodeposited ZnO layers are much higher than those of the SnO_2:F coating. Higher values of the haze ratio are observed for the samples deposited with higher quantities of the ZnO suspension added. It must be noted that the layers with higher values of the average roughness, as determined by AFM (table 1), demonstrate higher diffuse reflectance and haze ratio. Figure 4 b and c show the variation of the haze ratio with the roughness of the deposited films at two wavelengths – 550 and 1000 nm. It is very clearly seen that the value of the haze ratio increases with the increase of the surface roughness with maximal values of about 93%, 86% and 50% at 400 nm, 550 nm and 1000 nm, respectively.

Conclusions

The influence is studied of ZnO water suspension added to the electrolyte – an aqueous solution containing $ZnCl_2$, KCl, (pH 4.00), on the structure, the grain size, the diffuse reflection and the haze ratio of the electrochemically deposited ZnO films. The higher values of the diffuse reflectance and haze ratio in the spectral range of 400 - 1100 nm are evidence of ZnO layers with larger average grain size and higher average surface roughness. ZnO layers with similar properties could be used as light trapping structures in thin film solar cells.

Acknowledgements

The work has been supported by the Bulgarian National Science Fund under project DO0-207 and by the 7th European FP under project NanoPV No. 24331.

References

[1] Lo S, Haung D and Jan D-J 2010 *Optics Express* **18** 662
[2] Lupan O, Guerin V, Tiginyanu I M, Ursaki V V, Chow L, Heinrich H and Pauporte T 2010 *J. Photochem. Photobiol. A: Chem.* **211** 65
[3] Schropp R E I, Li H, Franken R H, Rath J K H, van der Werf C H M, Schüttauf J W A and Stolk R L 2008 *Thin Solid Films* **516** 6818
[4] Dimova-Malinovska D and Nikolaeva M 2002 *Vacuum* **69** 227
[5] Dimova-Malinovska D, Andreev P, Sendova-Vassileva M, Nichev H and Mikli V 2009 *Proc. 24 PVSECE* (Hambourg Germany 21-25 Sept 2009) pp 391-4
[6] Dimova-Malinovska D, Nichev H and Angelov O 2008 *Phys. Stat. Sol. C* **5** 10.1002/pssc.200778886

Electrical and optical properties of ZnO thin films doped with Nb deposited by rf magnetron sputtering

K Lovchinov[1], **O Angelov and D Dimova-Malinovska**

Central Laboratory of Solar Energy and New Energy Sources,
Bulgarian Academy of Sciences, 72, Tzarigradsko Chaussee, 1784 Sofia, Bulgaria

E-mail: lov4@abv.bg

Abstract. The optical and electrical properties are investigated of Nb doped ZnO (ZnO:Nb) thin films deposited by r.f. magnetron co-sputtering on glass substrates at different temperatures, T_s, between 50 °C and 500 °C. The optical transmittance and reflectance spectra are measured and used to determine the optical band gap. The values of the optical band gap are in the range 3.29- 3. 44 eV and are typical for ZnO thin films. The films have about 90 % transparency, while their resistivity increases in the range of 2.2×10^{-3} Ω cm - 6.4×10^{2} Ω cm as T_s is raised. The results demonstrate the potential of using Nb-doped ZnO films as a transparent conductive oxide in thin films solar cells and various optoelectronic devices.

1. Introduction

The thin film technologies and ZnO have been studied for the development of new electronic thin film devices [1], magnetic memories [2], transparent conductive oxides (TCO) for applications in thin film solar cells [3] etc. ZnO thin films doped with transition elements provide a variety of applications in gas sensors [4], diluted magnetic semiconductors (DMS) [5], etc. Studying the properties of doped ZnO with different metal (Al, V, Ga, Nb, Tb, etc.) [6-9] has led to the attractive idea of developing a new kind of highly conductive transparent thin films.

In this paper, a study is reported of the influence of varying the substrate temperature (50 °C – 500 °C) on the optical and electrical properties of Nbdoped ZnO thin films (ZnO:Nb) deposited by magnetron r.f. sputtering in Ar atmosphere.

2. Experimental

The ZnO thin films doped with Nb (ZnO:Nb) were prepared by r.f. magnetron co-sputtering of a ZnO target with pieces of Nb plates with area of 80 mm^2 in the maximum erosion zone on its surface in Ar atmosphere at pressure of 0.5 Pa and r.f. power of 160 W using ALCATEL DION 420 equipment. The films were deposited on glass substrates at different temperatures, T_s, in the range 50 °C – 500 °C. The thickness of the layers was between 700 and 950 nm, as measured by a Talystep Hobson profilometer.

Transmittance and reflectance spectra were obtaind by a Shimadzu UV.3600 spectrophotometer in the range of 320 - 1800 nm. The resistivity of the films was measured by the four-point probe method using a VEECO instrument.

[1] To whom any correspondence should be addressed.

3. Results and discussion

The values of the resistivity, ρ, of ZnO:Nb films are in the range 2.2×10^{-3} Ω cm – 6.4×10^{2} Ω cm (table 1) and are close to the ones reported in [10]. The values of ρ increase with T_s, particularly for $T_s > 350$ °C (figure 1). This is probably due to the changes in the stoichiometry of the films deposited at higher temperatures, as reported in [10]. At higher T_s, the deposition conditions favor a higher rate of reflection of ZnO atoms arriving at the substrate during the sputtering process from the substrate, thus increasing the concentration of O atoms in the lattice of ZnO films deposited (respectively decreasing the O vacancies). This results in an increased ZnO films resitivity of, as Zn in the ZnO lattice is a donor.

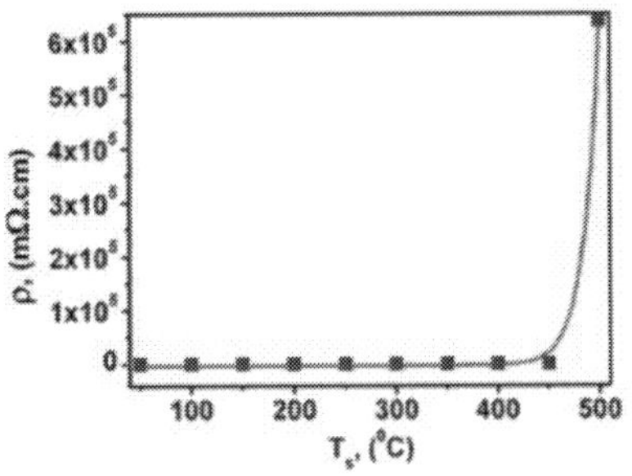

Figure 1. Dependence of ZnO:Nb films resistivity on the substrate temperature.

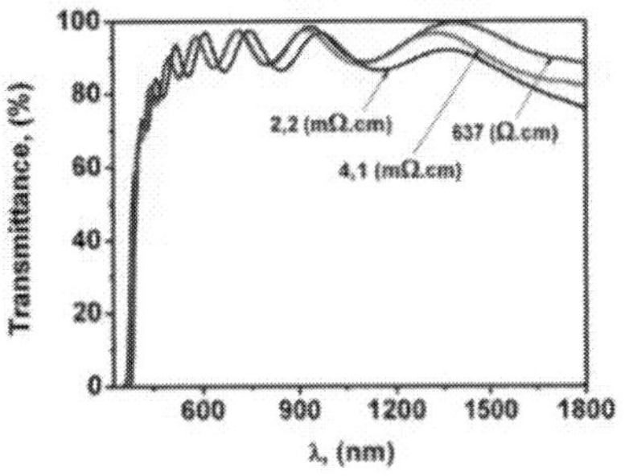

Figure 2. Transmittance of ZnO:Nb films with different resistivity.

Transmittance spectra of the ZnO:Nb thin films are presented in figure 2. The spectra are corrected for the transmittance of the glass substrate. All spectra demonstrate transmittance value about 90% in the range 550 – 1000 nm. In the IR region beyond 1200 nm, the transmission of the films with resistivity < 4.1×10^{-3} Ω.cm decreases due to the absorption of by free carriers (plasma resonance).

The values of n are calculated from the transmittance [11] in the range 500 – 1200 nm and are given in table 1. The value of n is about 1.99 and decreases slightly as T_s increases. This could be related to the oxidation of interstitial Zn in the films or to the decreasing concentration of O vacancies at higher T_s, which results in a decrease in the free carrier concentration. It is probable that the effect of any possible increase in the mobility, μ, upon ρ is less pronounced than the changes in the film stoichiometry [12].

The dependences of the absorption coefficient, α, on the energy, $h\nu$, for the ZnO:Nb films are presented in figure 3. The absorption coefficient was calculated as in [9].

The value of the energy band gap, E_g, the Urbach energy, E_0, and the coefficient B^2 (table 1) are calculated for direct inter-band electron transitions as described in [8, 13, 14]. The spectral dependence of α exhibits two regions: a power law at high photon energies and an exponential one at lower energies. The formula for direct allowed transitions can be used to obtain the optical gap, E_g,:

$$\alpha(h\nu) = B[(h\nu - E_g)^{1/2}/h\nu] \,. \qquad (1)$$

In the lower energy range $(h\nu < E_g)$, where α varies exponentially with the photon energy, the spectral dependence of the absorption edge follows the Urbach formula [10, 15, 16]:

$$\alpha(h\nu) = \alpha_0 \exp[(h\nu - E_1)/E_0] \,, \qquad (2)$$

where α_0 is the Urbach absorption at the edge (E_1, where $E_1 < E_g$), and E_0 is the Urbach energy.

Table 1. The values of the optical band gap, E_g, Urbach tail, E_0, the coefficient B (values of B^2 are given), refractive index, n, the thickness and the resistivity, ρ, of ZnO:Nb films deposited at different substrate temperature T_s.

sample	T_s, (^{0}C)	Thickness, (nm)	E_g, (eV)	E_0, (meV)	B^2, (cm^2.eV)	n, refractive index	ρ, (mΩ.cm)
D262 ZnO:Nb	50	860	3.42	77	8.42E11	1.99	2.2
D263 ZnO:Nb	100	950	3.44	76	7.79E11	1.97	2.5
D264 ZnO:Nb	150	760	3.43	77	9.84E11	2.00	3.95
D265 ZnO:Nb	200	710	3.37	72	1.15E12	1.96	4.1
D337 ZnO:Nb	250	810	3.37	71	1.14E12	1.95	4.2
D268 ZnO:Nb	300	750	3.38	69	1.22E12	1.99	4.9
D269 ZnO:Nb	350	850	3.38	69	1.25E12	1.99	3.5
D270 ZnO:Nb	400	750	3.34	67	1.32E12	1.95	13.5
D338 ZnO:Nb	450	860	3.32	62	1.44E12	1.94	15
D271 ZnO:Nb	500	770	3.29	59	1.52E12	1.93	6.37×10^5

The spectral dependence of the absorption coefficient of ZnO:Nb films is shown in figure 4. The values obtained of the band gap are in the range 3.29 – 3.44 eV and are typical for ZnO. The value of the band gap decreases as T_s increased, as seen in figure 4. The blue shift is observed for the films with lower resistivity which is attributed to the Burstein–Moss effect [17].

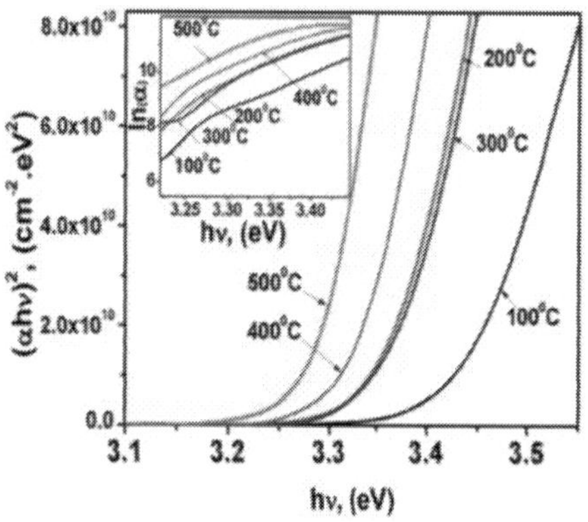

Figure 3. Dependence of $(\alpha h\nu)^2$ on $h\nu$. The inset shows the plot of lnα vs. $h\nu$.

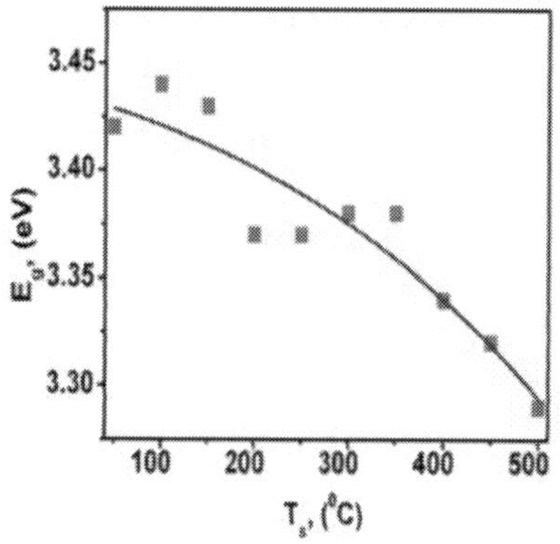

Figure 4. Dependence of the optical band gap on the T_s.

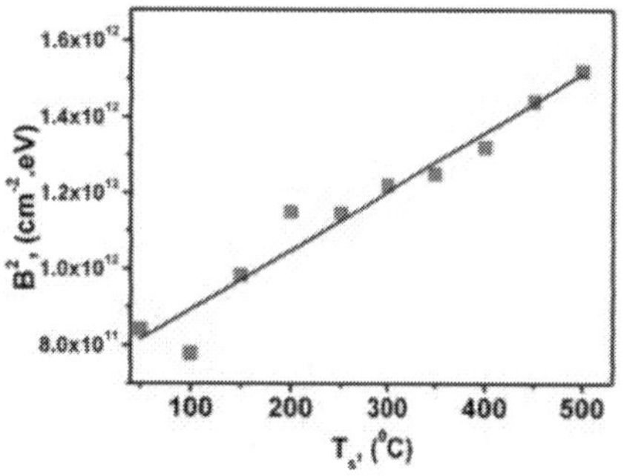

Figure 5. Dependence of the coefficient B^2 on the substrate temperature.

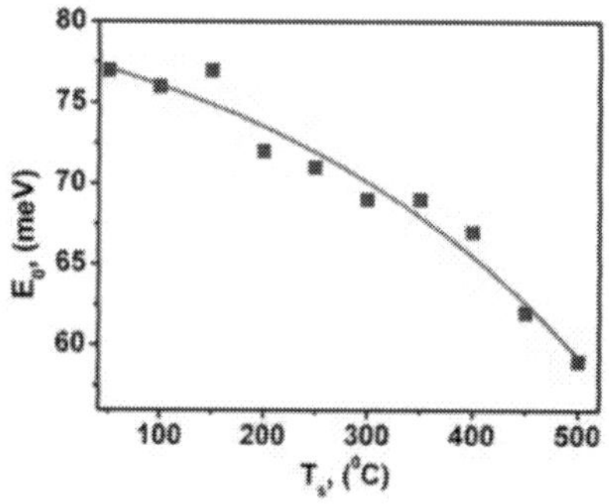

Figure 6. Dependence of the Urbach energy E_0 on the substrate temperature.

The variation of the coefficient B^2 and the Urbach energy, E_0, with the substrate temperature is shown in figures 5 and 6, respectively. The values of the coefficient B^2 increase with the substrate temperature. In contrast, the Urbach energy has an inverse dependence on T_s. The Urbach region ($h\nu < E_g$) is due to the perturbation of the parabolic density of the states at the band edge. The increase of the structural disorder results in an increase in Urbach energy. The coefficient B^2 and the Urbach energy are assumed as being parameters of the conduction and valence band tail states, respectively [18, 19]. A decrease in B^2 and an increase in E_0 implie an increase in the band tails' width. The dependences observed for ZnO:Nb demonstrate that the band tails width decreases with the increase of the substrate temperature, probably due to the better structural properties of the films deposited at higher T_s, as is reported in [10,13].

Conclusions

The influence was studied of the substrate temperature on the optical properties and electrical resistivity of ZnO thin films doped with Nb. The films have transmittance about 90% in the visible range. The resistivity of the deposited films decreases with T_s. The films deposited at substrate temperature $T_s = 50$ °C exhibited lower resistivity (2.2×10^{-3} Ω.cm). The values of the optical band gap of the ZnO:Nb films are in the range of 3.29 – 3.44 eV and decrease as T_s is raised. The refractive index is in the range 1.93 – 2.00 at about 500 nm. The ZnO:Nb film deposited at lower T_s (50 – 100°C) are transparent (~90%). They have low resistivity and demonstrate a potential for application as a transparent conductive oxide in thin film solar cells and LED.

Acknowledgements

The work was supported by the Bulgarian National Sciens Fund, project DO0-207 and by the 7th European Framework Program, project NanoPV No. 24331.

References

[1] Li Y, Gong J, McCune M, He G and Deng Y 2010 *Synthetic Metals* **160** 499
[2] Seshadri R 2005 *Current Opinion in Solid State and Materials Science* **9** 1
[3] Minami T 2005 *Semicond. Sci. Technol.* **20** S35
[4] Koshizaki N and Oyama T 2000 *Sensors and Actuators B* **66** 119
[5] Fukumura T, Jin Z, Ohtomo A, Koinuma H and Kawasaki M 2000 *Appl. Phys. Lett.* **75** 3366
[6] Rath J, Liu Y, de Jong M, de Wild J, Schuttauf J, Brinza M 2010 *Thin Solid Films* **518** 129
[7] Ekem N, Korkmaz S, Pat S, Balbag M, Cetin E and Ozmumcu M 2009 *Int. J. Hydrogen Energy* **34** 5218
[8] Lovchinov K, Nichev H, Angelov O, Sendova-Vassileva M, Dimova-Malinovska D and Mikli V 2010 *J. Phys.: Conf. Ser.* **253** 012030
[9] Su-Hua Y, Sheng-Yu H and Sneg-Hsun T 2010 *Japan J. Appl. Phys.* **49** 06GJ06
[10] Dimova-Malinovska D, Nichev H, Angelov O, Grigorov V and Kamenova M 2007 *Superlattices and Microstructures* **42** 123
[11] Swanepoel M 1983 *J. Phys. E: Sci. Instrum.* **16** 1214
[12] Roth A and Williams D 1981 *J. Appl. Phys.* **52** 6685
[13] Dimova-Malinovska D, Nichev H and Angelov O 2009 *Phys. Stat. Sol.* **5** 3353
[14] Dragoman D and Dragoman M 2002 *Optical Characterization of Solids* (Heidelberg: Springer-Verlag)
[15] Girtan M and Folsher H 2003 *Surf. Coat. Technol.* **172** 242
[16] Melsheimer J and Zielger D 1985 *Thin Solid Films* **35** 129
[17] Burstein E 1954 *Phys. Rev.* **93** 632
[18] Mott N and Davis E 1979 *Electronic Processes in Noncrystalline Materials* (Oxford: Clarendon) p 289
[19] Saito N 1985 *J. Appl. Phys.* **58** 3504

Journal of Physics: Conference Series **356** (2012) 012027 doi:10.1088/1742-6596/356/1/012027

NO_2 sensing properties of amorphous silicon films

V Georgieva[1], N Donkov[2,6], P Stefanov[3], M Sendova-Vassileva[4], A Grechnikov[5] and V Gadjanova[1]

[1] Georgi Nadjakov Institute of Solid State Physics, Bulgarian Academy of Sciences, 72, Tsarigradsko Chaussee, 1784 Sofia, Bulgaria
[2] Emil Djakov Institute of Electronics, Bulgarian Academy of Sciences, 72, Tsarigradsko Chaussee, 1784 Sofia, Bulgaria
[3] Institute of General and Inorganic Chemistry, Bulgarian Academy of Sciences, Acad. Georgi Bonchev Str., block 11, 1113 Sofia, Bulgaria
[4] Central Laboratory of Solar Energy and New Energy Sources, Bulgarian Academy of Sciences, 72, Tsarigradsko Chaussee, 1784 Sofia, Bulgaria
[5] Vernadsky Institute of Geochemistry and Analytical Chemistry, Russian Academy of Sciences, 19, Kosygin Str., 119991 Moscow, Russia

E-mail: nikolaidd@abv.bg

Abstract. The sensitivity to NO_2 was studied of amorphous silicon thin films obtained by e-beam evaporation. The process was carried out at an operational-mode vacuum of 1.5×10^{-5} Torr at a deposition rate of 170 nm/min. The layer's structure was analyzed by Raman spectroscopy, while its composition was determined by X-ray photoemission spectroscopy (XPS). To estimate their sensitivity to NO_2, the Si films were deposited on a 16-MHz quartz crystal microbalance (QCM) and the correlation was used between the QCM frequency variation and the mass-loading after exposure to NO_2 in concentrations from 10 ppm to 5000 ppm. A considerable sensitivity of the films was found in the interval 1000 ppm – 2500 ppm NO_2, leading to frequency shifts from 131 Hz to 208 Hz. The results obtained on the films' sorption properties can be applied to the development sensor elements.

1. Introduction

Nitrogen oxides (NO and NO_2) are typical air pollutants. They cause acid rain and smog and can also destroy the stratospheric ozone, affect the flora and fauna and lead to problems for human health [1].

In the past few decades, when new types of sensors appeared, a large number of techniques have been developed for environmental monitoring and control of the toxic gases. Among them, the QCM-based techniques attract particular interest due to their high mass-sensitivity. The analytical performance of such sensors strongly depends on the properties of a sensing coating deposited on the QCM electrodes. Semiconductor thin layers, such as SnO_2, WO_3, ZnO, MoO_3, and TiO_2 have been widely investigated for sensing different toxic gases [2]. Many forms of silicon are also interesting materials in the gas sensing field. In particular, porous Si can be successfully used as a sensing coating of chemical sensors for detection of different gases, including NO_2 [3]. The main disadvantage of

[6] To whom any correspondence should be addressed.

porous Si is that it cannot be prepared as a thin layer by different methods of deposition. A possible alternative is using amorphous α-Si, which can be prepared as thin layers by standard methods.

The aim of the work is to study the sensing properties to NO_2 of Si thin films by applying the QCM method.

2. Experimental

We used a 16-MHz QCM covered with thin Si films on both sides by e-beam evaporation. The experimental setup is described in detail in [4].The evaporator consisted of a heated tungsten filament (I_F=14.5 A AC); cathode voltage (U_c) was applied to the filament and to the screen. The evaporation power necessary for heating the crucible with Si powder is the product of the anode current I_a = 75 mA and U_c = −7 kV DC. The evaporation process was carried out at a pressure of $1{,}5\times10^{-5}$ Torr and calculated evaporation power of 525 W at a deposition rate of 170 nm/min. The as-deposited Si films were characterized by Raman spectroscopy, X-ray photoemission spectroscopy (XPS), and talystep techniques. The Raman spectra were recorded on a Horiba Jobin Yvon LabRAM HR800 spectrometer using a 600 l/mm grating and a HeNe laser for excitation. The thin film samples on glass substrates were placed under the 100X achromatic objective of a BX41 microscope and measured in back scattering configuration. The laser power on the sample was kept below 0.5 mW, so that no heating effects could be observed in the spectra or on the microscope image. The composition and the electronic properties of the Si films were studied by XPS measurements performed on an ESCALAB Mk II (VG Scientific) electron spectrometer with Al radiation (hν = 1486.6 eV).

The sensitivity to NO_2 of as-deposited Si films was measured on a laboratory setup. The methodology of measurements is described in [5]. Dry synthetic air and 1 % NO_2 gas were used. The NO_2 concentrations in the test camera were created by changing the ratio of the NO_2 gas and the diluting gas flow. The working temperature was 26±0.2 °C. The QCM frequency was registered by a counter connected to the QCM and to a computer. The measurements can be divided into three stages: purging the Si-QCM structure by dry air until the frequency reaches a constant value, *Fo*; creating a certain NO_2 concentration and reaching saturation of the frequency value (F); purging the camera with dry air thus restoring the Si-QCM system. The sensing proprties of the Si films were estimated by measuring the frequency-time characteristics (FTCs) of the Si-QCM at NO_2 concentrations ranging from 10 ppm to 5000 ppm. According to the Sauerbrey equation [6], the change in the resonant frequency shift ($\Delta F = Fo - F$) can be related to the mass change (Δm) of the adsorbed gas.

3. Results and discussion

3.1. Properties of as-deposited Si thin films

The spectrum in figure 1 is a typical Raman spectrum of amorphous silicon [7, 8]. The TO-like band is observed at 470 cm^{-1}; the longitudinal acoustic (LA) and the longitudinal optic (LO) regions between 420 and 200 cm^{-1} and the transverse acoustic (TA) region below 200 cm^{-1} are also clearly visible. The position of the TO band corresponds to non-hydrogenated amorphous silicon. The full width at half maximum of the peak is 84 cm^{-1}, which is consistent with a comparatively good short-range order. There are no microcrystalline grains, which would give rise to a narrow peak around 520 cm^{-1}. Figure 2 shows the XPS spectrum of the native oxide of amorphous silicon. The photoelectron spectrum exhibits a structure with two well-resolved peaks. The intensive peak at 99.4 eV corresponds to silicon, while the peak at ~102.8 eV corresponds to oxidized silicon. The latter peak is broad and is decomposed into two peaks using mixed Gaussian-Lorentzian curve fitting. The 103.2 eV peak shows a binding energy similar to that for silicon dioxide (103.4 eV); the peak at 101.6 eV falls in the binding energy range of suboxides (SiO_x with $x < 2$). Thus, the XPS study indicated the formation of a thin layer on the surface of the α-Si layer containing a mixture of SiO_2, SiO_x and Si. To estimate the thickness of the SiO_x layer, we used the quantitative relation proposed by J.M. Hill et al. [9]. The exact estimation is questionable due to the uncertainty in some of the parameters included in this relation, such as the escape depth of photoelectrons emitted from SiO_x and the intensity ratio of the signals from infinitely thick SiO_x and Si. There are significant discrepancies in the literature for these values. We used the value of 2.86 nm for the escape depth of Si 2p photoelectrons in SiO_2 excited by Al $K\alpha$ radiation as obtained by K. Takahashi et al [10]. Also, the value of 0.82 for the intensity ratio of infinitely thick SiOx and Si was taken from the work of M. F. Hochella et al. [11] for Si2p spectrum excited by Al $K\alpha$ radiation. Using these values we calculated the oxide layer thickness to be 2.1 nm.

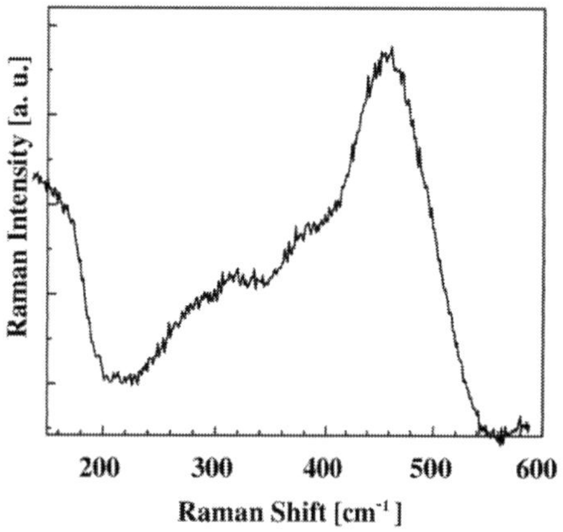

Figure 1. Raman spectrum of a Si film.

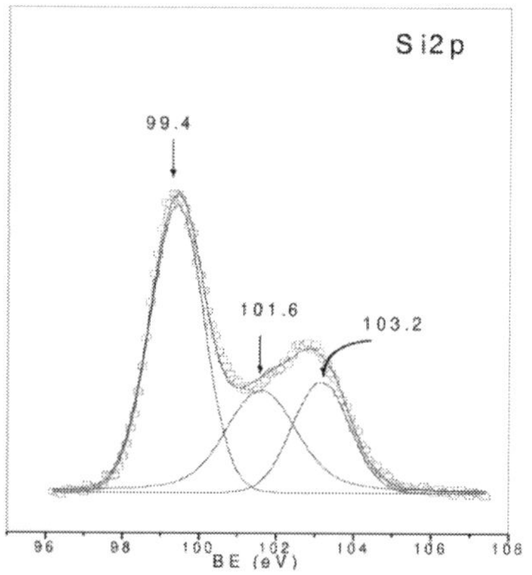

Figure 2. XPS spectrum of a Si film.

3.2. Gas sensing properties to NO_2 of the Si films

The gas sensing properties of α-Si were measured by a consecutive increase of the NO_2 concentration after the saturation process, because of the low unloading rate of the Si-QCM system. A considerable initial layer sensitivity was detected at 1000 ppm NO_2.

Figure 3 shows the response of a Si thin film to 1500 ppm NO_2 concentration. Three regions can be distinguished: the first one indicates a quick response, the second one, a slower response, and the third one is the unloading region after the saturation process. While in the first 52 s F changes by 21 Hz, in the next 68 s it decreases by only 6 Hz, i.e. the sorption rate decreases from 24 Hz/min to 5.4 Hz/min.

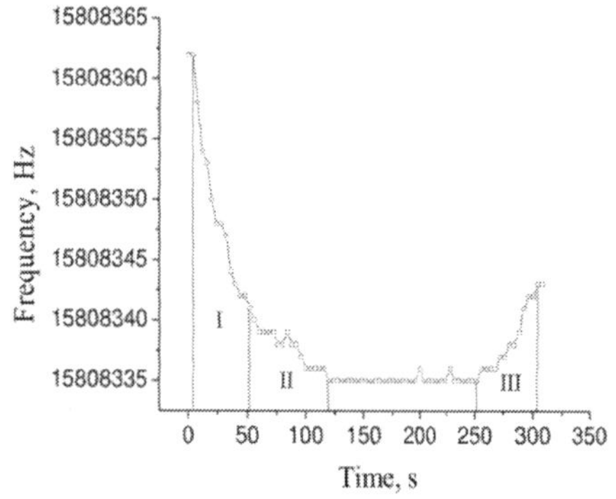

Figure 3. FTC of a Si-QCM structure. Si thickness – 174 nm, NO_2 – 1500 ppm.

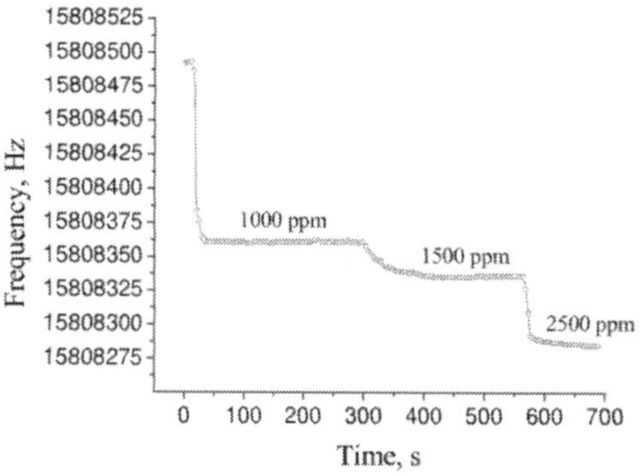

Figure 4. FTC of a Si-QCM structure vs NO_2 concentration.

The presence of two sorption stages proves that the process is limited by the layer's surface and its thickness. The desorption process in the third region proceeds slowly. Comparing the first region of sorption with the desorption region for a period of 52 s points to a desorption rate lower by a factor of three than the sorption rate (9 Hz/min). The low desorption rate is caused by the quantity of sorbed mass on the layer and its depth.

Recovering the frequency by scavenging the system showed that the sorption has a physical character. Figure 4 illustrates the response of the Si films to different (1000 – 2500 ppm) NO_2 concentrations. After treating the system by NO_2 (1000 ppm), a rapid decrease of the frequency shift was registered. The consecutive increase of the NO_2 concentrations in the flow after the processes of saturation showed a considerable decrease in the frequency shift. The ΔF for each NO_2 concentration was calculated.

Figure 5a demonstrates the linear character of the dependence obtained. In the interval studied of 1000 – 2500 ppm, ΔF changed (131 Hz - 208 Hz) and the sorbed mass of NO_2 was calculated correspondigly. The dependence is shown in figure 5b. The sorbed mass is increasing linearly from 29,13 ng to 46,26 ng when ΔF changes from 131 Hz to 208 Hz.

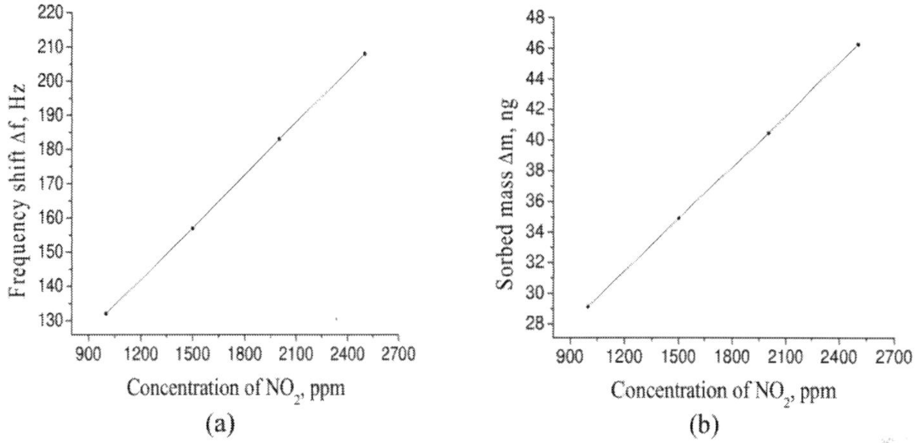

Figure 5. Frequency change (a) and mass-loading (b) of Si-QCM vs NO_2 concentration.

Conclusions

Si thin films were prepared by e-beam evaporation. The results obtained showed that the layers are amorphous with a thin SiOx films existing on their surface. The layers possess high sensitivity in the interval 1000 – 2500 ppm NO_2. The process of sorption is physical, so that the response times are short. The Si investigated in the Si-QCM structure could be used as sensor element for NO_2 detection.

Acknowledgements

The work is supported partly by the bilateral project (2009-2011) IFTT, BAS and GEOKHI, RAS and by Bulgarian Science Fund Project No. DO02-167/2008.

References

[1] Zhuikow S and Miura N 2007 *Sens. and Actuators* B **121** 639
[2] Ibrahim A, Al-Homoudi J, Thakur R, Naik and Auner G 2007 *Appl. Surf. Sci.* **253** 8607
[3] Bisi O, Ossicini S, Pavesi L 2000 *Surf. Sci. Reports* **38** 1
[4] Georgieva V, Donkov N, Spassov L 2008 *J. Phys.: Conf. Series* **113** 012053
[5] Georgieva V, Stefanov P, Raicheva Z 2009 *J. Optoelectr. Adv. Mater.* **11** 1363
[6] Sauerbrey G 1959 *J. Physik* **155** 206
[7] Street R 1991 *Hydrogenated Amorphous Silicon* (Cambridge Cambridge University Press)
[8] Lanin J, Pilione L, Kshirsagar S, Massier R and Ross R 1982 *Phys. Rev. B* **26** 3506
[9] Hill J, Royce D, Fadley C, Wagner L and Grunthaner F 1976 *Chem. Phys. Lett.* **44** 225
[10] Takahashi K, Nohira H, Hirose K and Hattori T 2003 *Appl. Phys. Lett.* **83** 3422
[11] Hochella M and Carim A 1988 *Surf. Sci. Lett.* **197** L260

Compositional dependence of the optical properties of silver containing As_2Se_3 thin films

V Ilcheva[1,2,3], P Petkov[2], T Petkova[1], V Boev[1] and B Monchev[1]

[1] Institute of Electrochemistry and Energy Systems, Bulgarian Academy of Sciences, Acad. G. Bonchev Str., Bl. 10, 1113 Sofia, Bulgaria
[2] Department of Physics, University of Chemical Technology and Metallurgy, 8, St. Kl. Ohridski blvd, 1756 Sofia, Bulgaria

E-mail: vania_ilcheva@yahoo.com

Abstract. We obtained amorphous thin films by evaporation and condensation in three-component systems based on As, Se and Ag. The aim was to investigate the influence of the third component on the thin film structure and optical properties.
The refractive index and the film thickness were determined from the upper and lower envelopes of the optical transmission spectra measured in the spectral range 400 – 2500 nm.
The absorption coefficient (α) was determined after extrapolating the values of n in the region of strong absorption (where $\alpha \geq 10^4$ cm^{-1}); its spectral distribution is discussed. The dispersion of the refractive index was analyzed in terms of the single-oscillator Wemple-DiDomenico model. The shift of the optical absorption edge was described using the non-direct transition model proposed by Tauc. The optical band gap was calculated from the Tauc plot $\alpha h\nu = B(Eg^{opt} - h\nu)^2$; the compositional dependence of the optical gap is also discussed.

1. Introduction

Chalcogenide glasses based on sulfide, selenide and telluride alloys in binary or multi-component systems are very promising materials for various optical and photonic applications in the spectral range 0.6 to 15 μm. Up to now, they have been studied mostly for applications as passive devices, such as lenses, windows, fibers; however, these glasses are also attractive for preparation of active devices, such as laser fiber amplifiers and non-linear components [1].

The glassy chalcogenide semiconductors' properties are affected by the addition of a third element. Experimental results reported by various workers have shown that the additives in various chalcogenide glasses change their optical properties [2].

In this paper, we present studies of the distribution of the optical parameters of VTE-layers as a function of the composition. The variation of the optical band-gap is discussed in terms of the local ordering modification due to Ag incorporation.

2. Experimental

Three compositions of the glassy $(As_2Se_3)_xAg_x$ alloys with x = 0, 15 and 20 mol. % Ag, synthesized by melt-quenching as described elsewhere [3], were used for thin film preparation by vacuum thermal

[3] To whom any correspondence should be addressed.

evaporation (VTE).

The thermal evaporation process was carried out at a source-substrate distance of 0.12 m, evaporation source temperature of 700-800 K and a residual gas pressure of 1.33×10^{-4} Pa. In order to avoid thickness non-uniformities, the substrates were rotated during the evaporation process. The thickness of prepared films varied from 627–1120 nm. The optical transmission spectra of the thin films were recorded in the wavelength range of 400 to 2500 nm at room temperature using a double-beam computer-controlled Cary 5E UV–VIS–NIR spectrophotometer with an accuracy of ±0.5 nm. The optical constants were derived by applying the Swanepoel method.

3. Results and discussion

The optical transmission spectra of the thin films were recorded in the wavelength range of 400 to 2500 nm. The films are transparent in the visible and near infrared spectral regions, with the spectra exhibiting interference maxima and minima and approaching the transmission of the glass substrate (figure 1).

The absorption edge is between 550 and 700 nm depending on the film composition. It gradually shifts to the longer wavelengths when Ag is added into the As-Se amorphous matrix. The red shift of the absorption edge observed in the spectra after addition of silver is probably caused by formation of defect states localized in the band gap.

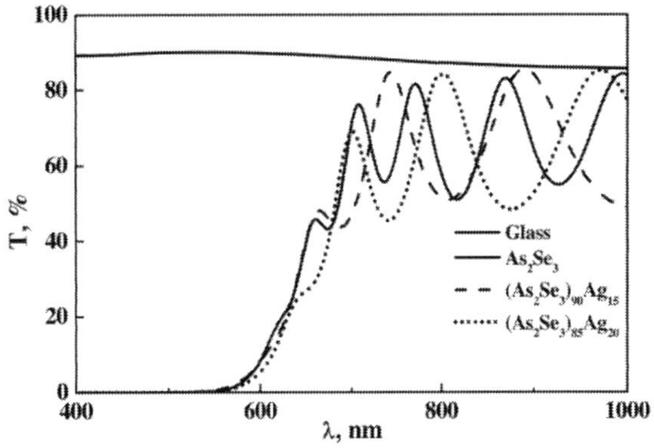

Figure1. Transmission spectra of the $(As_2Se_3)_{1-x}Ag_x$ thin films.

The optical constants and the thickness of the thin films were calculated from the transmission spectra using the method suggested by Swanepoel [4] for the case of uniform thin layers, which is based on the upper and lower envelopes of the transmission spectrum at normal incidence.

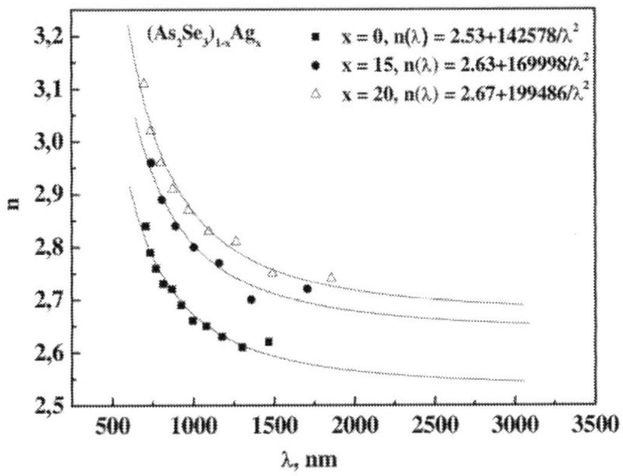

Figure 2. Spectral distribution of the refractive index of $(As_2Se_3)_{1-x}Ag_x$ films.

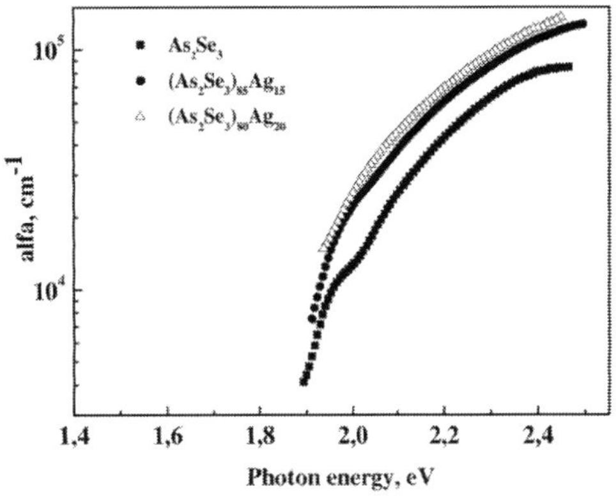

Figure 3. Spectral distribution of the absorption coefficient in $(As_2Se_3)_{1-x}Ag_x$ films.

Figure 2 shows the spectral dependence of the refractive index *n* of $(As_2Se_3)_{1-x}Ag_x$ films. The overall tendency is an increase in *n* with the silver content in the films, which is probably related to an increase in the number of heteropolar bonds. These bonds replace part of the homopolar bonds, present in the structural units, leading to an increase in the effective polarizability [5].

The dispersion of the refractive index was analyzed in terms of the Wemple–Di Domenico model [6,7], which is based on the single-oscillator formula $n^2(h\nu)-1=\frac{E_dE_0}{E_0^2-(h\nu)^2}$. By plotting $(n^2-1)^{-1}$ versus $(h\nu)^2$ and fitting a straight line, the single-oscillator energy (E_0) and the dispersion energy (E_d) can be determined from the slope, $(E_0E_d)^{-1}$, and the intercept, E_0/E_d, on the vertical axis, respectively. The values of the dispersion parameters E_d and E_0 are presented in table 1. The proportion between E_0 and Eg^{opt} will be discussed later, taking into account the relation $E_0=2Eg^{opt}$, found by Tanaka [8].

The absorption coefficient α for all films investigated was determined in the region of strong absorption ($\alpha \geq 10^4$ cm^{-1}), which involves optical transitions between the valence and conduction bands [5]. For this purpose, the values of n obtained were extrapolated in the strong-absorption region and α was estimated using the equation given in [4].

The spectral dependences of the absorption coefficient (α) of films with different Ag content are plotted in figure 3. It is obvious that introduction of silver leads to an increase in the absorption coefficient.

As we mentioned above, the strong-absorption region ($\alpha \geq 10^4$ cm^{-1}) corresponds to transitions between extended states in both valence and conduction bands where Tauc law [9] is valid. Thus, above the exponential tail, the absorption coefficient of amorphous semiconductors can be described by the relation $\alpha h\nu = B(h\nu - Eg)^m$, where $h\nu$ is the photon energy; Eg, the optical band gap; B, a constant that depends on the transition probability; m, an index depending on the nature of electronic transitions. For amorphous materials, non-direct optical transitions ($m = 2$) are observed.

The optical band gap Eg^{opt} was determined from the intercept on the energy axis of the linear fit in the strong-absorption region ($\alpha \geq 10^4$ cm^{-1}) of the plot $(\alpha h\nu)^{1/2}$ versus $h\nu$ (where α is the absorption coefficient and $h\nu$ is the incident photons energy), known as Tauc extrapolation [9].

The optical band gaps of the films derived from the Tauc plots are presented in table 1. The dependence $(\alpha h\nu)^{1/2}$ vs. $h\nu$ corresponds to a straight line giving a proof that non-direct transitions are responsible for the optical absorption in the high-energy spectral region [5]. The values obtained of Eg^{opt} for the initial As_2Se_3 chalcogenide matrix are in good agreement with the results published earlier [10].

Table 1. The optical band gaps of the films derived from the Tauc plots.

Glass composition	E_d, eV	E_0, eV	Eg^{opt}, eV
As_2Se_3	20,934	3,82	1,79
$(As_2Se_3)_{85}Ag_{15}$	21,618	3,59	1,74
$(As_2Se_3)_{80}Ag_{20}$	21,354	3,42	1,72

The decrease of the band gap with addition of silver is related to the structural transformation in the material as a result of silver incorporation. According to [11], the basic structural units in As-Se glasses are $AsSe_3$ pyramids. The addition of Ag in to As-Se matrix leads to a decrease in the $AsSe_3$ pyramids, which are replaced by Ag_3AsSe_3 structural units. The appearance of new structural units causes an increase of the disorder and the number of defects in the material structure.

The decrease in the optical band gap can be discussed in the framework of Mott and Davis model [12] relating the degree of disorder and defects in an amorphous structure to the width of the localized states near the mobility edges. As a result of insufficient number of atoms, unsaturated bonds together with some saturated bonds (such as dative bonds) are produced. These unsaturated bonds cause the formation of some defects in the material, responsible for the presence of localized states in the band gap of the amorphous solids. As we mentioned above, the introduction of silver results in an increase of the number of defects, due to formation of structural units with greater and/or smaller coordination than the regular structural units, which leads to an increase in the density of the localized states and, consequently, a decrease in the optical band gap of the material. We can conclude from the results listed in Table 1 that the values of the gaps E_0 and Eg^{opt} satisfy the already introduced relationship $E_0 = 2Eg^{opt}$.

Conclusions

$(As_2Se_3)_{100-x}Ag_x$ amorphous thin films with three different compositions were successfully deposited by vacuum thermal evaporation; their optical properties were determined from the transmission spectra measured at normal incidence.

The refractive index calculations, carried out by the modified Swanepoel method, show the tendency of an increase in n with the silver content over the entire spectral range studied, which is probably related to an increase in the effective polarizability after structural transformations in the films.

The optical band gap Eg^{opt}, determined by Tauc extrapolation, is found to decrease after silver addition. The decrease is probably due to the formation of defects in the amorphous structure giving rise to an increase in the localized states density and, consequently, a decrease in the optical band gap of the material.

The dispersion of the refractive index was analyzed in terms of the Wemple–Di Domenico model in order to determine the values of the dispersion parameters – single-oscillator energy E_0 and dispersion energy E_d. The values of E_0 and Eg^{opt} satisfy the Tanaka relationship $E_0 = 2Eg^{opt}$, as is typical for amorphous materials.

References

[1] Krbal M, Wagner T, Vlcek Mil, Vlcek Mir and Frumar M 2006 *J. Non-Cryst. Solids* **352** 2662
[2] Pandey V, Tripathi S K and Kumar A 2007 *J. Ovonic Res.* **3**/2 29
[3] Ilcheva V, Petkova T, Petkov P, Boev V, Socol G, Sima F, Ristoscu C, Mihailescu C N, Mihailescu I N, Popov C and Reithmaier J P 2009 *Appl. Surf. Sci.* **255**/24 9691
[4] Swanepoel R 1983 *J. Phys. E: Sci. Instrum.* **16** 1214
[5] Ma′rquez E et al. 2001 *Thin Solid Films* **396** 183
[6] Wemple S H and Di Domenico M 1971 *Phys. Rev. B* **3** 1338
[7] Wemple S H 1973 *Phys. Rev. B* **7** 3767
[8] Tanaka K 1980 *Thin Solid Films* **66** 271
[9] Tauc J 1974 *Amorphous and liquid semiconductors* (Plenum Press New York)
[10] Fayek S A, Elocker M, Foad S S, El-Fouly M H and Amin G A 1995 *J. Phys. D: Appl.Phys.* **28** 2150
[11] Němec P, Jedelský J, Frumar M, Štábl M and Černošek Z 2005 *Thin Solid Films* **484**/1-2 140
[12] Mott N F and Davis E A 1971 *Electronic Proc. in Non-Cryst. Mat.* (Clarendon Press, Oxford)

Journal of Physics: Conference Series **356** (2012) 012029 doi:10.1088/1742-6596/356/1/012029

Preparation and protection of iron and iron compounds

N Koprinarov[1,4], M Konstantinova[1], G Avdeev[2], T Ruskov[3] and Tz Tzacheva[2]

[1] Central Laboratory of Solar Energy and New Energy Sources,
Bulgarian Academy of Sciences, 72, Tsarigradsko Chaussee, 1784 Sofia, Bulgaria
[2] Institute of Physical Chemistry, Bulgarian Academy of Sciences,
Acad. G. Bontchev Str., Bl. 11, 1113 Sofia, Bulgaria
[3] Institute for Nuclear Research and Nuclear Energy,
Bulgarian Academy of Sciences, 72, Tsarigradsko Chaussee, 1784 Sofia, Bulgaria

E-mail: koprin@phys.bas.bg

Abstract. Iron, iron carbide and iron oxide nano- and micro-particles were synthesized in a hermetically sealed container using ferrocene and a mixture of ferrocene, xylene and water. The particles produced possess well expressed magnetic properties and are wrapped in a protective carbon cover. Carbon provides excellent protection against moisture and chemical influences and insures a long-lasting stability. Structural changes in the particles and their covers were examined at up to 1000 ^{0}C in vacuum and 800 ^{0}C in air, as were their stability under the influence of acids. The particles morphology was examined by scanning (SEM) and transmission electron microscopy (TEM); their chemical composition and crystal structure were studied by X-ray diffraction (XRD), Mössbauer spectroscopy and electron probe X-ray micro analysis and energy dispersive X-ray spectrometry (EDS).

1. Introduction

Iron is one of the most widely spread elements in nature, as it is a part of the elemental series H, He, C, O, Fe with which the star matter evolves. It forms compounds with a wide spectrum of chemical elements; its compounds with C and O represent a substantial part of this series. Because of its chemical activity under normal conditions on Earth, it cannot be found in a pure form. The good properties of its compounds, however, has made it the most often used metal in practically all indistrial sectors and in everyday life, but the problem still remains to be solved of the changes of its properties and those of its compounds under ambient influence. This is especially important in the cases when usage in micro- and nanostructures is intended [1, 2, 3], since the number of atoms on the surface of a very small iron particle becomes commensurate with that in the particle volume. The surface atoms undergo transformation to an unwanted compound, which leads to a drastic change in the particle's properties. Logically, the idea comes to mind to protect iron and its compounds from contact with the surrounding ambient by depositing a protective coating. This idea is not new [4, 5]. The search for implementation of reliable protection still continues, as no sufficiently good techniques for solving the problem have so far been found. In this respect, carbon seems to be a promising candidate. It is chemically stable when a graphite-type lattice is created and possesses good mechanical properties.

[4] To whom any correspondence should be addressed.

Journal of Physics: Conference Series **356** (2012) 012029 doi:10.1088/1742-6596/356/1/012029

Moreover, when carbon is deposited from a gaseous phase, iron act as a catalyst in the formation of the graphite-type carbon covering on its surface [6 - 9].

The studies presented here were aimed at establishing whether different iron particles can be obtained and coated during the decomposition of carbon and iron containing compounds.

2. Experimental

Ferrocene ($Fe(C_5H_5)_2$) (Ventron) is used as a source of atoms for forming iron particles. It has the advantage that, when decomposing, except iron atoms it also supplies carbon atoms necessary for the protective coating in even greater numbers than needed. The pyrolytic decomposition proceeds in a stainless steel container (figure 1). Its opening is hermetically sealed by a sphere made also of stainless steel, which is continuously pressed against the container by a hydraulic press thus creating a pressure of 80 MPa inside the container. The temperature of decomposition is reached at a rate of 200 °C/min by heating by a gas burner, while the cooling rate is 30 °C/min. The temperatures of synthesis are within the 300 °C – 600 °C interval and are controlled by a thermocouple mounted in the container wall. The prepared material is studied without any additional treatment. Some of the experiments are carried out with a ferrocene, xylene and water mixture in weight ratio 3:4.5:1 as a starting material. The water is used to stimulate the iron oxidation, while xylene is chosen as an additional source of carbon atoms.

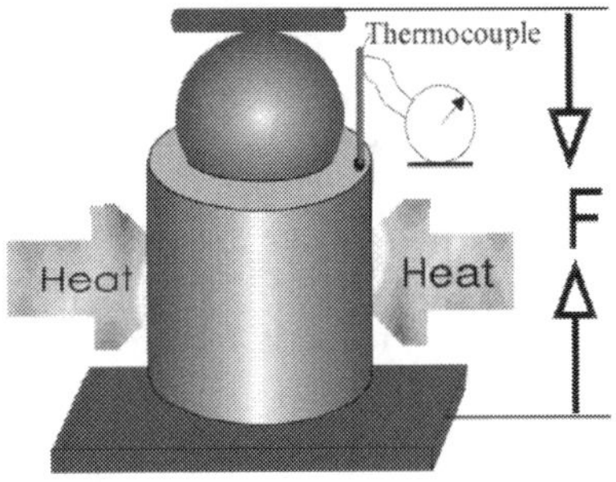

Figure 1. Schematic of the equipment used.

3. Results and discussion

The experiments on pyrolytic decomposition of ferrocene in the temperature interval 300 °C - 600 °C indicate that the temperatures in this interval are sufficiently high for ferrocene decomposition. The XRD spectra of the material obtained demonstrate that the iron and iron compounds prepared differ in composition and structure depending on the synthesis temperature, the process duration and the presence of additional starting substances in the container. For instance, the spectrum of the material prepared at 300 °C confirms the decomposition of ferrocene (figure 2) and the formation of Fe, Fe_3C, and Fe_2O_3. This is supported by Mössbauer spectroscopy of the materials synthesized at 300 °C for different times: 1 min, 5 min, and 30 min (figure 3). The Fe-containing phase ratio

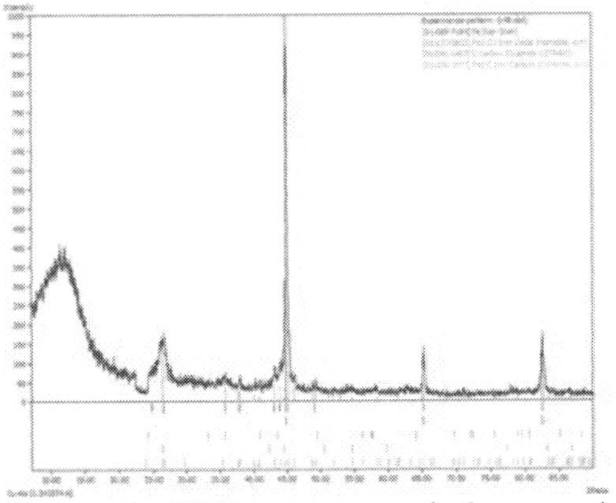

Figure 2. XRD spectrum of the material prepared.

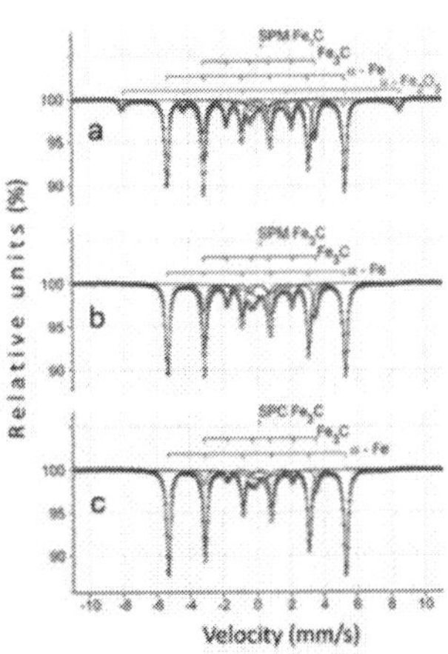

Figure 3. Mössbauer spectra of materials synthesized at 300 °C for: (a) 1 min; (b) 5 min; (c) 30 min.

α-Fe /Fe_3C/superparamagnetic Fe_3C/α-Fe_2O_3 obtained from the Mössbauer spectra is, respectively, 56/25/11/8; 64/21/15/0; 78/11/11/0. One should conclude that the synthesis duration influences not only the material type produced but its quantity as well.

The number of carbon atoms released by the decomposed ferrocene exceeds by a factor of ten that of the iron atoms; when the iron nuclei are of small size, this number is sufficent to cover them by protective coatings [10]. The case of low temperature pyrolysis is similar. The TEM and SEM observations and their analysis indicate that the iron atoms produced by low-temperature ferrocene decomposition (300 °C; 1-5 min) agglomerate and form nuclei of pure iron or of iron compounds covered by carbon (see figure 4a). Because of the small atomic mass of carbon, the carbon layer in the TEM and SEM photos is semitransparent. The number of available carbon atoms rises together with the nucleus size growth when the process lasts longer (20 - 60 min). Conditions appear for growth of carbon nanotubes inside which iron nuclei often remain enclosed (see figure 4b).

The nuclei become larger with morphology typical of magnetite (figure 5a) when the synthesis duration is extended (30 - 60 min) at high temperatures (600 °C - 700 °C) in the presence of xylene and water, as confirmed by the EDS spectra (figure 5b). Under these conditions, only a part of the carbon atoms are spent for the development of coatings. The remaining carbon atoms do not come in contact with the nuclei and build perfect carbon spheres. In the SEM image in figure 5a, the crystallized iron nuclei are seen in white and their carbon coating and the pure carbon spheres are darker.

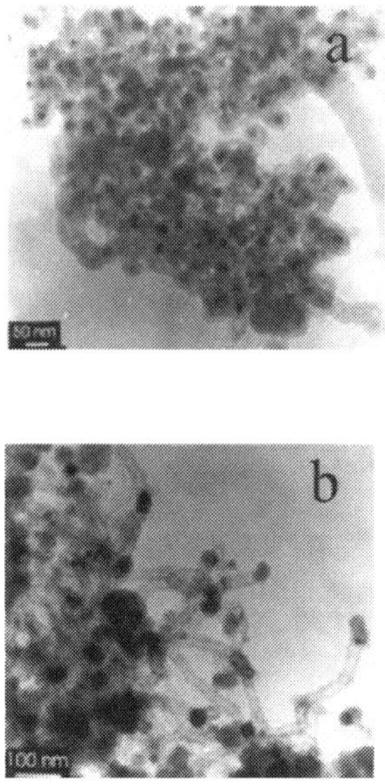

Figure 4. TEM images of material synthesized at low temperature for: (a) 1 min., (b) 30 min.

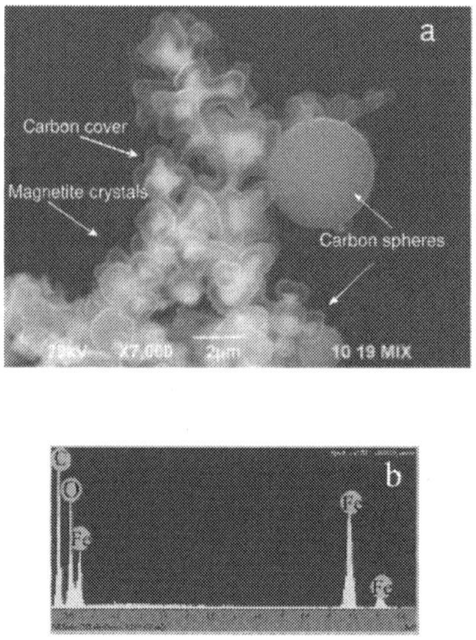

Figure 5. Material synthesized at 600^0 C:
(a) SEM image;
(b) EDS spectrum.

Figure 6. Magnetite crystals: (a) before etching; (b) after etching in HNO_3; (c) after etching in HCl.

The ability of the coatings obtained to protect the iron nuclei was tested by placing the synthesized material in 30 % HCl and in 70 % HNO_3 at 80 °C. After a 100 hour treatment, no variation in the particle morphology was registered (figure 6). Moreover, no changes in the materials morphology were noticed following thermal treatment at 1000 °C in vacuum and at 600 °C in air.

Conclusions

Experiments with ferrocene pyrolytic decomposition in a hermetically sealed container show that nano- and micro-particles of various morphology and chemical composition can be synthesized. The protective coatings formed are of different types. At low temperatures and a short synthesis process, the iron containing nuclei are small in size and tightly wrapped in a carbon layer. As the temperature and the synthesis duration are increased, carbon starts to form nanotubes using an iron nucleus as a catalyst. In most of the cases, the nucleus remains enclosed inside the tube. At still higher temperatures, besides coating iron nuclei, carbon starts building pure carbon spheres.

Acknowledgements

This work was supported by the Bulgarian National Science Fund (contract No. DOO2-241/18.12.2008 and contract No. DTK 02-77/22.12.2009).

References

[1] Jung C W and Jacobs P 1995 *Magn. Res. Imaging* **13** 661
[2] Zhang L, Gu F X, Chan J M, Wang A Z, Langer R S and Farokhzad O C 2008 *Clinical Pharmacology & Therapeutics* **83/5** 761
[3] Artemov D, Mori N, Okollie B and Bhujwalla Z M 2003 *Magnetic Resonance in Medicine* **49** 403
[4] Hayashi K, Ohsugi M, Kamigaki M, Xia B and Okuyama K 2002 *Solid State Lett.* **5** J9
[5] Sun G, Li X, Wang Q and Yan H 2010 *Mater. Res. Bulletin* **45** 519
[6] Endo M, Kim Y A, Matusita T, Hayashi T 2001 Carbon Filaments and Nanotubes: Common Origins, Differing Applications *NATO Sci. Series E* eds L P Biro, C A Bernardo, G G Tibbetts and P Lambin **372** 51 (Springer)
[7] Liu B, Wei L, Ding Q and Yao J 2004 *J. Cryst. Growth* **277** 293
[8] Kim H and Sigmund W 2005 *J. Cryst. Growth* **276** 594
[9] Ando I, Zhao X, Sugai T and Kumar M 2004 *Mater. Today* **22**
[10] Koprinarov N, Konstantinova M and Marinov M 2010 *J. Solid State Phen.* **159** 105

Transformation of the coherent population trapping resonance into enhanced fluorescence in a paraffin coated Rb vacuum cell

E Taskova, E Alipieva[1] and G Todorov

Emil Djakov Institute of Electronics, Bulgarian Academy of Sciences,
72, Tsarigradsko Chaussee, 1784 Sofia, Bulgaria

E-mail: alipieva@ie.bas.bg

Abstract. The ultra-narrow resonances achieved in coherent spectroscopy of alkali atoms originate from the destruction of the laser-induced coherence in the ground state. An anti relaxation cell coating preserves the coherence created and, consequently, the resonance narrows by almost two orders of magnitude. Coherence resonances are used in magnetic field sensors, atomic clocks, quantum informatics, ets. In this work, we investigate the features of the CPT resonance on the D1 87 Rb line obtained in fluorescence in paraffin coated cells. The transformation from a black to a bright resonance is found to depend on he laser power and the experimental geometry.

1. Introduction

Coherent population trapping (CPT) is an interference effect arising in a three-level atomic system after interaction with two coherent fields which connect the ground levels to a common upper level [1]. In a Hanle-configuration, the CPT resonance is due to the interference between the Zeeman sub-levels of a single hyperfine level. The hyperfine transition is excited by linearly polarized single frequency laser light which connects the Zeeman sub-levels in Λ-systems through its σ^+ and σ^- components. The coherent fields connect the lower levels of the system in a new, non-absorbing state. The effect manifests itself as a dark resonance (lack of fluorescence), or as a peak of transmission. The resonance width is determined by the lifetime of the coherent state. As the ground states are long lived, the CPT resonances are ultra narrow. Widths of 1-10 Hz have been measured.

Alkali atoms in vapor phase depolarize after collision with the cell walls thus limiting the coherence lifetime of the non-absorbing state. In order to prevent such depolarization, vapor cells are covered with an anti-relaxation coating. The most popular is a paraffin coating which has been shown to allow up to 10 000 collisions before destroying the atom coherence [2]. In a Hanle configuration experiment the signal is very sensitive to a stray magnetic field. An advantage of the covered cells is that magnetic field gradients are reduced which narrows the resonance.

The CPT resonances in coated cells have some new features with respect to the uncoated ones. The long lifetime of the coherent state makes possible the transfer of coherence by multi-photon processes because of mixing of the contributions of polarization moments with different ranks. The competition

[1] To whom any correspondence shoul be addressed.

Journal of Physics: Conference Series **356** (2012) 012030 doi:10.1088/1742-6596/356/1/012030

between processes with different rates leads to new structures in the signal observed. They cannot be connected with one definite polarization moment [3] and may change in sign. Additional "bright" resonances appear in the Λ-level system, where only dark resonances are registered in an uncoated cell [4].

CPT in a coated cell has been investigated mainly in absorption [5]. In this work we investigate the coherent population trapping in a paraffin covered cell by means of the fluorescence signal.

2. Experimental

2.1. The set up

The experimental set-up is shown in figure 1(a). It consists of a single-mode diode laser oscillating at $\lambda = 794.76$ nm, a Rb cell and a system for registration and control of the experiment. The laser beam passes through an optical isolator to avoid the optical feedback and a polarization rotator to change the laser polarization direction when necessary. The laser power is attenuated by means of grey filters. Then the laser beam passes through the vapor cell enclosed in a 3-layer magnetic shield which screens the laboratory magnetic field The cell is paraffin coated, containing ^{87}Rb. The frequency of the laser is controlled by monitoring the fluorescence from a second Rb vapor cell. All measurements are performed at room temperature (25°C). A dc magnetic field **B**, created by a solenoid, is applied collinearly to the laser beam. The current through the solenoid is controlled by a PC.

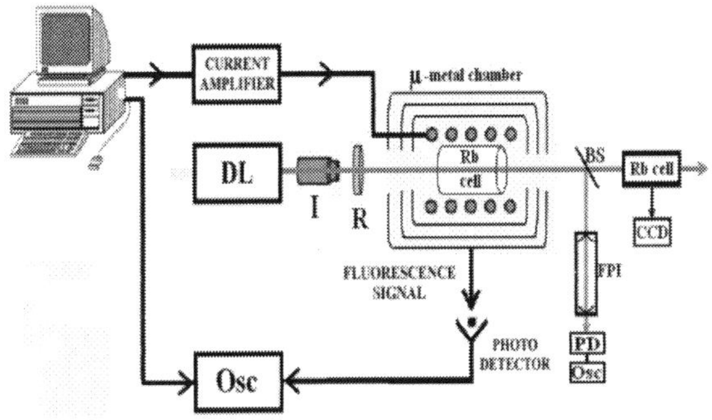

(a)

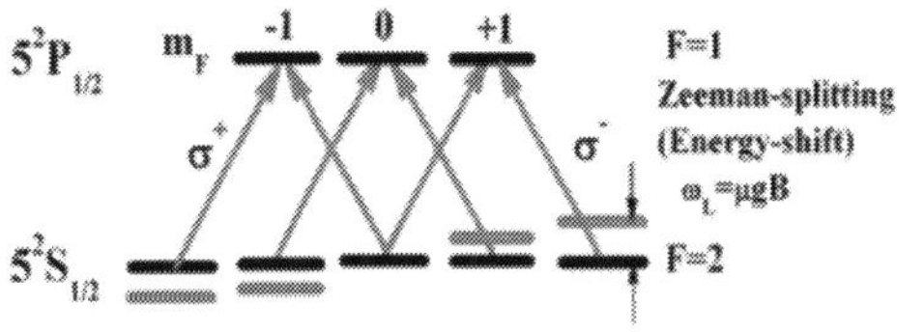

(b)

Figure 1. The experimental set-up (a); Λ-system in 5S1/2→5P1/2 87Rb D1 formed by a linear polarized laser (b).

The fluorescence from the cell is detected by a photodiode. Its dependence on the dc magnetic field **B** is stored in a two channel digital oscilloscope. Our experiment was performed on the $5S_{1/2}\rightarrow 5P_{1/2}$ transition ($F_g=2\rightarrow F_e=1$) of the ^{87}Rb D_1 line shown in figure 1(b). The linearly polarized laser beam through its σ^+ and σ^- components connects all sub-levels of the transition investigated into 3 Λ-systems, creating CPT resonance when the levels are degenerate, i.e. at zero magnetic field.

2.2. Experimental results

The CPT resonance in coated cells has a dual structure – a narrow resonance superimposed on a broad pedestal [5]. The pedestal is due to the time of flight and is determined by the laser beam radius. Interesting in view of applications is the narrow resonance part whose width is determined by the relaxation of the laser-created coherent state. Due to a difference of two orders of magnitude in the two widths they are recorded separately. In this work we investigate the narrow resonance. The magnetic field was scanned from -1 mG to +1 mG. The laser power P_{las} was varied from 5 μW to 500 μW. The fluorescence signal was detected in two orthogonal directions. The resonance detected in the direction parallel to the E preserves its sign. At low P_{las} (less than 100 μW), the resonance detected in a direction orthogonal to E is "bright" indicating enhanced fluorescence around the zero point of the scanned magnetic field. Above this value, the resonance sign changes and a "dark" resonance is detected in both directions. The CPT resonance sign and width in an uncoated cell does not depend on the observation direction and the laser power and is "dark" [6]. The width of the CPT resonance in fluorescence detected along the E_{las} is twice as narrow. The resonances obtained in the orthogonal directions are shown in figure 2.

The possibility of controlling the transformation of CPT resonances in Λ-systems from "dark" to "bright" have been recently investigated theoretically [7,8]. Experimental results have been reported using two counter-propagating laser beams [9] in the presence of a magnetic field applied in a transverse direction.

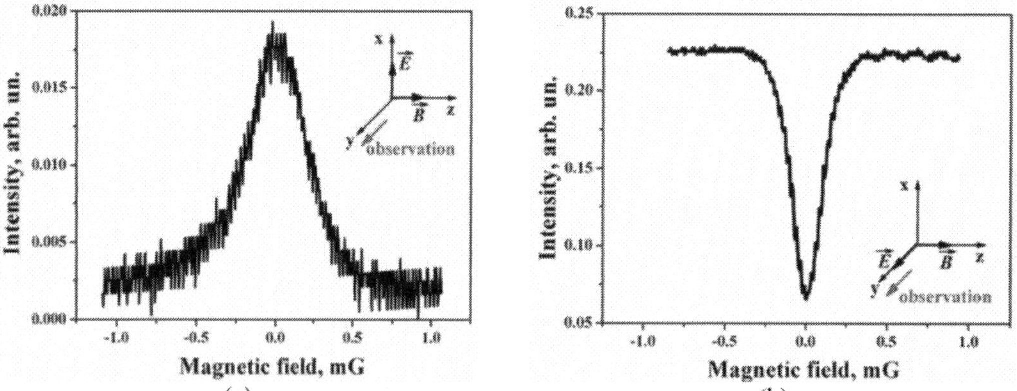

(a) (b)

Figure 2. CPT resonance detected in direction perpendicular to E (a); CPT resonance detected in direction parallel to E (b).

3. Theoretical

The theoretical analysis of the CPT resonances obtained by means of a Hanle effect configuration and observed in fluorescence is based on the standard semiclassical description of the atomic system by the statistical operator $\hat{\rho}$ in density matrix representation. The irreducible tensor operator (ITO) formalism is used. The numerical calculations are based on the model presented in [10]. The task is to estimate the influence of the relaxation constants of the tensor components with different ranks on the shape and sign of the resonance detected in two directions, namely, parallel and orthogonal to the laser polarization vector.

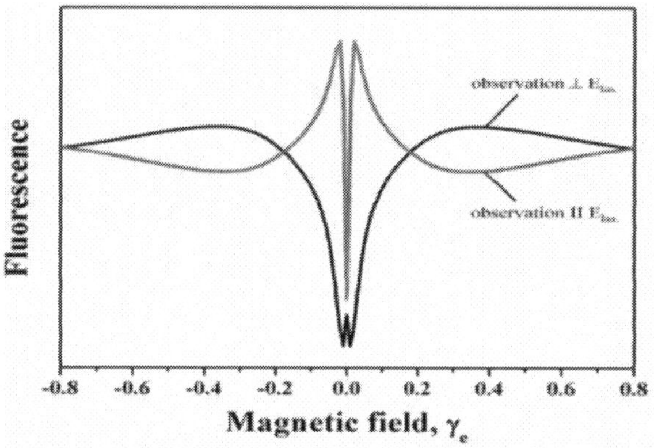

Figure 3. Numerically calculated dependence of the fluorescence detected in two orthogonal directions, i.e. parallel and perpendicular to the laser polarization.

The relaxation constants of the ground state $\gamma_g(\kappa)$ and the Raby frequency were varied taking into account that in cells with an anti-relaxation coating the transverse relaxation rate is much higher than the longitudinal relaxation rate $\gamma_g(2) >> \gamma_g(0)$. Figure 3 shows the signals in fluorescence for two orthogonal directions of observation – along and perpendicular to the vector of laser polarization E. They show qualitative agreement with the experiment. The resonance obtained in a direction perpendicular to E is twice as narrow as the one obtained parallel to E. The sign of this resonance changes as the laser power is increases, while the resonance in a collinear observation preserves its

Journal of Physics: Conference Series **356** (2012) 012030 doi:10.1088/1742-6596/356/1/012030

sign. These predictions were experimentally confirmed in our experiment. The numerical calculations will be improved by taking into account the velocity distribution of the atoms, the Gaussian shape of the laser beam and the stray magnetic fields.

Conclusions

The features were investigated of the CPT resonance of the D1 ^{87}Rb line obtained in fluorescence in a paraffin-coated cell. Transformation from black to bright resonance was observed depending on laser the power and the experimental geometry. The resonances obtained in fluorescence in two directions – parallel and perpendicular to the laser polarization vector, differ by a factor of two. The numerical calculations show qualitative agreement with the experiment.

Acknowledgements

This work was supported by the Bulgarian National Science Fund (Grant No: DO-02-108/2009).

References

[1] Arimondo E 1996 *Prog. Opt.* **35** 257-354
[2] Bouchiat M A and Brossel J 1966 *Phys. Rev.* **147** 41
[3] Balabas M V et al 1999 *JETP Lett.* **70** 198-202
[4] Gateva S, Alipieva E, Taskova E and Todorov G 2011 *Proc. SPIE* **7747** 77470G
[5] Klein M, Novikova I, Phillips D F and Walsworth R L 2006 *J. Mod. Opt.* **53**/16-17 2583-91
[6] Alipieva E, Gateva S, Taskova E and Cartaleva S 2003 *Opt. Lett.* **28**/19 1817-9
[7] Chou H-S and Evers J 2010 *Phys. Rev. Lett.* **104** 213602
[8] Braznikov A, Taichanachev and Yudin V 2011 *Eur. Phys. J. D* 10.1140/epjd/e2011-20112-6
[9] Kim S K, Moon H S, Kim K and Kim J B 2003 *Phys. Rev. A* **68** 063813
[10] Huss A, Lammegger R, Windholz L, Alipieva E, Gateva S, Petrov L, Taskova E and Todorov G 2006 *JOSA B* **23**/9 1729-36

Annealing of Si surface region modified by plasma immersion implantation of nitrogen

S Alexandrova[1], E Halova[1,4], A Szekeres[2], E Vlaikova[2], M Gartner[3] and M Anastasescu[3]

[1] Technical University of Sofia, 8, Kl. Ohridski Blvd., 1797 Sofia, Bulgaria
[2] Institute of Solid State Physics, Bulgarian Academy of Sciences, 72, Tsarigradsko Chaussee, 1784 Sofia, Bulgaria
[3] Institute of Physical Chemistry, Romanian Academy, Spl. Independentei 202, 060021 Bucharest, Romania

E-mail: ehalova@tu-sofia.bg

Abstract. In the present work, the formation of a nano-scale Si surface layer is studied after high-temperature annealing of Si modified by shallow plasma immersion implantation of nitrogen with fluences of 10^{16} - 10^{18} cm^{-2}. The implanted profiles of the atomic (N^+) and molecular nitrogen (N_2^+) are modeled by SRIM for different annealing durations taking into account the diffusion process. The presence of Si-O and Si-N bonds is established by Fourier (FTIR) spectral analysis and spectroscopic ellipsometry (SE). The refractive index value measured at 632.8 nm varies between 1.46 and 1.59, corresponding to a low y/x ratio. The models using VIS and IR ellipsometric data reveal formation of nanostructured SiO_xN_y layer with Si inclusions.

1. Introduction

Silicon oxynitride films find different applications as an important dielectric material. It exhibits low absorbance and adjustable refractive index between 1.46 (SiO_2) and 1.9 (Si_3N_4) [1], being thus suitable for optoelectronic devices. Low-energy ion-implantation of nitrogen for formation of silicon nitride layers offers an alternative to the traditional CVD deposition methods with the advantage of ensuring clean conditions. By adjusting yhe implantation parameters and the post-implantation conditions, high-quality nanoscaled layers with easily controlled composition and thickness [2] can be produced.

The aim of the present study was to synthesize ultrathin SiO_xN_y layers on the silicon surface with controllable nitrogen concentrations by using shallow nitrogen plasma implantation and subsequent annealing in oxidizing ambient.

2. Experimental details

The experiments were carried out on Cz-grown (100)p- and (111)n-type silicon substrates with resistivity of 5-8 Ohm.cm. The first step of the processing was the incorporation of nitrogen ions into the near surface Si region through plasma-beam ion immersion (PII) implantation technology. The N^+ ion beam with energy of 4 keV was directed at normal incidence with fluencies ranging from

[4] To whom any correspondence should be addressed.

10^{16} to 10^{18} cm^{-2}. It should be noted that the implanted species are atomic nitrogen (N^+) and molecular (N_2^+) ions from the plasma. The implanted Si substrates were subjected to high-temperature furnace annealing at 1050 °C up to 20 min in dry oxygen O_2 ambient at atmospheric pressure.

The optical and structural properties of the synthesized SiO_xN_y layers were studied in detail with spectroscopic ellipsometry (SE) in the visible (VIS-SE) and in the infrared (IR-SE) range and Fourier transform infrared (FTIR) transmission spectroscopy. The ellipsometric measurements were performed in the range 190–1700 nm at an angle of incidence of 70° on a J.A. Woollam Co. variable-angle spectroscopic ellipsometer. The infrared (IR) spectroscopic measurements were carried out using a Perkin-Elmer 1430 IR spectrophotometer in transmission mode at normal incidence. The spectra were collected in the spectral range from 400 to 4000 cm^{-1} with a resolution of 4 cm^{-1} at room temperature. The depth profiles of implanted ions and vacancies were derived using the SRIM simulation code based on a Monte-Carlo method for different annealing durations taking into account the diffusion of the implanted species.

3. Results and Discussion

The thickness and the refractive index of the layers synthesized on Si substrates were estimated from the VIS-SE ellipsometric data. Figure 1 shows the thickness and the refractive index values at a wavelength of 632.8 nm of the layers for implantation with different N^+ fluencies and different annealing durations. The refractive index reaches 1.59, corresponding to a low y/x ratio.

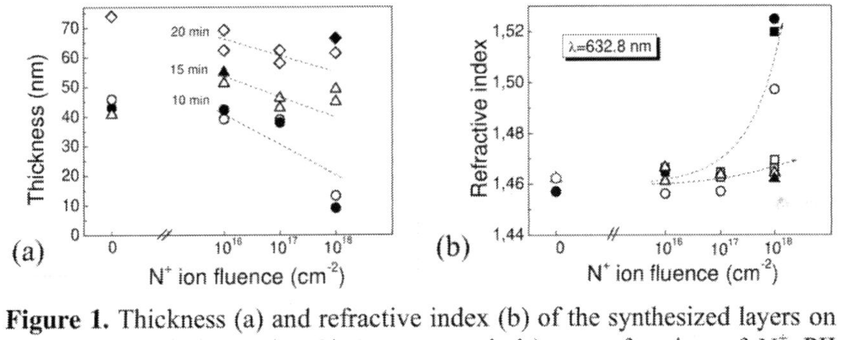

Figure 1. Thickness (a) and refractive index (b) of the synthesized layers on n-Si (full symbols) and p-Si (empty symbols) as a function of N^+ PII implantation fluence.

The process of SiO_xN_y formation is determined to a high extent by the profile of the implanted nitrogens and the induced defects, such as vacancies and interstitials, N and Si. Taking into account that the plasma implantation ambient provides N^+ and N_2^+ ions; figure 2 presents the ion and the vacancy profiles obtained by the SRIM modeling. It is seen that the concentration of N atoms available for SiO_xN_y formation changes substantially with the depth. For both N^+ and N_2^+, the concentration of vacancies is much higher than the concentration of the implanted nitrogen atoms. The results of SRIM modeling, given in figure 2, show the expected depth of nitrogen ion penetration with a maximum concentration located at 14.4 nm for N^+ and 8.3 nm for N_2^+ below the Si surface. It should be stated that for 10^{18} cm^{-2}, the concentrations corresponding to the profiles are close to a nitrogen content of $5{,}8\times10^{22}$ cm^{-3} of stoichiometric Si_3N_4 over a substantial depth in the Si substrate [3].

The SiO_xN_y layer formation is accomplished by formation of Si-N and Si-O bonds. The vacancies play an important role for diffusion of the oxidizing species and as available sites for N and/or Si atoms. It can be suggested that the local concentration ratio of the available atoms and vacancies will substantially determine the reactions leading to formation of SiO_xN_y, i.e., the degree of nitridation through the formed layer.

The high temperature synthesis can be alter the nitrogen implantation profile due to diffusion of N atoms. During annealing under oxidation conditions, nitrogen atoms already present in the Si network will compete with the diffusing oxygen to form Si-bond. The presence of oxygen can influence the impurity diffusion. The implanted nitrogen can be found mostly at interstitial position forming directly interstitial pairs, i.e., N-N dimers [4]. Only a very small amount of N goes to substitutional Si sites [4].

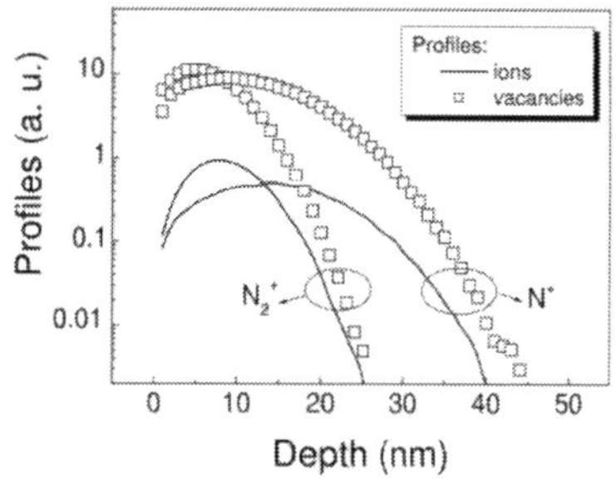

Figure 2. Ion and vacancy profiles for N^+ and N_2^+ implants.

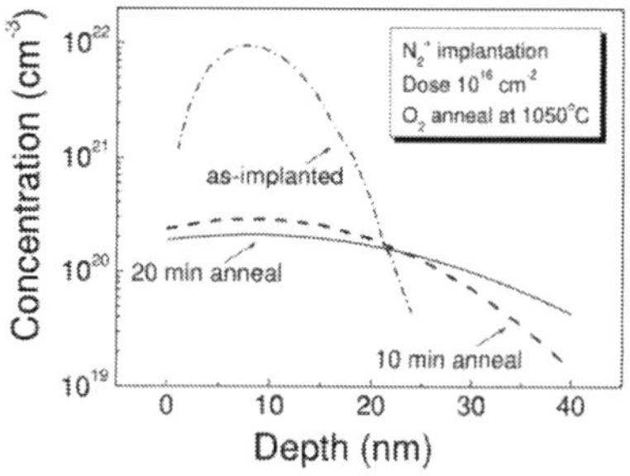

Figure 3. Ion profiles before and after high temperature treatment.

The N-N dimer was found to move with an activation energy of ~3 eV, a value very close to the experimentally found energy for diffusion of nitrogen in Si. We suggest that the N_2^+ implants serve as a N-N dimmer precursor. Therefore, we examined the N_2 migration at the annealing temperature of 1050°C. The results from the simulation of the implantation profile in the Si substrate are plotted in figure 3 for annealing times of 10 and 20 min. For comparison, the as-implanted ion profile is also given. The simulation was performed using the implantation profile and standard diffusion equations with a diffusion coefficient $D = D_o \exp(-E_a/kT))$, where $D_o = 0.87 x 10^{-4}$ m^2s^{-1} and the activation energy E_a = 3.9 eV. The broadening of the profile is evident and the nitrogen concentration decreases in the whole implanted region, becoming substantially lower than the value of 5.8×10^{22} cm^{-3} for stoichiometric Si_3N_4 [3], substantiating the assumption for formation of SiO_xN_y with low N content. Moreover, a contribution to this effect comes from the high outdiffusion of the implanted atoms because of the shallow plasma implantation. No results are shown in figure 3 for the N^+ migration since the diffusion was found to proceed rapidly with a barrier of only 0.4 eV leading to flattened N profile at an insignificant low concentration level.

Information that SiO_xN_y synthesis terminated with formation of Si-N and Si-O bonds was obtained from the IR spectra analysis. No bands connected to the Si–H (2170 cm^{-1}), N–H (3350 cm^{-1}) or O–H (3640 cm^{-1}) bonds were detected. The IR transmission spectra are shown in figure 4 in the range below 2000 cm^{-1} since the main bands for SiO_xN_y appear up to about 1400 cm^{-2}. Evidence of SiO_xN_y formation is the broad band in the range of 650-1000 cm^{-1}, marked in figure 4, attributed to the stretching mode of Si-N bonds. This band is observed in all spectra; its intensity increases with the N fluence. A band related to Si-N stretching bonds in silicon nitride is known to appear within the 800-900 cm^{-1} range with position and intensity depending on the y/x ratio in Si_xN_y [5-7]. One should note that as the N fluence increases, the intensity of the 1096 cm^{-1} band decreases because of the increased implantation damage and enhanced SiOx formation [8, 9]. The band becomes less pronounced and the full width at half height (FWHH) increases with the fluence as seen in the inset of figure 4. This is an indication that the structure of the synthesized layer becomes less ordered. The rocking mode of Si–O–Si bonds at 460 cm^{-1} is also present, obviously independent of the ion fluence. This is reasonable, since this mode is not being influenced by structural changes [8]. The bending mode of the Si–O bond at 800 cm^{-1} cannot be distinguished because it overlaps with the broad Si-N band. The band at 614 cm^{-1} is typical for the

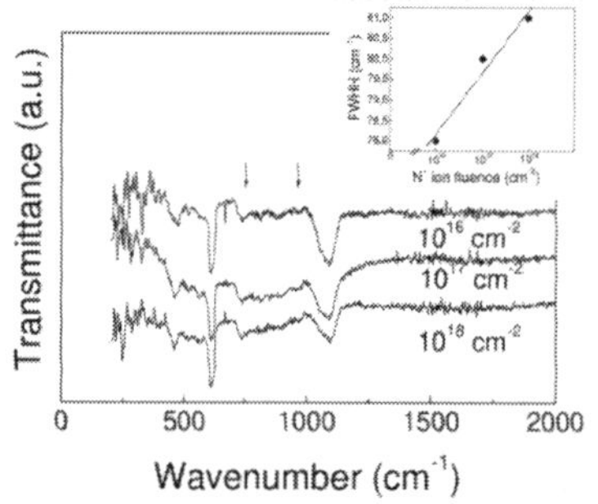

Figure 4. IR spectra of the SiO_xN_y layers synthesized at different nitrogen fluences.

Si–Si bond vibration and is an indication for the presence of Si-Si clusters. These considerations allow estimation of the synthesized layer as silicon-rich SiO_xN_y with Si inclusions.

Additional information about the chemical composition and molecular vibrations is gained from ellipsometric measurements in the IR region. Figure 5 shows the spectra of the Ψ angle for the studied layers. The Ψ spectra in figure 5 a for 10 min annealing contain a feature at 1250 cm^{-1}. The intensity of this peak decreases steadily with the ion fluence without the characteristic shift. It can be ascribed to the stretching LO-phonon vibration mode of Si-O-Si bonds. Due to the Berreman effect, this peak appears in the IR reflectance spectrum recorded at a proper incidence angle but it is not present in the normal incidence transmittance IR spectrum. The peak around 1250 cm^{-1} was found in native oxides and in thermal silica, as well as in SiO_x [10]. A shoulder at about 1150 cm^{-1} is an indication of Si-N bond formation, as is typical for Si_3N_4 [11]. Upon prolonged annealing for 20 min, the Ψ angle spectra change, as can be seen in figure 5 b. The shoulder at about 1150 cm^{-1} has become clearly pronounced, which evidences formation of Si_3N_4.

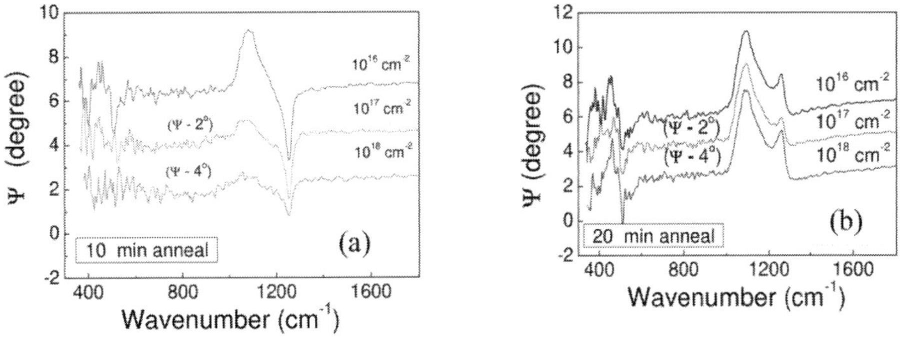

Figure 5. IR-SE spectra of the ellipsometric angle Ψ for films synthesized on N^+ ion implanted Si with different implantation fluence (shifted along Y-axis) for 10 min (a) and 20 min (b).

Conclusions

SiO_xN_y layers were synthesized by high temperature annealing in O_2 ambient of plasma-beam nitrogen ion implantation for different durations. The FTIR analysis revealed the presence of Si–N bonds and SiO_x depending on the implantation fluence. The refractive index was found to increase with the ion fluence up to 1.59, implying a low y/x ratio. The IR-SE analysis revealed a gradual increase of the Si_3N_4 fraction. The FTIR and VIS-SE experimental results, as well as the SRIM–based modeling, allowed us to identify the layers as nanostructured silicon-rich SiO_xN_y films with low N content.

References

[1] del Prado A, San Andrés E, Martínez F L , Mártil I, González-Díaz G, Bohne W, Röhrich J, Selle B and Fernández M 2002 *Vacuum* **67** 507
[2] Pan J S, Wee A T S, Huan C H A, Tan H S and Tan K L 1997 *Appl. Surf. Sci.* **115** 166
[3] Bachurin V I, Churilov A B, Potapov E V, Smirnov V K, Makarov V V and Danilin A B 1999 *Nucl. Instrum. Meth. B* **147** 316-9
[4] Jones R, Hahn I, Goss J P, Briddon P R and Oberg S 2004 *Solid State Phenomena* **95-96** 93-8
[5] Honda R Y, Mota R P, Batocki R G S, Santos D C R, Nicoleti T, Kostov K G, Kayama M E, Algatti M A, Cruz N C and Ruggiero L 2009 *J. Phys.: Conf. Series* **167** 012055
[6] Larijani M M and Haidari G 2010 *World Appl. Sci. J.* **8** 972-9
[7] Lange P, Schnakenberg U, Ullerich S and Schliwinski H-J 1990 *J. Appl. Phys.* **68** 3532-7
[8] Rui Y, Chen D, Xu J, Li W, Cen Zh, Huang X and Chen K 2007 *App. Surf. Sci.* **253** 8647-51
[9] Kaneta C, Yamasaki T and Kosaka Y 2003 *Fujitsu Sci. Techn. J.* **39** 106-18
[10] Cigal J C 2002 A Novel Spectroscopic Ellipsometer in the Infrared *Thesis* (Technische Universiteit Eindhoven The Netherlands)
[11] Korte E H and Roseler A 1998 *Analyst April* **123** 647-51

Investigation of the optical properties of InAs/InGaAs/GaAs quantum dot in quantum well multilayer structures for infrared photodetectors

Ts Ivanov[1,5], V Donchev[1], K Germanova[1], Ts Tellaleva[1], K Borissov[1], V Hongpinyo[2], P Vines[3], J P R David[3] and B S Ooi[4]

[1] Faculty of Physics, St. Kliment Ohridski University of Sofia, 5, J. Bourchier Blvd., 1164 Sofia, Bulgaria
[2] Department of Electrical and Computer Engineering, Lehigh University. Bethlehem, PA 18015, USA
[3] Department of Electronic and Electrical Engineering, University of Sheffield, Sheffield S1 3JD, UK
[4] Division of Physical Sciences and Engineering, King Abdullah University of Science and Technology, Saudi Arabia

E-mail: tsvetanivanov@lycos.com

Abstract. A detailed study of InAs/InGaAs quantum dots in quantum well (DWELL) structures grown on GaAs substrates for infrared photodetectors was performed using surface photovoltage (SPV) spectroscopy. Three types of samples were investigated: as-grown, and annealed with dielectric coating SiO_2 or SiN. The annealing resulted in intermixing of the material components. The amplitude and phase SPV spectra were measured at room temperature under various experimental conditions. The comparison of the SPV with the photoluminescence (PL) spectra allows one to conclude that the spectral features are due to optical transitions in the DWELL structure. The blueshift observed of these features in the intermixed samples implies that the energy levels responsible for the transitions change correspondingly due to the intermixing process. The interface band-bending in the samples and the mechanisms of the carrier dynamics were determined by a comparative analysis of the SPV amplitude and phase spectra, using our vector model for representation of the SPV signal.

1. Introduction

The infrared photodetectors (IRPD) based on quantum dots in quantum well (DWELL) structures have several advantages compared to other architectures, for example, higher operating temperature [1, 2], multicolour detection [3] and better optical quality of the QDs due to strain relaxation [4]. They also allow tuning of the detection wavelength by changing the parameters of the QW (composition, thickness), thus changing the position of the excited state, without modifying the bound state in the QD. An alternative post-growth approach is interdiffusion of the material components, by performing a rapid thermal annealing [5, 6].

[5] To whom any correspondence should be addressed.

The SPV spectroscopy is an advanced semiconductor characterization technique, successfully applied in the study of complex nanostructures in recent years [7, 8]. It is contactless, nondestructive, and provides information for the optical properties of the samples even at room temperature. Compared to the commonly used PL, the SPV spectroscopy gives information for the electronic transitions between ground and excited states. In this work we used the SPV and PL spectroscopy to investigate the optical properties of multi-layer self-assembled InAs/InGaAs/GaAs DWELL structures for IRPD, annealed with SiO_2 and SiN dielectric coatings.

2. Experimental

The samples are grown by MBE technique on semi-insulating GaAs substrate and start with an n^+ GaAs (2×10^{18} cm^{-3}) contact layer [figure 1(a)]. After a GaAs buffer layer (49 nm), a Si delta doping plane is introduced. The active region consists of 5 periods, each containing an InAs quantum dot (QD) layers (0.7 nm) embedded in an $In_{0.15}Ga_{0.85}As$ quantum well (QW) (1 nm – bottom, 7 nm – top) and a GaAs (1 nm) barrier layers. The top two layers are a GaAs (50 nm) and an n^+ GaAs ($2x10^{18}$ cm^{-3}, 400 nm) contact. During the growth process, a slight n-type residual doping is introduced. The QD density is approximately 6×10^{10} cm^{-2}. The samples coated by SiO_2 or SiN are processed by rapid thermal annealing at 800°C for 2 minutes under N_2 ambient for material intermixing.

The SPV measurements were performed at room temperature using the metal-insulator-semiconductor operation mode of the SPV technique [9]. A 250 W halogen lamp along with a SPEX grating monochromator (f = 0.25 m, 600 gr/mm) and an optical chopper (353 Hz) were used to illuminate the sample. The probe signal with respect to the ground was fed to a high-impedance unity gain buffer and then measured by an SR830 lock-in amplifier. The photon flux was kept constant ($\approx1.5\times10^{14}$ $cm^{-2}s^{-1}$) within ± 0.5 % for all the wavelengths, by positioning a neutral density filter with variable optical density. The reference signal from the optical chopper defined the zero value of the SPV phase [10]. More details about the SPV experimental set-up and measurement procedure can be found in [11]. The PL measurements were performed at 77 K by a 532-nm diode-pumped laser as the excitation source, and an InGaAs detector was used for signal detection.

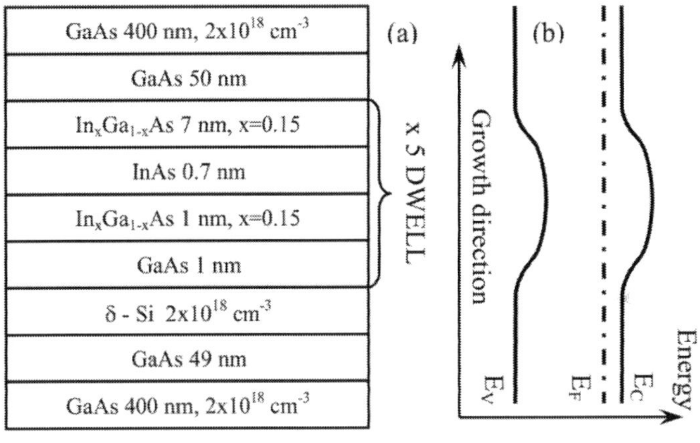

Figure 1. a) Layer sequence in the samples; b) Band diagram of the DWELL structure.

3. Results and discussion

Figure 2 presents typical SPV amplitude and phase spectra of the three samples in the range 0.95 – 1.35 eV. For comparison purposes, the corresponding PL spectra are also shown. They are shifted to the lower energies by 55 meV to account for the temperature dependence of the InAs bandgap between 77 K and 300 K.

The SPV amplitude spectrum of the as-grown sample reveals a broad bump in the range 1.0 - 1.25 eV. In the same range, a broad PL peak is observed centered at 1.086 eV with a shoulder at ~ 1.17 eV [figure 2(a)]. These spectral structures are ascribed to optical transitions in the QDs. Their broad shape indicates a relatively large inhomogeneity in the QD sizes.

The alignment of the Fermi levels in the layers with different doping level implies that the energy bands in the nominally undoped DWELL structure and the 50 nm thick GaAs layers are bent upward with respect to the surrounding highly-doped 400 nm thick GaAs layers, as represented schematically in figure 1(b). The SPV signal from the QDs is due to photocarriers, which thermally escape from the InAs potential wells of the DWELL structure into the surrounding 50 nm thick GaAs layers, where they are separated by the built-in electric field of the space charge regions (SCR). It is important to note that the SCRs at the top and the bottom interfaces of the DWELL structure have directions opposite to the built-in electric field [figure 1(b)]. Therefore, their contributions to the SPV signal have (nearly) opposite phases [10] and are, consequently, subtracted. The SPV phase for this sample is in the IInd quadrant [see figure 2(a)]. This indicates that for some reasons the contribution of the top DWELL interface, where the electrons (holes) move towards the surface (the bulk), dominates. A simple reason for this could be the exponential decrease of the photon flux in depth of the sample.

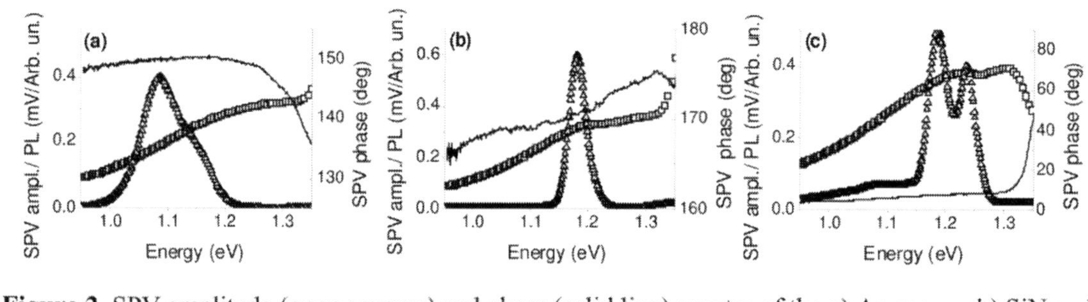

Figure 2. SPV amplitude (open squares) and phase (solid line) spectra of the a) As-grown, b) SiN and c) SiO_2 samples. The PL spectra (triangles), redshifted by 55 meV, are shown for comparison.

In the SPV amplitude spectrum of the SiN annealed sample [figure 2(b)], the bump mentioned above is also present, but it is seen as a well-resolved step at 1.20 eV. The corresponding PL peak is sharp and positioned at 1.182 eV in agreement with the SPV step. Therefore, the annealing has improved the QD size homogeneity and resulted in a blueshift of the optical transition energy. The latter is due to the blueshift of the QD electron and hole levels as a result of the material intermixing. To assess the magnitude of the blueshift it is easier to employ the PL results - from the PL peak positions one obtains 96 meV.

A smaller step is observed at ~1.27 eV in the SPV spectrum [see figure 2(b)]. This energy is slightly above the band-gap of bulk $In_{0.15}Ga_{0.85}As$ (1.21 eV) [12] and, therefore, the step is ascribed to optical transitions in the $In_{0.15}Ga_{0.85}As$ QW of the DWELL structure. A similar, although much weaker, step can be seen at ~1.26 eV in the SPV spectrum of the as-grown sample [figure 2(a)] upon a careful look after magnification. These results show that the transition energy of the QW is less influenced by the intermixing, which is explained by its lower surface-to-volume ratio. It is to be noted that the PL spectra do not reveal any particular structures in this energy range. Thus, the SPV spectra complement the information obtained by PL, where the lowest energy transition is predominantly observed. The SPV phase behavior [figure 2(b)] of the SiN annealed sample is similar to that of the as-grown sample [figure 2(a)] and can be explained in a similar way.

In the SPV amplitude spectrum of the SiO_2 annealed sample [figure 2(c)], the QD related step is clearly resolved at 1.22 eV. As compared to the SiN annealed sample, it is at a higher energy and is slightly broader. This may be understood with the help of the PL spectrum, which shows a sharp doublet of peaks situated at 1.184 and 1.234 eV implying two types of optical transitions, which are not resolved in the SPV spectrum. These observations could be explained assuming a non-uniform intermixing in the structure, e.g. two different planes of QDs blueshifted by different degrees. However, the sharp PL peaks indicate an improvement of the QD homogeneity inside each QD plane similar to that in the SiN annealed sample.

The SPV phase spectrum of this sample is different from those of the other two samples. In the QD spectral range, the phase remains in the Ist quadrant near 0 degrees. According to the vector model [11], such SPV vector results from the addition of two SPV vectors – one in IVth quadrant corresponding to the contribution of the bottom interface of the DWELL structure and one in IInd quadrant corresponding to its top interface. For some reasons, the contribution of the bottom DWELL interface is more pronounced, which pushes the resulting SPV vector closer to 0 degrees.

In the range 1.26-1.30 eV, the SPV amplitude increases, while the phase remains nearly constant. This amplitude step corresponds to the QW transitions in the DWELL structure. However this step is not completed, because above 1.30 eV the phase suddenly changes counter-clockwise towards the IInd quadrant, while the amplitude decreases sharply. Such behavior could be explained by the inclusion and growth of a new SPV generation process with a phase in IInd quadrant. Its SPV vector makes an obtuse angle with the already present vector and, as a result, their addition rotates the overall vector counter-clockwise, while decreasing its magnitude. In view of its phase, the new process should be related to free carrier generation at the top interface of the DWELL structure, but its exact origin is still unknown.

Further, an external voltage was applied to the sample surface in order to change the band bending at dark. Preliminary SPV spectral measurements have confirmed our expectations that with applying a voltage with the correct polarity we can change the direction of the slope of the potential along the structure and thus alter the relative contributions of the top and bottom interfaces to the SPV signal.

Conclusions

This is the first SPV observation of QD optical transitions in InAs/InGaAs/GaAs DWELL structures for IRPD. It is found that the carrier dynamics is governed by the different energy band-bending on the opposite interfaces of the DWELL structure. The RTA has two major effects: a blue shift of the optical transitions and improved size homogeneity of the QDs.

This investigation has a practical aspect since it provides information for the design optimization, growth parameters and post-growth treatment of the structures which is useful for high quality IRPD production. It also has fundamental research aspects, since it gives information for the physical properties of nanosized multilayer structures.

Acknowledgements

This work was supported by the research fund of the University of Sofia (contract 38/2011).

References

[1] Lim H, Tsao S, Zhang W and Razeghi M 2007 *Appl. Phys. Lett* **90** 131112
[2] Barve A V, Montaya J, Sharma Y, Rotter T, Shao J, Jang W-Y, Meesala S, Lee S J and Krishna S 2011 *Infrared Phys. Technol.* **54** 215
[3] Barve A V and Krishna S 2011 *Semiconductors and Semimetals (Advances in Infrared Photodetectors)* eds S Gunapala, D Rhiger and C Jagadish (San Diego: Elsevier) **84** 153
[4] Ye Z, Campbell J C, Chen Z, Kim E-T and Madhukar A 2002 *J. Appl. Phys.* **92** 7462
[5] Krishna S, Raghavan S, Gray A L, Stintz A and Malloy K J 2002 *Appl. Phys. Lett.* **80** 3898
[6] Gargallo-Caballero R, Miguel-Sanchez J, Guzman A, Hierro A and Munoz E 2008 *J. Phys. D* **41** 065413
[7] Chan C H, Chen H S, Kao C W, Hsu H P, Huang Y S and Wang J S 2006 *J. Appl. Phys.* **100** 064301
[8] Ivanov T, Donchev V, Germanova K, Gomes P F, Iikawa F, Brasil M J S P and Cotta M A 2011 *J. Appl. Phys.* **110**
[9] Kronik L and Shapira Y 1999 *Surf. Sci. Rep.* **37** 1
[10] Donchev V, Kirilov K, Ivanov T and Germanova K 2006 *Mat. Sci. & Eng. B* **129** 186
[11] Ivanov T, Donchev V, Germanova K and Kirilov K 2009 *J. Phys.D: App.Phys.* **42** 135302
[12] http://www.ioffe.rssi.ru/SVA/

Room temperature humidity sensors based on co-evaporated TeO_2 and Sn

B Georgieva[1], I Podolesheva and G Spasov

Acad. J. Malinowski Institute of Optical Materials and Technologies, Bulgarian Academy of Sciences, Acad. G. Bonchev Str., Bl. 109, 1113 Sofia, Bulgaria

E-mail: biliana@clf.bas.bg

Abstract. Thin layers prepared by vacuum co-deposition of TeO_2 and Sn are studied as humidity sensors working at room temperature. They are amorphous and, due to chemical interaction between the two substances, are built up of a mixed Sn- and Te-oxide matrix and a dispersed phase that at $R_{Sn/Te} < 2.0$ consists of elemental Te. The optimal composition of the layers is obtained by varying the atomic ratio $R_{Sn/Te}$. Auger analysis is used to follow the in-depth distribution of the components.

It is shown that the layers have a homogeneous composition over their entire depth. For $R_{Sn/Te} = 0.8 - 0.9$, they exhibit very high sensitivity to humidity – the electrical resistance changes by more than 4 decades in the range of 20 - 90% RH. The sensor response is very fast and the recovery period, very short. A very high long-term stability is observed as well.

1. Introduction

Over the past ten years, thin vacuum-deposited layers prepared by co-evaporation of Sn and TeO_2 [1] have been the object of intensive investigation in our laboratory since they form the basis for the development of gas sensors, conductive layers, etc. The as-deposited layers are amorphous with high porosity and large specific area. They are built of a fine oxide (SnO_2, SnO, and TeO_2) matrix (grains with dimensions of several nm) and a dispersed phase (Te, Sn and/or SnTe) depending on the atomic ratio Sn/Te ($R_{Sn/Te}$). Upon thermal treatment, the layers become polycrystalline [2].

We have already reported that the amorphous layers exhibit very good characteristics as humidity sensors working at room temperature, namely, high sensitivity and selectivity [2].

In this paper we present results from a systematic study of the preparation reproducibility and the homogeneity of the amorphous layer deposited, as well as of the composition range where the layers possess high sensitivity to humidity. Data are also included about the sensor response and recovery time, the long-term stability, etc.

2. Experimental

The humidity sensing thin layers (50-60 nm) were prepared by thermal vacuum co-evaporation of Sn and TeO_2 from two independent Knudsen-type cells with different atomic ratios $R_{Sn/Te}$. The fluxes of both substances were controlled continuously during the evaporation process using quartz crystal monitors. The chemical composition, i.e. the amount of both substances (the ratio $R_{Sn/Te}$, respectively)

[1] To whom any correspondence should be addressed.

Journal of Physics: Conference Series **356** (2012) 012033 doi:10.1088/1742-6596/356/1/012033

and the thickness of the layers were calculated on the basis of the measured evaporation rates using computer programs as described in [3]. Analogous calculation procedures were used for selecting the experimental conditions necessary for obtaining the desired composition and thickness of the samples [3]. Polycrystalline layers were produced by thermal treatment up to 360 °C (15 min heating at each step of 40 °C). The layers' chemical composition and $R_{Sn/Te}$ were controlled by energy dispersive spectrometry (EDS) in SEM. Auger electron spectrometry (AES) was used to determine the in-depth distribution of Sn, Te and O. The Auger spectra are derivative and obtained by a spectrometer with an energy resolution of ~0.3% (single-pass CMA and coaxial electron gun). Primary electron energy of 3 keV and modulation voltage of 4 Vpp were employed. The peak for O (510 eV) and the doublets for Sn and Te were monitored. The profiling was done by sputtering by 3 keV argon ions at a current of 2 μA. The quantification was made through a modified method of the elemental sensitivity factors. SnO_2 was used as own standard. The intensity of the tellurium doublet was measured from the maximum of the low-energy peak to the minimum of the high-energy peak.

Cr comb-like electrodes patterned by a photolithographic technique were used [4] to perform the electrical measurements at different temperatures and relative humidity (RH [%]). The sensor's response to water vapor was studied as a function of the substrate temperature in a test chamber equipped with temperature and RH controllers. The measurement of the electrical resistance R [Ohm] of the films was performed by a National Instruments multichannel ohmmeter. The data acquisition and processing, as well as the control of the ohmmeter, were computerized using Lab View software. The kinetic measurement was made with the same ohmmeter using saturated solutions that maintain constant relative humidity in the atmosphere over them – $(NH_4)_2SO_4$ keeps the humidity at 82% and NaOH, at 6% RH. The samples were hold above the solution and the change of R [Ohm] was measured at intervals of 2 s.

3. Results and discussion

3.1 Preparation reproducibility and homogeneity of the layers

As it will be shown later, the sensing properties of the layers are strongly dependent on their composition. To check the reliability of the computer calculations described above and the preparation reproducibility, we compared the calculated $R_{Sn/Te}$ with the respective data obtained by the EDS-analyses in SEM. As seen in figure 1, the agreement is very good – the deviation is ~2%.

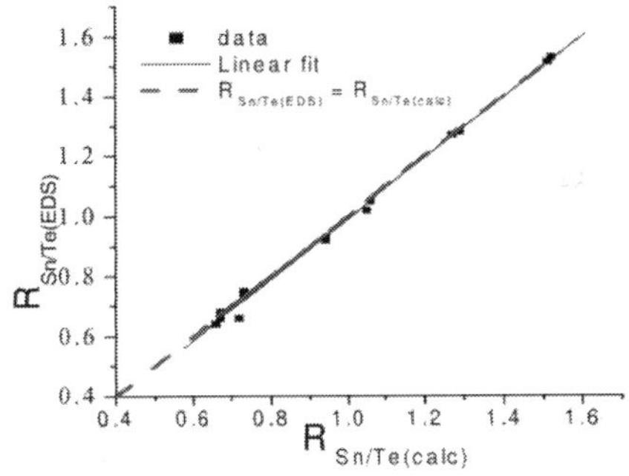

Figure 1. EDS control of the calculated layer composition.

As already mentioned, we used AES profiling to investigate the homogeneity of the as-deposited layers, i.e. the elements' depth distribution. Unfortunately, the direct application of the method is not possible because of the mobility of elemental Te upon heating and electron and ion beam bombardment, which could compromise the evaluation of its original distribution in the layer [1]. On the other hand, it is known that the other components of the layer do not exhibit mobility. Therefore, if the mobile Te atoms are removed from the layer by the electron beam prior to the AES - analysis, the distribution of the other components could be used to assess the homogeneity. The layers used for the analysis were with $R_{Sn/Te} \approx 0.8$ (confirmed by EDS analysis in SEM), i.e. they contained 20 parts of oxygen, 10 parts of Te and 8 parts of Sn (respectively 52.6, 26.3 and 21.1 at.%).

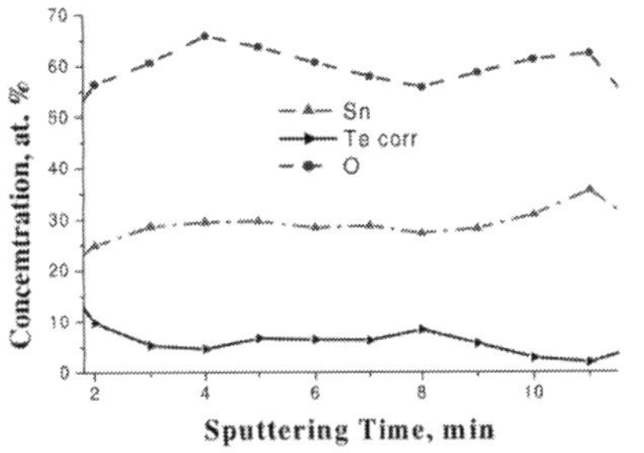

Figure 2. AES depth profiling of films with $R_{Sn/Te} \approx 0.8$ deposited on Si substrates.

The solid-state reaction between TeO_2 and Sn during their co-deposition has resulted in 8 parts of SnO_2, 2 parts of TeO_2 and 8 parts of elemental Te. After removal of the elemental Te, the composition has become O:Te:Sn = 66.7:6.7:26.7. Their distribution should be homogeneous if the original as-deposited layer has been homogeneous. The dose used for removing the unbound Te exceeded by a factor of 5×10^4 the dose applied for EDS. As can be seen in figure 2, the profiles are quite homogeneous, the average plateau concentrations are O:Te:Sn = 61(±1):6.4(±0.4):28.6(±0.3). After correction accounting for the contamination with C (4%), the atomic concentration of O becomes approximately 64%.

3.2. Humidity sensing properties

In our previous investigations, we have shown that layers with $R_{Sn/Te}$ = 0.7÷0.8 exhibit excellent characteristics as humidity sensors. Figure 3 presents the sensor response kinetics under step-like changes in RH. As can be seen, the sharp rise in relative humidity (from 6% to 82%) causes a very fast response of the sensor, the recovery period being very short as well. This leads to the conclusion that the water molecules are physically adsorbed. Results from the ageing tests are demonstrated in figure 4. They indicate that measurements performed over a long period of time (2002-2011) do not show significant differences, i.e. the layers exhibit a very good long-term stability.

We studied the sensitivity of the layers to water vapor over a wide range of atomic ratios and, as seen in figure 5, the best humidity sensing properties are exhibited by layers with $R_{Sn/Te} \approx$ 0.4 - 0.9. As shown in the same figure (inset), after crystallization (thermal treatment up to 360 °C), the layers with $R_{Sn/Te}$ = 0.7 lose their sensing properties to water vapor.

3.3 Temperature dependence

Figure 6 presents the dependence of *R* [Ohm] on the relative humidity measured at different temperatures – 25, 40, 60 and 80 °C. The sensor layers are practically not sensitive to water vapor above 40 °C. The cyan curve was obtained after heating the sample up to 80 °C (which does not lead to compositional and structural changes of the layers) and cooling

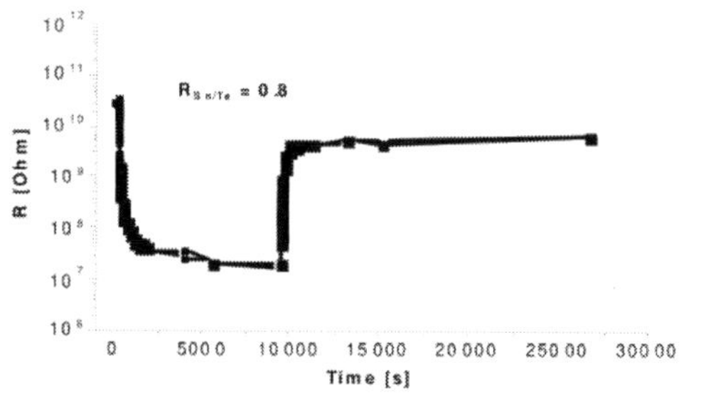

Figure 3. *R* [Ohm] as a function of time [s] at RH = 82% and 6%.

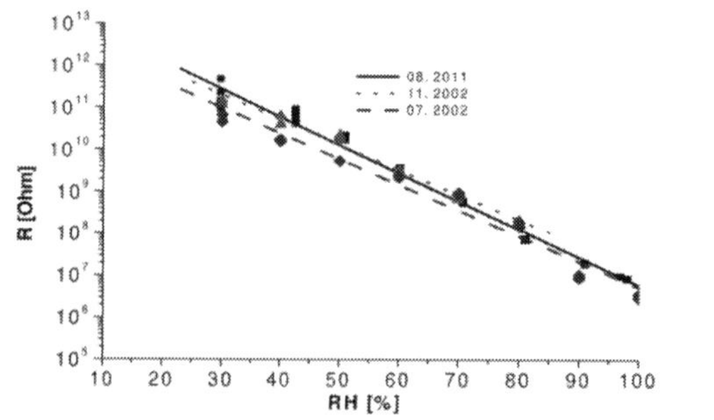

Figure 4. Long term stability of the sensing layers.

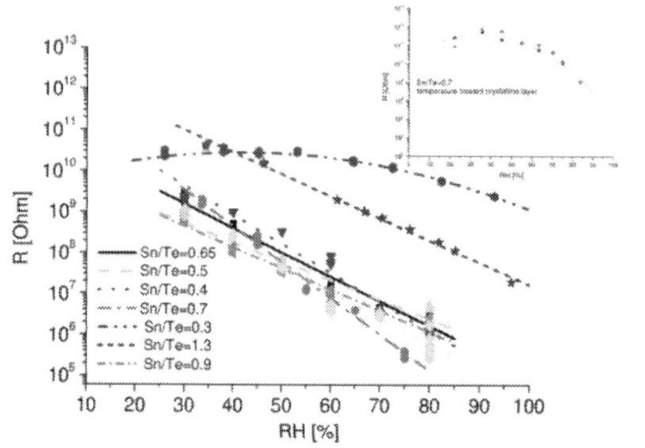

Figure 5. *R* [Ohm] as a function of RH [%] for layers with different $R_{Sn/Te}$. Inset - *R* [Ohm] as a function of RH [%] of a thermally treated polycrystalline layer with $R_{Sn/Te}$ = 0.7.

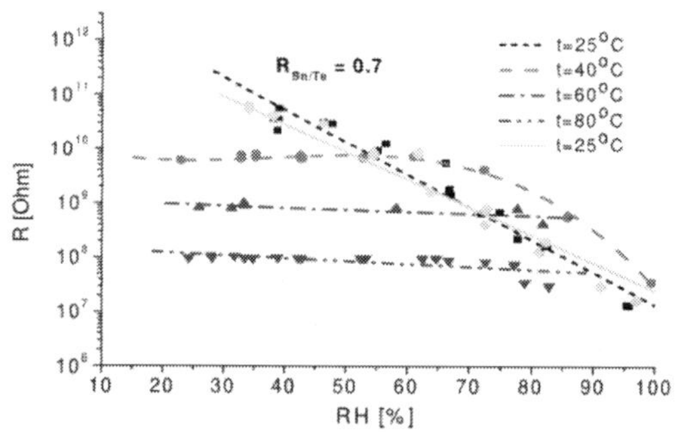

Figure 6. *R* [Ohm] as a function of RH [%] at different temperatures.

back to room temperature. It almost coincides with the black curve that was obtained before heating, with the as-deposited layer. This means that temperature variations in the range 20-80 °C do not result in changes in the sensing properties of the layers.

Figure 7 shows the dependence of *R* [Ohm] on the temperature at different RH values. As seen, at low humidity values, the dependence is linear and the activation energy is 0.44 eV (figure 7 - inset). At high humidity values (80-85%), a maximum appears at about 50 °C, probably due to desorption of the water vapor off the layer's surface.

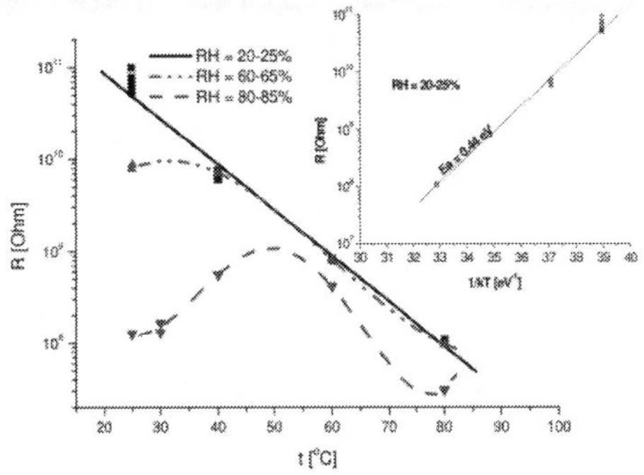

Figure 7. Temperature dependence of the sensing layer at different RH [%]. Inset – *R* [Ohm] as a function of $1/kT$ [eV^{-1}].

Conclusions

Our investigation shows that reproducible and homogeneous thin layers are obtained using the method of co-evaporation of Sn and TeO_2. In the as-deposited state, the amorphous layers with $R_{Sn/Te} \approx 0.4 \div 0.9$ possess good humidity sensing properties at room temperature. No change in the sensing behavior is observed if the layers are heated up to 80 °C and cooled back to room temperature. If they are heated up to 360 °C, however, and become polycrystalline, they lose their humidity sensing ability. The amorphous sensing layers show fast response, short recovery time and very good long-term stability. The activation energy is calculated using the temperature dependence in low-humidity atmosphere. Based on the kinetics studies, the conclusion can be drawn that the water molecules are physically adsorbed on the layer's surface.

Acknowledgements

This work was supported by the Bulgarian Ministry of Education and Science under Contract DO 02-148/16.12.2008.

References

[1] Podolesheva I, Dimov V, Tarassov M, Platikanova V and Konstantinov I 1998 *J. Vac. Sci. Technol. A* **16**
[2] Georgieva B, Podolesheva I, Pirov J and Platikanova V 2005 *J. Optoelectr. Adv. Mater.* **7** 2595
[3] Podolesheva I, Gushterova P, Platikanova V and Konstantinov I, 1998 *J. Vac. Sci. Technol. A* **16**, 674
[4] Mednikarov B and Sahatchieva M 2005 *J. Optoelectr. Adv. Mater.* **7** 1415

Journal of Physics: Conference Series **356** (2012) 012034 doi:10.1088/1742-6596/356/1/012034

Study of reticulated vitreous carbon surface treated by plasma immersion ion implantation for electrodes production

L L G Silva[1,5], D A S Conceição[2], S S Oishi[2], A Toth[3] and M Ueda[4]

[1] Technological Faculty of Pindamonhangaba – FATEC,
Vereador Abel Fabricio Dias Road, 4010, Pindamonhangaba SP, 12455-010, Brazil
[2] Engineering Faculty-FEG, State University of São Paulo – UNESP,
333 Dr. Ariberto Pereira da Cunha Av., Guaratinguetá/SP, 12516-410, Brazil
[3] Institute of Material & Environmental Chemistry, Chemical Research Center, Hungarian Academy of Sciences, Budapest, Hungary
[4] Associated Plasma Laboratory, National Institute for Space Research – INPE, Astronautas Av., 1758, São José dos Campos/SP, 12227-010, Brazil

E-mail: leide.lili@feg.unesp.br

Abstract. RVC samples were treated by nitrogen plasma immersion ion implantation (N-PIII) for electrodes production. High-voltage pulses with amplitudes of −3.0 kV or −10.0 kV were applied to the RVC samples while the treatment time was 10, 20 and 30 min. The samples were characterized by scanning electron microscopy (SEM), X-ray photoelectron spectroscopy (XPS) and electrochemical measurements. The SEM images present an apparent enhancement of the surface roughness after the treatment probably due to the surface sputtering during the PIII process. This observation is in agreement with the specific electrochemical surface area (SESA) of RVC electrodes. An increase was observed of the SESA values for the PIII-treated samples compared to the untreated specimen. Some oxygen and nitrogen containing groups were introduced on the RVC surface after the PIII treatment. Both plasma-induced process: the surface roughening and the introduction of the polar species on the RVC surface are beneficial for the RVC electrodes application.

1. Introduction

The reticulated vitreous carbon (RVC) is a form of macroporous polymeric carbon, which is produced by carbonization of polymers in the expanded form or by voids generating materials. The RVC has semiconducting properties, low density (0.05 g/cm^3), high corrosion resistance and low thermal expansion. The RVC is a kind of material that is frequently applied for electrodes production due to its high surface area and good electrical properties, such as high current density, low electrical resistance and chemical inertness [1]. RVC electrodes are applied in flow cells for elements detection in aqueous solutions. Additionally, it is used in the electro-analytical field for removing heavy metals (Hg, Cu, Pb) and for measuring the toxicity of water and solutions [1]. RVC has been recently applied as a stack material in thermo-acoustic devices [2], in the preparation of PbO_2 coatings for the electro-oxidation of organic pollutants [3] and in biofuel cells as an electrode [4].

[5] To whom any correspondence should be addressed.

RVC electrodes can be applied as-received or modified by some kind of treatment to improve further certain desired properties. This work describes the modification of RVC by plasma ion immersion implantation (PIII) in view of production of electrodes. The RVC samples were characterized by scanning electron microscopy (SEM), X-ray photoelectron spectroscopy (XPS), corrosion tests, cyclic voltammetry and open-circuit potential measurements.

2. Experimental

The RVC was purchased from Sanko Foams Ltda in the form of a polyurethane foam with 70 ppi (pores per inch). As-received RVC substrates were treated by PIII using nitrogen as gas precursor. A detailed description of the PIII system was given elsewhere [5]. The samples were immersed in nitrogen glow discharge plasma and pulsed at high negative voltage. Ions extracted from the plasma were implanted at normal incidence into all sides of a 3D sample. The samples were treated for 10, 20 and 30 minutes using high-voltage pulses with amplitudes of 3.0 and 10.0 kV. The other process parameters, namely, working pressure of 6.0×10^{-3} mbar, 200 Hz pulse repetition rate and pulse width of 20 μs were kept fixed during the treatment. The morphological changes on the surface of untreated and treated samples were examined by a JEOL model JSM 5310 microscope. The chemical characterization of the surface was carried out by X-ray photoelectron spectroscopy using a Kratos XSAM 800 spectrometer, using Mg Kα radiation and fixed analyzer transmission mode. The electrochemical measurements were obtained by means of an Autolab 302N potenciostat using a conventional three-electrode glass cell.

3. Results and discussion

3.1.Scanning electron microscopy (SEM)

The SEM technique was used to obtain information about the surface morphology of the RVC; the images obtained are presented in figure 1. Figures 1 (a-b) show the general view and the detailed view of a macroporous RVC sample. The SEM image of RVC treated at 3kV and 10 minutes shown in figure 1(c) exhibits an apparent roughening of the surface when compared to the untreated sample in figure 1(b). This effect is caused by the RVC surface sputtering during the PIII process. This observation is in agreement with the specific electrochemical surface area (SESA) values of RVC electrodes, which will be discussed later.

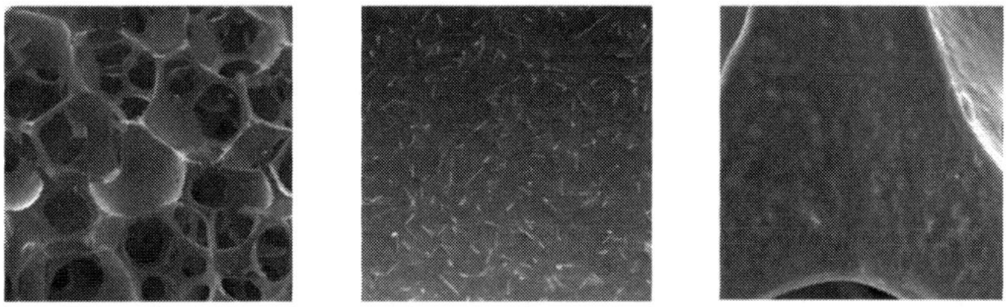

(a) (b) (c)

Figure 1- SEM images of untreated samples: (a) magnification of 50 x and (b) 1000X, (c) treated sample at PIII-3kV-10min - magnification of 1000 x.

3.2. X-ray photoelectron spectroscopy (XPS)

The chemical composition of surface of the RVC samples was examined by XPS analysis (table 1). For both treatments, the PIII induced surface changes on the RVCs in the form of nitrogen and oxygen containing polar groups formed on the RVC surface. The presence of oxygen groups in the as-received RVC was attributed to the furfuryl resin used in the manufacture of RVC. A gradual incorporation of nitrogen takes place upon increasing the treatment time at 3.0 kV, while at 10.0 kV an abrupt increase of the N content is observed. However, for the treatment time of 30 min, the N content tends to saturate due to sputtering that occurs during the PIII process. Considering similar treatment times, at higher voltage (10 kV) the amount of incorporated N scales from 4% up to 11 at.% and at lower voltage (3 kV), the N content ranges from 2% up to 5 at %. This behavior can be explained by the fact that nitrogen is incorporated on the RVC surface during the N-implantation, as confirmed by PIII treatment in other materials [6]. The evolution of the O-content is not monotonic. At 3 kV after 10 min treatment, it increases from 20 to 39 at%, but at longer treatment times the increase of O-content is smaller (33 at% after 20 min and 28 at% after 30 min). A similar behavior was noted for the samples treated at 10.0 kV pulse voltage but with different O-contents. The effect can be ascribed to the presence of oxygen impurities on the PIII chamber wall, so that at lower treatments times (10 min), a higher O-content is observed. However, as the treatment time increases, the O impurities are gradually removed from the system and less O is implanted. At a higher pulse voltage (10.0 kV), a similar process takes place, with the O-content decreasing down to 18-20 at% for treatment times longer than 20 min, which is the same as for the untreated sample.

Table 1. Surface composition of RVC samples.

Treatment		Composition		
U [kV]	t [min]	O [at%]	N [at%]	C [at%]
0	0	20	0	80
3	10	39	2	59
3	20	33	4	63
3	30	28	5	67
10	10	34	4	62
10	20	18	11	72
10	30	20	10	71

3.3. Electrochemical measurements

The open-circuit potential method consists in measuring the potential versus the time until a transition in the potential profiles is observed. Measurements of the open-circuit potential of RVC samples were performed to verify the stability of the electrodes in a NaCl 3.5% w/w and pH= 6.0 solution aiming to obtain information about the electrodes behavior in a corrosive environment. The untreated sample quickly reaches stabilization potential around 0.3 V, while the treated samples requires a longer time to achieve stabilization. The untreated sample shows a more positive potential (0.30 V) than the treated electrodes. However, the electrodes treated at 3.0 kV during 10 min and 10.0 kV during 30 min showed stabilization potential around 0.28 volts, which is closer to the untreated one indicating electrodes with higher corrosion resistance. On the other hand, the other treated electrodes exhibited a less noble behavior.

Polarization curves were investigated for analyzing the corrosion behavior of the electrodes in the NaCl 3.5% w/w solution. They were obtained immediately after the measurements of the open-circuit potential. It can be seen in figure 2 the untreated RVC samples have higher corrosion potentials (E_{corr} = 0.23 V) compared to the treated samples. For these samples the corrosion potential vary between 0.0 and –0.26V. The results also show that the samples treated at 10 kV and 30 minutes have corrosion current density higher by a factor of about 2.0 than the untreated sample (3.0×10^{-9} A/cm^2), which indicates a slight decrease of the corrosion resistance after the implantation. However, this increment in the current densities is not significant because the corrosion currents still remain quite low (10^{-9} A/cm^2) which indicates a good corrosion resistance.

Using cyclic voltammetry it is possible to determine the specific electrochemical surface area (SESA) using the Randles-Sevcik equation [7]. These tests were performed in a solution of 1 mM $K_4Fe(CN)_6.3H_2O$ in 0.1 M KCl using a scan rate of 5 mV/s. They are used to determine the peaks of anodic and cathodic current densities that are used in the SESA determination. It was observed that the anodic current densities peaks for treated samples increased compared to the untreated one (2.0×10^{-5} A/cm² up to 3.6×10^{-5} A/cm²). The increase of these current densities results in an increase of the SESA values as can be seen in Table 2. Increasing the implantation dose (RVC electrode treated at 10 kV and 30 min) resulted in a smaller anodic current density peak. The

Journal of Physics: Conference Series **356** (2012) 012034 doi:10.1088/1742-6596/356/1/012034

highest current density (3.6×10^{-5} A/cm^2) was exhibited ny the electrode treated at 3kV for 10 min. This current density is almost twice as high as the value of the untreated sample, which suggests an approximately twofold increase in the surface area of the RVC electrodes (table 2). The increase of the SESA values for the treated samples probably occurs due to the surface attack (sputtering) during the implantation with nitrogen, which promotes a general increase of the surface roughness [6]. However, as can be seen in table 2, as the implantation dose increases the SESA area and the anodic peak current density decrease. This is probably due to the enhanced sputtering of the RVC samples at higher implantation dose, which smoothen the material surface.

Table 2. SESA values for RVC samples.

RVC Samples	*SESA [cm²/cm³]*
Untreated	38.0
3kV-10 min	71.5
3kV-20min	45.4
3kV-30min	46.4
10kV-10min	56.5
10kV-20min	47.7
10kV-30min	36.6

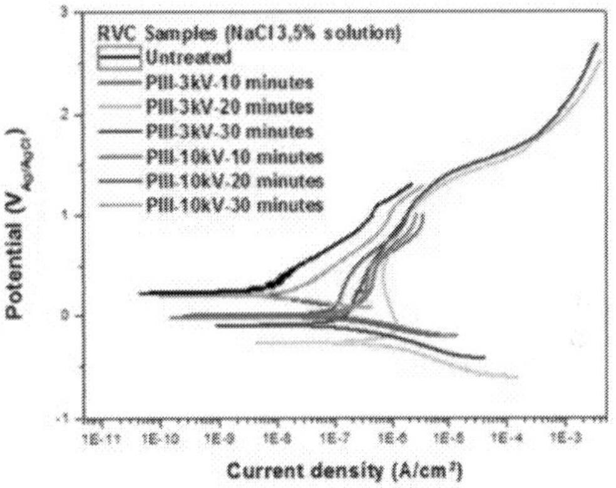

Figure 2. Polarization curves for RVC samples.

Conclusions

RVC samples were treated by nitrogen plasma immersion ion implantation. The PIII treatment caused a slight decrease in the corrosion resistance of the RVC samples. However, they maintained their current densities in the same order of magnitude as the untreated sample. There was up to twofold increase of the active area (SESA) for the electrodes treated by PIII compared with the untreated sample. These results are in agreement with the SEM images, as the PIII-treated samples appear to be rougher than the untreated sample. The results of the XPS analysis for all treatments show that the N-PIII treatment introduced active oxygen and nitrogen-related polar groups on the RVC surface, thus improving their wettability. This fact, together with the increase of the surface roughness, would contribute to the RVC application as electrodes.

References

[1] Friededric J M, Ponce-De-León C, Reade G W and Walsh F C 2004 *J. Electroanal. Chem.* **561** 203
[2] Tasnim S H, Mahmudb S, Fraser R A and Pop I 2011 *Int. J. Heat Mass Transfer* **54** 3811
[3] Recio F J, Herrastia P, Sirés I, Kulaka A N, Bavykina D V, Ponce-De-León C and Walsha F C 2011 *Electrochimica Acta* **56** 5158
[4] Rincon R A, Lau C, Garcia K E, Atanassov P 2011 *Electrochem. Acta* **56** 2503
[5] Ueda M, Berni L A, Castro R M, Beloto A F, Rossi J O, Barroso J J and Lepienski C M 2002 *Surf. Coat. Technol.* **156** 71
[6] Silva L L G, Ueda M, Mello C B, Codaro E N and Lepienski C M 2008 *J. Mater. Sci.* **43** 5989
[7] Wang J 2000 *Analytical Electrochemistry* (New York: John Wiley & Sons Inc.)

Journal of Physics: Conference Series **356** (2012) 012035 doi:10.1088/1742-6596/356/1/012035

Surface modification of aluminium alloys using hybrid treatment techniques

P Petrov

Emil Djakov Institute of Electronics, Bulgarian Academy of Sciences,
72, Tsarigradsko Chaussee, 1784 Sofia, Bulgaria

E-mail: pitiv@ie.bas.bg

Abstract. The application of electron and laser beams to the modification of metal and alloys opens up new possibilities of improving the materials properties. In recent years, the studies aimed at developing new materials have brought about the development of hybrid treatment techniques. These techniques perform an additional alloying in the zone treated, which has a substantial effect on the physical and mechanical properties of the materials processed. In this work we report results on the microstructural changes and mechanical properties of heat-treated $AlSi_{12}CuNiMg$ alloys resulting from additional alloying with Fe, Co, Ni, Cr by hybrid electron-beam techniques. It is established that the mechanical properties of Al-Si alloys can be improved by means of additional alloying with Fe, Co, Ni, Cr by hybrid electron-beam techniques. The specimens' properties practically do not change after heat-treatment (aging) for up to 200 hours at 250 °C.

1. Introduction

Using electron beams to modify metals and alloys offers new possibilities of improving the materials properties [1-2]. The processes of surface modification of metals and alloys can be roughly divided into two groups depending on whether the electron beam acts on the material in a solid or in a liquid phase. Both types of techniques have found practical applications. In recent years, the studies aimed at developing new materials have resulted in the development hybrid treatment (HEB) techniques that perform an additional alloying in the zone treated, thus affecting substantially the physical and mechanical properties of the materials processed [3-7].

In this work we report results on the changes of the microstructure and the mechanical characteristics of heat-treated Al-Si alloys resulting from additional alloying with Fe, Co, Ni, Cr by HEB techniques.

2. The process of electron-beam hybrid treatment techniques

The experiment on applying electron-beam hybrid treatment techniques is shown schematically in figure 1. The alloying elements were deposited on the specimens' surface by plasma spraying in argon atmosphere with the thickness of the layer formed being about 0.1 mm. The experiments were performed on Leybold Hereaus EWS300-15/60 equipment. Electron beam remelting was achieved by scanning electron beam with power P = 1.5 - 3 kW, frequency f = 0.5 - 10 kHz, amplitude A = 2 - 5 mm and sample motion speed V = 0.5 - 2.5 cm/sec.

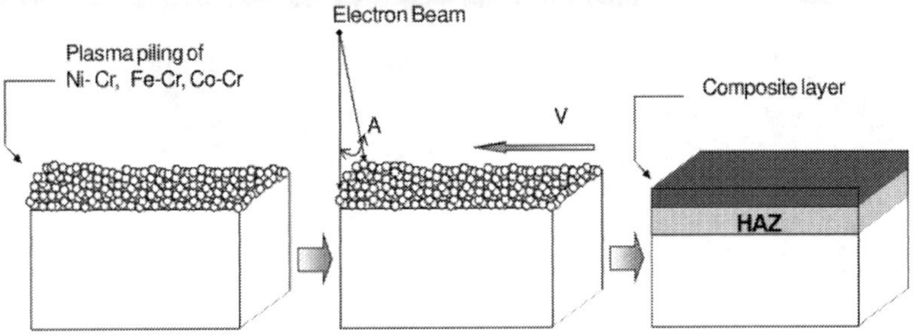

Figure 1. Schematic representation of hybrid electron-beam treatment of aluminium alloys.

The heat-treated Al-Si alloys used in the experiments reported were AlSi12CuNiMg (wt% - 11.45% Si, 1.25% Cu, 0.96% Mg, 1.41% Ni, 0.38% Mn). The alloying was done using Fe-Cr-Ni-Co powders, table 1.

Table 1. Chemical composition of the Fe-Cr-Ni-Co powders, wt%.

Composition	Ni	Fe	Co	Cr	Si	B	C
Ni-Cr	base	11.76	-	16.30	3.27	3.20	0.23
Fe-Cr	2.34	base	-	21.69	3.63	2.10	0.88
Co-Cr	3.33	1.2	base	22.16	-	-	-

The starting specimens with dimension of 50×50×18 mm were prepared by bulk thermal processing in the following sequence: (i) homogenization at 505 °C for 4 h and at 515 °C for another 4 h; (ii) water quenching; (iii) aging at 200 °C for 15 h. Prior to the analyses by optical and electron microscopy, the specimens were cut into appropriate pieces and a standard procedure was carried out, which included developing in Keller's solution and electrode etching. The samples' microhardness was assessed using a LEITZ measuring instrument with loading force of 1 N. X-ray structural analysis was conducted by DRON-1 facilities with Co-K_α emission. The strength measurements were performed after hybrid electron beam processing, as well as after a furnace heat treatment at 250 °C for up to 200 h.

3. Results and discussion

A transverse cross-section of the samples' zone treated by the HEB technique at $P = 3$ kW, $V = 2$ cm/s, $f = 10$ kHz and $A = 5$ mm is shown in figure 2a; the longitudinal cross-section is shown in figure 2b. Some defects, such as cracks in the case of alloying with F-Cr powder, are observed in alloyed zone. The microstructure observed is quite different from the initial one, figure 3 and figure 4. Electron beam treatment without alloying results in the formation of an α-hard solution and a highly dispersed intermetallic compound, as shown in figure 3a, b. The crystallization is of a dendritic nature.

The data of the X-ray diffraction analyses of the layers produced by hybrid electron beam treatment show: (i) Fe-Cr powder - $FeSiAl_6$, Al_3Fe, $CrAl_7$, $(CrFe)_4Si_4Al_{13}$, $(CrFe)_5Si_8Al_2$ phases; (ii) Ni-Cr powder - $NiAl_3$, Cu_4NiAl_7 and $(CuFeNi)_2Al_3$ phases and (iii) Co-Cr powder - Co_2Al_9 and $CoCu_2Al_7$, $(CoFeNi)_2Al_9$ phases. After heat treatment for 24 h at 250 °C, the alloy's phase composition in the layer does not change, as proven by X-ray diffraction.

The layer formed following the application of the HEB technique (curves 3, 4, 5 in figure 5) exhibits hardness substantially exceeding that of the base metal (curve 1 in figure 5), as well as that

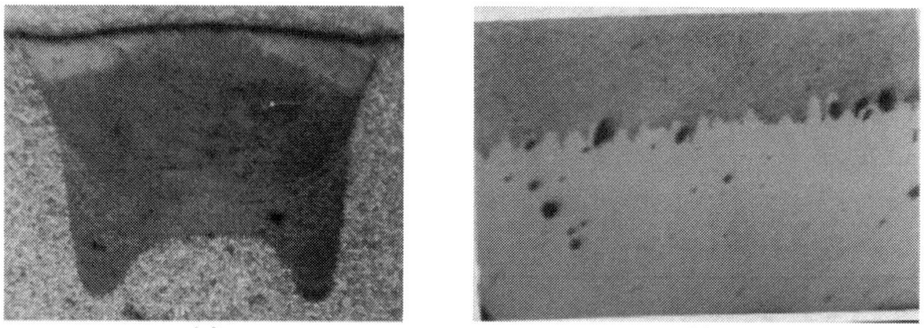

(a) (b)

Figure 2. Transverse cross-section of the alloyed zone (×4) - a); longitudinal cross-section of the alloyed zone (×2) - b).

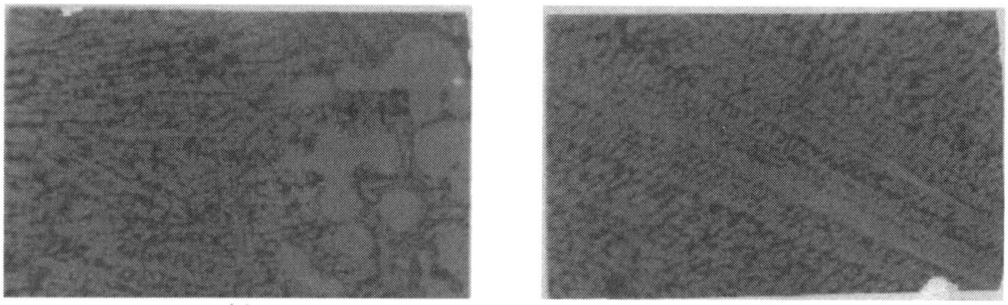

(a) (b)

Figure 3. Microstructure of the treaded zone without alloying (×160): a) near the border with initial material; b) in the remelted zone.

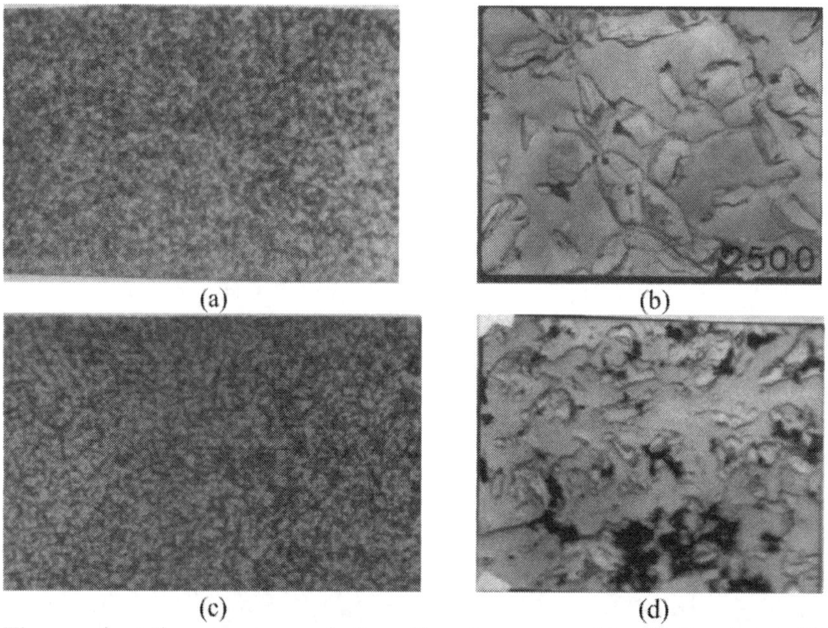

(a) (b)

(c) (d)

Figure 4. Microstructure of the alloyed zone: with Ni-Cr a)-(×160), b)-(×2500); with Fe-Cr c)-(×160), d)-(×2500).

after electron-beam remelting without alloying (curve 2 in figure 5). Aging for up to 200 hours at 250 C leads to a decrease of the hardness of all types of layers, especially in the first 24 hours. However, the hardness of the layers subjected to HEB treatment (alloying and electron beam remelting) remains the highest.

Figure 6 shows the variation of the tensile strength *R* as a function of the type of dopant powder. The highest tensile strength is measured in the treated samples in the case of doping with Co-Cr powders. The results obtained by the microhardness and tensile strength measurements demonstrate the advantages of applying hybrid electron-beam treatment of aluminium alloys as compared with the standard electron-beam remelting. The hardness and the tensile strength at normal and higher temperatures increase due to the higher dispersivity of the structure, the appearance of new structural components and the formation of a stable frame of inter-metallic compounds.

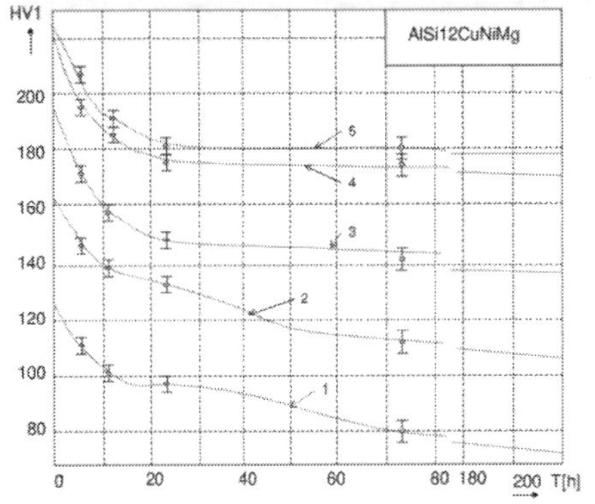

Figure 5. Microhardness variation during aging for up to 200 hours at 250 °C: curve 1 - base material; curve 2 - after electron-beam remelting; curve 3 - (Ni-Cr), curve 4 - (Fe-Cr), curve 5 - (Co-Cr).

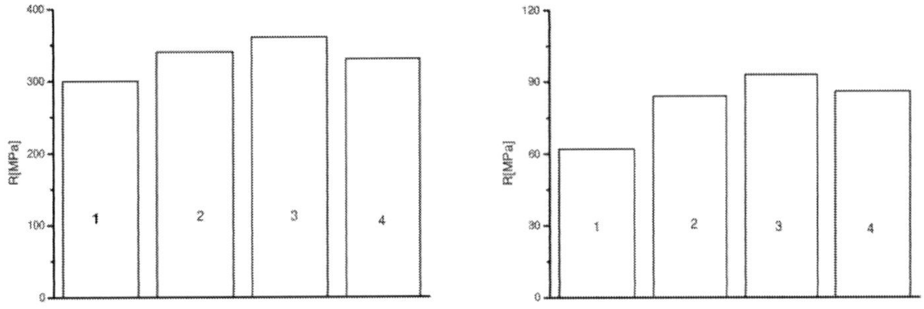

(a) (b)

Figure 6. Tensile strength *R* after HEB processing: a); RT b) elevated temperature at 250 °C ; (1 - base material; 2 – alloying with Ni-Cr; 3 - alloying with Co-Cr, 4 - alloying with Fe-Cr).

Conclusions

As a result of the systematic studies carried out, it is established that the mechanical properties of Al-Si alloys can be improved by means of additional alloying with Fe, Co, Ni, Cr by hybrid electron-beam techniques. The specimens' properties practically do not change after heat-treatment (aging) for up to 200 hours at 250 °C.

References

[1] Schiller S and Pancer S 1988 *Ann. Rev. Mater. Sci.* **18** 121
[2] Surla V and Ruzic D 2011 *J. Phys. D: Appl. Phys.* **44** 1740
[3] Petrov P and Dimitrov D 1993 *Vacuum* **44** 857
[4] Petrov P *Vacuum*1997 **48** 49
[5] Zhou Q 2001 *Ordinance Mater. Sci. Eng.* **24** 28
[6] Song R, Zhang K and Chen G 2002 *Surf. Coat. Technol.* **157** 1
[7] Dalke A, Buchwalder A, Zenker R and Biermann H 2009 *Int. Heat Treatm. Surf. Eng.* **3** 147

Journal of Physics: Conference Series **356** (2012) 012036 doi:10.1088/1742-6596/356/1/012036

XPS study of ZnO nanostructures prepared by laser ablation

G Atanasova[1], A Og Dikovska[2,4], M Stankova[3], P Stefanov[1] and P A Atanasov[2]

[1] Institute of General and Inorganic Chemistry, Bulgarian Academy of Sciences, Acad. G. Bonchev Str., Bl. 11, 1113 Sofia, Bulgaria
[2] Emil Djakov Institute of Electronics, Bulgarian Academy of Sciences, 72, Tsarigradsko Chaussee, 1784 Sofia, Bulgaria
[3] Serví de Recursos Científicas i Técnicas, Campos Sescelades, Universitad Rovira i Virgili, 26, Av. Paisos Catalans, 43007 Tarragona, Spain

E-mail: dikovska@ie.bas.bg

Abstract. The subject of the present work is different types of nanostructured ZnO films produced by pulsed laser deposition. The influence was investigated of the surface morphology on the physicochemical and optical properties of the nanostructured films. X-ray photoelectron spectroscopy was used to study the composition and chemical state of the nanostructured ZnO. The changes occurring on the surface of the films after prolonged exposure in air are considered in the light of practical applications of ZnO films. The change in the number of adsorbed species on the surface with the time affects the surface conductivity, the latter in turn modifying the optical properties of the samples.

1. Introduction

In recent years, the synthesis and modification of nanosized ZnO have been the object of comprehensive studies due to the wide range of possible applications in gas sensors, photocatalysts, chemical absorbents, solar cells, electrical and optical devices.

However, in the case of nanostructured materials, undesired effects could be expected due to surface inhomogeneities, such as non-stoichiometry or selective adsorption of foreign species, in addition to the abrupt termination of the lattice periodicity. ZnO is known to adsorb readily atoms and, particularly, oxygen-containing species [1-3], which affect the surface conductivity of the nanostructures. The recent study of Yang et al. [4] indicated that the presence of OH^- and H^- species play a dominant role in facilitating surface recombination. Also, Bai et al. reported that a cover of hydroxyl groups on the surface of ZnO can inhibit the adsorption of ethanol [5], thus affecting the sensing properties. Besides, the possibility of interaction between chemical groups on the ZnO surface and the molecules that are to be detected can also influence the performance of a ZnO sensor. Our previous investigations showed that the gas sensing properties of the as-prepared sensor element may differ from those after a prolonged period of time. Therefore, investigating the influence of the adsorbed hydroxyl groups on the electronic and physical properties is necessary in view of the practical applications of ZnO nanostructures.

In the present work, the surface morphology and the electronic and optical properties of different types of ZnO nanostructured films were studied with the aim of monitoring the changes in the surface

[4] Towhom any correspondence should be addressed.

composition, electronic configuration and optical properties after a prolonged exposure in air, depending on the nanostructure of the ZnO layers.

2. Experimental

Thin ZnO films were prepared on amorphous SiO_2 substrates by pulsed laser deposition (PLD). The third harmonic of a Nd:YAG laser (λ = 355 nm, τ = 18 ns full-width at half-maximum (FWHM), and repetition rate of 10 Hz) was used for ablation of a ZnO ceramic target. The target-to-substrate distance was fixed at 40 mm in a standard on-axis configuration. All experiments were performed in oxygen atmosphere. The substrate temperature was kept at 300 °C during deposition.

The nanostructured films were fabricated via a two step process [6,7]. As a first step, a thin ZnO film was prepared at high oxygen pressure in order to form growth nuclei. Oxygen pressures of 20, 50, and 100 Pa were used for preparation of three different types of nuclei. The laser fluence was kept constant at 2 J/cm^2. The second step consisted in the deposition of a ZnO film on the as-created nuclei. At this stage, the oxygen pressure was decreased to 5 Pa and kept constant to the end of the experiment. During the second step, laser fluence of 3.5 J/cm^2 was used.

The surface morphology of the samples was characterized by atomic force microscopy (AFM) using an Agilent 5500 microscope. The film composition and electronic structure were investigated by X-ray photoelectron spectroscopy (XPS). The measurements were performed on a VG ESCALAB II system using MgK_α radiation with energy of 1253.6 eV. The binding energies (BE) were determined with an accuracy of ± 0.1 eV utilizing the C 1s line at 285.0 eV (from adventitious carbon) as a reference. The optical transmission spectra were obtained by illuminating the samples by a white-light source - halogen and Xe lamps. The light transmitted through the samples was detected by an HR 4000 Ocean Optics Spectrometer with optical resolution of 0.02 nm.

3. Results and discussion

3.1. AFM studies

The AFM images of the as-deposited ZnO nanostructured films are presented in figure 1.

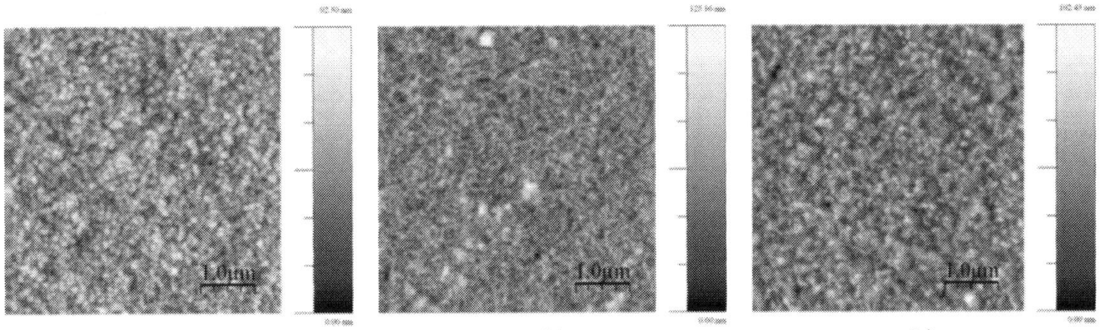

(a) (b) (c)

Figure 1. AFM images of ZnO nanostructured films prepared on the basis of different nuclei obtained at: (a) 20 Pa, (b) 50 Pa, and (c) 100 Pa oxygen pressure.

The surface morphology of the sample in figure 1a shows grains with clearly expressed spherical shape and mean grain-size in the range 60 - 110 nm. The grains are packed closely and are well distributed on the substrate with RMS surface roughness of about 10 nm. The scale bar in the AFM image (figure 1a) indicates a high peak-to-valley ratio. It is worth pointing out that the ZnO buffer layer deposited at 20 Pa oxygen pressure has the crystal structure as reported in our previous investigations [8]. The growth of the ZnO nanostructure follows the crystal nuclei.

The surface morphology of the sample prepared by a ZnO buffer layer deposited at 50 Pa is presented in figure 1b. The increase of the oxygen pressure applied during the buffer layer deposition leads to a change of the surface morphology, namely, a non-uniform structure with a larger grain-size in the range of 100 – 200 nm, in contrast with the sample shown in figure 1a. Additionally, the increase of the oxygen pressure worsens the crystal structure of the buffer layer. The further increase of the oxygen pressure applied during the buffer layer deposition (100 Pa) preserves the non-uniform structure of the surface morphology (figure 1c). In this case, the sample grain-size is ranging from 100 to 200 nm. Furthermore, at this high oxygen pressure the as-prepared buffer layer is practically amorphous. Again, one should point out that the difference in the surface morphology appears mainly between the film prepared on the basis of growth nuclei produced at 20 Pa (figure 1a) and samples prepared with growth nuclei obtained at higher oxygen pressures (figure 1b and c).

3.2. XPS studies

The composition and chemical state of the nanostructured film surfaces were studied by XPS. The asymmetric peak observed in the O1s region for all as-prepared samples (not shown) is resolved into two peaks by a Lorentzian - Gaussian spectral fitting. The peak with low binding energy (530.2 eV) corresponds to O^{2-} ions in a normal ZnO wurtzite structure, while the second one centered at 531.7 eV may be attributed to different kinds of species, such as adsorbed hydroxyl or/and carbonyl groups [9,10]. Also, some contribution to this peak may originate from O^{2-} ions in the oxygen deficient regions within the ZnO matrix [11]. However, it should be mentioned that the fitting of the O1s peak in the present study is in close agreement with that for a nanostructured ZnO film on a silica substrate [12], where the actual oxide stoichiometry is given as $ZnO_x(OH)_y$.

Further on, we followed the changes in the intensity of the oxygen peak at 531.7 eV depending on the time of exposure in air. The XPS data are presented in table 1. Variation should not be expected in the oxygen vacancies concentration on the surface of the ZnO nanoparticles during a prolonged exposure. Therefore, the changes in the intensity of this peak are mainly related to variation in the concentration of the adsorbed oxygen containing species, most probably OH^- groups. In addition, the atomic ratio of crystal lattice oxygen to zinc on the surface of the distinct oxide nanostructures is lower than the stoichiometric value of 1.0. It is evident that there is pronounced oxygen deficiency in the films. This fact indicates that there are large concentrations of vacancies and unsaturated chemical bonds that can interact with other atoms.

Table 1. Curve fitting data of Ols photoelectron spectra and Auger parameter of different ZnO samples.

Samples with growth nuclei prepared at:	Exposure [days]	O_{lat}/Zn	O_{OH^-}/O_{total}	ά [eV]
20 Pa O_2	as-dep	0.57	0.29	2010.6
	7	0.63	0.32	2010.4
	21	0.63	0.44	2010.8
	42	0.59	0.46	2010.9
50 Pa O_2	as-dep	0.62	0.34	2010.9
	7	0.68	0.29	2010.7
	21	0.72	0.40	2011.1
	42	0.66	0.41	2010.9
100 Pa O_2	as-dep	0.61	0.30	2010.8
	7	0.63	0.28	2010.6
	21	0.69	0.36	2011.1
	42	0.70	0.40	2011.0

The chemical state of Zn in the ZnO thin film was analyzed in detail by investigating the Zn 2p core level and Auger Zn *LMM* peaks. These peaks show irregular shifts whose origin is associated with the electrostatic charging of the film surfaces during the acquisition of the spectra. In order to avoid these discrepancies, we used the modified Auger parameter ά for more exact determination of the chemical state of Zn, thus eliminating the surface effects of the electrostatic charging. This parameter is defined as ά = KE (Zn $L_3M_{4,5}M_{4,5}$) + BE (Zn $2p_{3/2}$), where KE (Zn $L_3M_{4,5}M_{4,5}$) is the kinetic energy of the Auger transition and BE (Zn $2p_{3/2}$) is the binding energy of an electron in the core level [13].

The values presented in table 1 for the Auger parameter are in the range 2010.4 – 2011.1 eV. It is known that the ά values located at 2010 and 2014 eV are characteristic of the oxidation state Zn^{2+} and metallic state Zn^0, respectively [14,15]. Our values are consistent with the existence of Zn^{2+}, but the positive shift observed as the air exposure is increased is explained by charge transfer during the continuous adsorption of oxygenated species on the surface of the oxide nanostructures.

Table 1 also shows the values for the O_{OH^-}/O_{total} intensity ratio depending on the time of air exposure. This dependence of the O_{OH^-}/O_{total} intensity ratio is plotted in figure 2a for the three nanostructured films prepared on the basis of different nuclei. Clearly, the intensity ratio tends to saturate at prolonged exposure, suggesting that the chemical adsorption of the surface of the ZnO films is nearly saturated.

According to theoretical predictions, surface defects, such as oxygen vacancies, can dominate the electronic/chemical properties and adsorption behavior of metal oxide surfaces [16]. The film formed at the lowest oxygen pressure during the growth of nuclei exhibits the fastest increase compared with the films formed at higher oxygen pressures, indicating that more oxygen vacancies exist in the surface layers of the film.

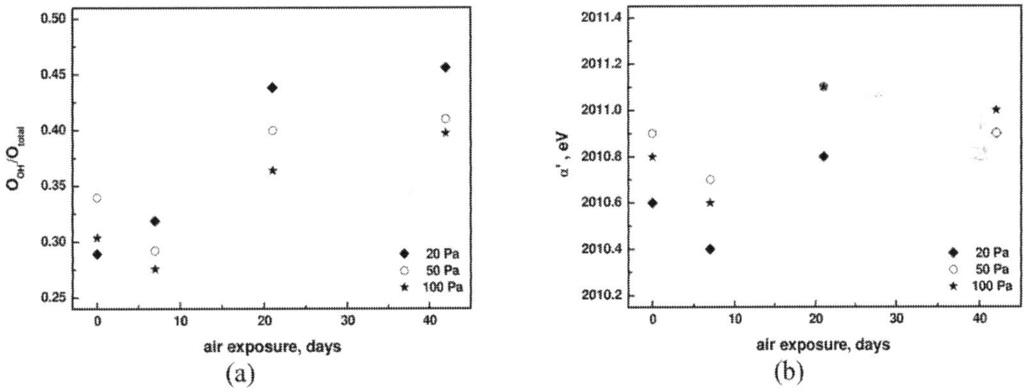

Figure 2. Dependence of: (a) the amount of adsorbed OH^- groups and (b) Auger parameter on the time of exposure to air.

The adsorption behavior of the films is compared with the variation of the modified Auger parameter in figure 2b. The Auger parameter has been recognized as a very powerful tool in acquiring information on the electronic interactions and the polarizability of the medium around the photoemitting atom [17]. In this plot, the film formed with growth nuclei at 20 Pa shows a lower degree of increase of the Auger parameter than the ones formed at higher oxygen pressures. As mentioned above, adsorption occurs at vacancy defect sites on the oxide surface. Charge transfer takes place, resulting in a change of the surface conductivity with the absorbed species gaining a negative charge; this causes the formation of a depletion layer on the surface of the *n*-type ZnO. The formation of this layer affects the photoemission process by increasing the relaxation energy, which leads to an increase of the Auger parameter values for the three samples in figure 2b. Thus, the variation of the Auger parameter reflects the variation of the surface conductivity as a result of increased number of adsorbed species. The nanostructured film with the highest number of oxygen vacancies exhibits the lowest conductivity change.

3.3. Optical studies

The optical transmission of the samples was measured during air exposure in order to investigate how the surface composition affects the optical properties of the samples. We followed the behavior of one fixed maximum in the transmission spectra as a function of the time of exposure in air; its shift is presented in figure 3. The shift in the maximum's position means that the optical density of the samples has changed [18]. Assuming that the samples thickness remains constant during exposure, this shift is probably due to the change in the refractive index of the samples. It should be mentioned that for all samples the behavior of the optical maximum shift follows the behavior of the O_{OH-}/O_{total} intensity ratio, presented in figure 2a. The change of the surface conductivity of the samples results in a variation of the optical properties, as it was previously reported [19]. The least change of the surface conductivity provokes the smallest shift in the maximum position.

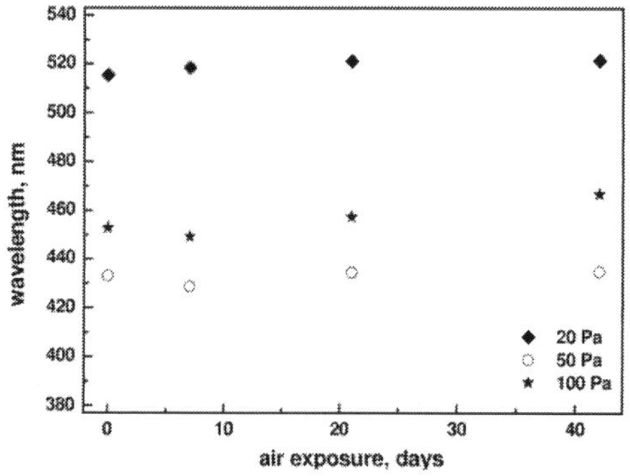

Figure 3. Shift of the optical transmission maximum versus time exposure.

Conclusions

We showed that the physicochemical properties of ZnO nanostructured films depend on the initial stages of growth. The analysis of the XPS data demonstrates that the amount of surface oxygen vacancies decreases with the increase of the oxygen pressure used for preparation of the initial growth nuclei. As the grain size of the zinc oxide films increases, their adsorption capability with respect to the hydroxyl groups is reduced. The changes in the intensity of the oxygen peak in the surface composition of the samples were observed depending on the time of exposure in air. The intensity ratio O_{OH-}/O_{total} tends to saturate at a prolonged exposure, suggesting that the chemical adsorption of the surface of the ZnO films is approximately saturated. The change of the number of adsorbed species on the surface during the exposure leads to a variation of the surface conductivity which affects the optical properties of the samples.

Acknowledgment

The authors acknowledge the financial support of the National Science Fund of Bulgaria (National Centre for New Materials UNION, Contract No DCVP-02/2/2009). The samples preparation was supported by the Bulgarian Ministry of Education and Science under Contract DO 02-293/08.

References

[1] Li Q H, Liang Y X, Wan Q and Wang T H 2004 *Appl. Phys. Lett.* **85** 6389
[2] Fan Z, Wang D, Chang P C, Tseng W Y and Lu J G 2004 *Appl. Phys. Lett.* **85** 5923
[3] Schmidt O, Geis A, Kiesel P, van de Walle C G, Johnson N M, Bakin A, Waag A and Döhler G H 2006 *Superlattices Microstruct.* **39** 8
[4] Yang L L, Zhao Q X, Willander M, Liu X J, Fahlman M and Yang J H 2010 *Appl. Surf. Sci.* **256** 3592
[5] Bai Z K, Xie C S, Hu M L, Zhang S P and Zeng D W 2008 *Mater. Sci. Eng. B* **149** 12
[6] Sun Y, Fuge G M and Ashfold M N R 2006 *Superlattices and Microstruct.* **39** 33
[7] Conley J F, Stecker L and Ono Y 2005 *Nanotechnol.* **16** 292
[8] Dikovska A Og, Atanasov P A, Vasilev C, Dimitrov I G and Stoyanchov T R 2005 *J. Optoelectr. Adv. Mater.* **7** 1329

[9] Tanaka K, Miyahara K and Toyoshima I 1984 *J. Phys. Chem.* **88** 3504
[10] Dupin J-C, Gonbeau D, Vinatier P and Levasseur A 2000 *Phys. Chem. Chem. Phys.* **2** 1319
[11] Fan J C C and Goodenough J B 1977 *J. Appl. Phys.* **48** 3524
[12] Armelao L, Fabrizio M, Gialanella S and Zordan F 2001 *Thin Solid Films* **394** 89
[13] Futsuhara M, Yoshioka K and Takai O 1998 *Thin Solid Films* **322** 274
[14] Xue M, Guo Q, Wu K and Guo J 2008 *Langmuir* **24** 8760
[15] Moulder J F, Stickle W F, Sobol P E and Bomben K D 1992 *Handbook of X-ray Photoelectron Spectroscopy* (Eden Prairie, Perkin- Elmer)
[16] Schaub R, Wahlstrom E, Ronnaus A, Laegsgaard E, Stensgaard I and Besenbacher F 2003 *Science* **299** 377
[17] Moretti G 1998 *J. Electron. Spectrosc. Relat. Phenom.* **95** 95
[18] Swanepoel R 1983 *J. Phys. E: Sci. Instrum.* **16** 1214
[19] Mazingue T, Escoubas L, Spalluto L, Flory F, Jacquouton P, Perrone A, Kaminska E, Piotrowska A, Mihailesscu I and Atanasov P 2006 *Appl. Opt.* **45** 1425

Thin dye layers – vacuum deposition and structure

D Dimov[1,3], J Assa[1], A Georgiev[2], R Kazakov[1], D Karashanova[1], G Danev[1] and E Spassova[1]

[1] Institute of Optical Materials and Technologies, Bulgarian Academy of Sciences, Acad. G. Bonchev Str. Bl. 109, 1113 Sofia, Bulgaria
[2] University of Chemical Technology and Metallurgy, 8, St. Kl. Ohridski Blvd., 1756 Sofia, Bulgaria

E-mail: dean@clf.bas.bg

Abstract. Vacuum-deposited layers of zinc phthalocyanine and perylene-3,4,9,10-tetracarboxylic dianhydride were obtained with regard to the development of converters of solar radiation into electricity, namely, organic solar cells. It was shown that there were no differences in the composition of the initial dyes and the deposited ones, which is a proof of their thermal stability and a prerequisite for the reproducibility of the composition of the single and mixed layers under study. The results obtained are interpreted in view of the possible applicability of the layers studied for the fabrication of organic solar cells in which the processes of creation, the lifetime and the diffusion paths of the current carriers depend significantly on the film composition and the defects formed at the time of deposition of the active dye layers.

1. Introduction

In recent decades, the progress related to the utilization of renewable energy sources, such as organic (OSCs) or inorganic solar cells, wind parks, the heat of the earth, etc,. has been substantial [1]. The major drawback of the OSCs is their comparatively low (up to 5-6 %) efficiency [2] with a predicted limit for some experimental donor-acceptor pairs of 11 % [3]. The growth of each vacuum deposited (VD) layer in an OSC affects the characteristics of the films deposited subsequently – roughness, defectlessness and granularity, thus determining to a large extent the efficiency of the OSCs.

The best parameters of a thin-layer OSC were reported by the group of Forest [2,4]. This group used copper phthalocyanine (CuPhTc) as a donor and 3,4,9,10-perylen-tetracarboxy bis-benzimidazol (PTCBI) or fullerene (C_{60}) as an acceptor. The efficiency reached was in the range of 3-5 % per 1 cm^2. [2,4]. Obviously, these parameters are considerably lower than those of the inorganic SC [5], which, together with their lower stability, are the reasons why the OSCs will not replace the SCs based on silicon in the near future.

As a result from the absorption of solar light in an OSC, a predominant formation of bound electrons and holes (photogenerated excitons) takes place [5]. The excitons are destroyed in the comparatively strong electrical field between the electrodes to charge carriers – electrons and holes. Their transformation is influenced by the presence of defects in the volume of the layers, their microstructure and the existence of traps. The latter are to a large extent a consequence of the

[3] To whom any correspondence should be addressed.

fabrication conditions – purity of the initial substances, reproducibility of the composition and structure of the active layers, etc. This is why the aim of the present study was to investigate the impact of the parameters for preparing films of VD "small molecules" organic substances on the composition and structure of the deposited layers with regard to their utilization as active layers in OSCs.

2. Experimental

Most of the experiments were carried out with VD films consisting of successive or mixed condensed ZnPhTc or PTCDA. The substrate used was a flexible polymer band – polyethylene-terephtalate (PET) covered by electroconductive films – 30 Ω/sq or 60 Ω/sq – of indium tin oxide (ITO). Al films of thickness 10-12 nm were used as upper electrodes. The ZnPhTc and PTCDA were thermally evaporated from specially designed sources. The latter allow one to heat carefully the organic dyes thus overcoming the problems of their low thermal conductivity and insufficient thermal stability. The deposition rates were 0.1 – 0.4 nm/sec at 550°C for ZnPhTc and 430-460°C for PTCDA at a basic pressure of 4×10^{-2} Pa. The rates and the final thicknesses were strictly controlled by quartz oscillators (Rate-thickness Monitor Inficon SQM 160, USA). The VD layers on a KBr substrate were characterized by Fourier transform infrared spectroscopy (FTIR). The FTIR absorbance spectra were recorded on a Bruker Tensor 27 spectrometer in the range of 4400 - 450 cm^{-1}. The surface morphology of the layers was investigated using Philips 515 SEM, and the microstructure, by means of a high-resolution transmission electron microscope (HRTEM JEOL 2100).

3. Results and discussion

3.1. FTIR investigations

The results from the FTIR investigations are presented in figures 1, 2. The spectra are normalized with respect to band 720 cm^{-1}. Figure 1 shows the FTIR spectra of the initial ZnPhTc dispersed in a KBr pill. The spectra of the VD layers of ZnPhTc – as deposited and after thermal treatment for 1 h at 200 °C, are compared in figure 2. The peaks at 720 cm^{-1}, 1120 cm^{-1}, 1165 cm^{-1} and 1333-1335 cm^{-1} in figures 1 and 2 are typical for the existence of ZnPhTc [6]. The band at 720 cm^{-1} is typical of an α-polymorphic modification which is characteristic of a thermally untreated material. The bands at 1287 cm^{-1}, 1070-1170 cm^{-1} and 720-770 cm^{-1} correspond to the existence of different polymorphic modifications. The important results from our study are that a constant composition is preserved upon

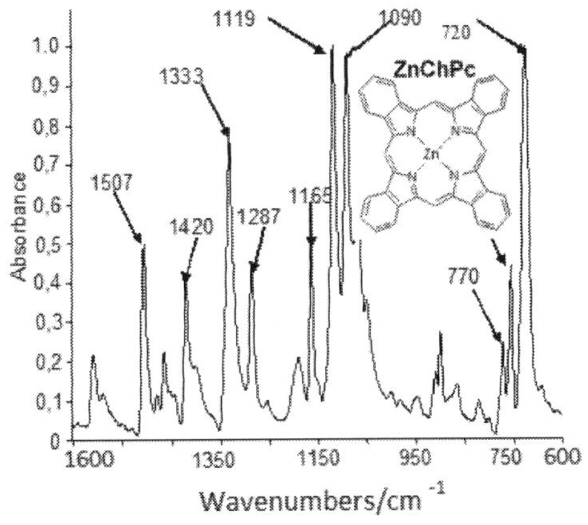

Figure1. FTIR of initial ZnPhTc in KBr pill.

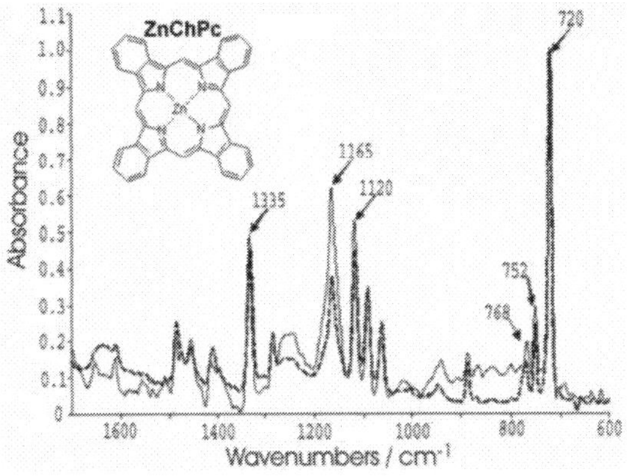

Figure 2. FTIR spectra of thin vacuum deposited ZnPhTc layers: dotted line – as deposited; solid line – after thermal treatment for 1 h at 200° C.

vacuum deposition of the ZnPhTc and negligible changes take place after thermal treatment at 200 °C of the condensed films whose structural defects can be reduced by annealing. Based on the spectra obtained and in accordance with the data in the scientific literature [6], we established a full correspondence between the initial substances and the composition of the VD layers prepared.

3.2. SEM and TEM investigations

Figure 3 presents electron microscope images of the surface morphology of ZnPhTc of thickness 63 nm and 190 nm.

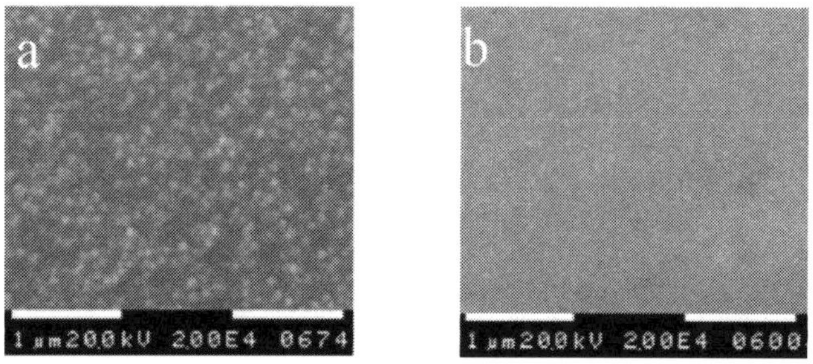

Figure 3. Dependence of the surface morphology of ZnPhTc films on the thickness: a) d = 63 nm b) d = 190 nm; PET/ITO substrate 60 Ω/sq.

Figure 4. SEM of profile of consecutively deposited ZnPhTc/PTCDA, d ≈ 1.2 μm.

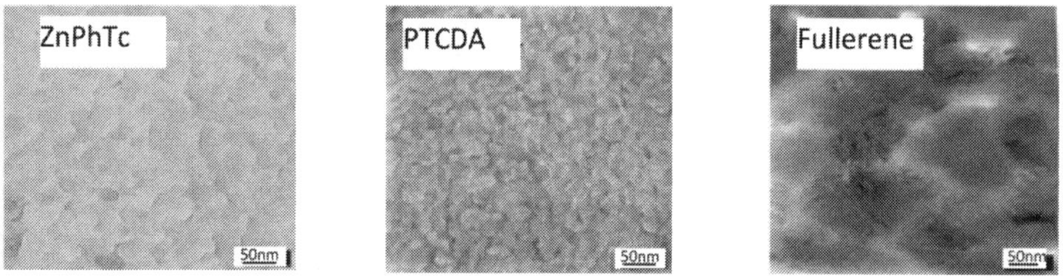

Figure 5. TEM of the different films with thicknesses in the range of 200 nm. Magn. 100k.

The PET/ITO (60Ω/sq) flexible substrate proved to be the most suitable one with regard to the smoothness of the surface, fine surface morphology and defectlessness of the active layer – figure 3. The thin ZnPhTc layers (d = 63 nm) are more sensitive to the type and structure of the substrate and replicate their defects. The thickness of the VD layer is also important. In the case of thicknesses of over 100 nm, no substrate-related effects were registered. In summary, the condensation of ZnPhTc on the substrates with conductivity 60Ω/sq determines the possibility of preparing dye layers (≥ 80 nm) with uniform fine structure and an almost defectless surface, which was one of the aims of the work.

In contrast to the typical crystalline films, in the ones studied by us (figure 4), column-like growth is only slightly marked, but the results observed are not sufficient for a final conclusion. It is important to note the absence of a phase boundary in the consecutive deposition of both dyes – ZnPhTc and PTCDA – figure 4.The results from the microscopy studies – figures 3, 4, as well as for fullerene – figure 5, show that the films obtained, similarly to the copper phthalocyanine ones [7], have an even surface, fine microstructure and lack considerable defects. This is an important prerequisite for a foreseeable structure of the active films in the multi-layer or tandem OSCs.

3.3. Electrical measurements

The structures ITO/ZnPhTc:PTCDA/PTCDA/Al show a typical diode behavior. The *I/U* curves – figure 6, cross the ordinate below the origin, which is a promising result for their applicability as a SC. The current increase in daylight and upon illumination by a lamp (60 W) is clearly discernible (approximately by one order of magnitude). In spite of the weak response – around pA to nA, registered as a photovoltaic effect, a low J_{sc} is measurable.

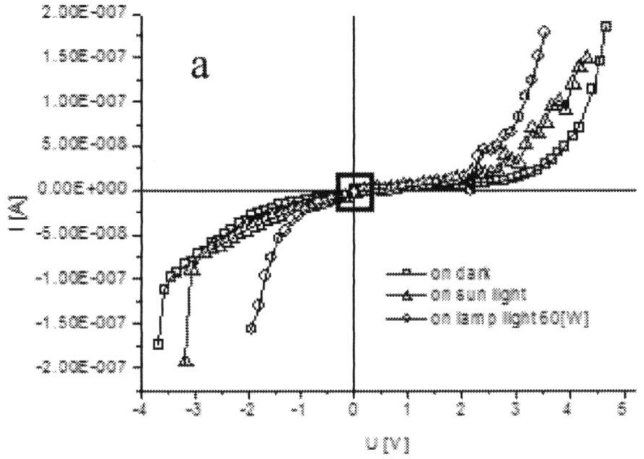

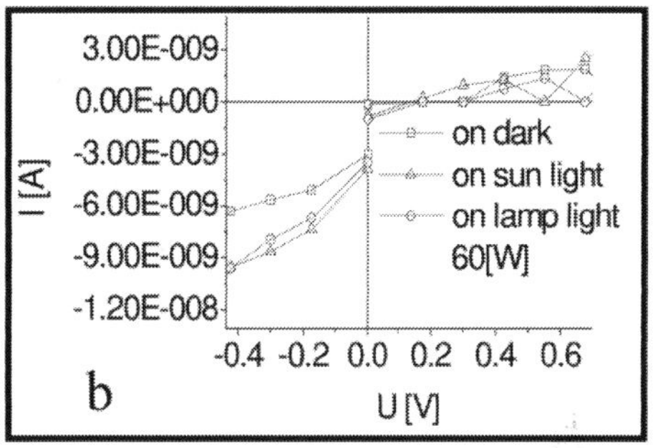

Figure 6. a) Volt-ampere curves of ITO/ZnPhTc:PTCDA/PTCDA/Al layers depending on the type of exposition, b) zoom of the square area in figure 6 a).

Conclusions

The experimental conditions are established for preparation of VD semiconductor "small molecules" organic films, consisting of pure or mixed dye layers in view of their application as active elements in OSCs. The analyses of the FTIR spectra of initial and VD materials show that they remain unchanged, which is a confirmation of the presence of ZnPhTc in the layers deposited. The results from the microscopy studies show that the films obtained have an even surface without considerable defects. This is an important prerequisite for the growth of the upper layers in the formation of multi-layered or tandem OSCs. Because of the relatively low resolution of the SEM Philips 515, confirming the presence of column-like growth needs additional experimental studies. The initial electrical measurements give us grounds to expect registering a better response to illumination by light similar to the solar light after the improvement of the experimental set-up.

Acknowledgements

This work was supported by the Bulgarian National Science Fund at the Ministry of Education, Youth and Science under contracts BY-TH- 205/06 and DO 02-254/18.12.2008 are gratefully acknowledged.

References

[1] Riede M, Mueller T, Tress W, Schueppel R and Leo K 2008 Small-molecule solar cells-status and Perspectives *Nanotechnol.* **19** 424001

[2] Xue J, Uchida S, Rand B and Forrest S 2004 Asymmetric tandem organic photovoltaic cells with hybrid planar-mixed molecular heterojunctions *Appl. Phys. Lett.* **84** 3013-5 **98** 5757-9

[3] Schuppel R, Schmidt K, Uhrich C, Schulze K, Wynands D, Bredas J, Maennig B, Pfeiffer M, Leo K, Brier E 2007 Tailored heterojunctions for efficient thin-film organic solar cells: a photoinduced absorption study *Proc. SPIE* **6656** 66560G1-11

[4] Peumans P and Forrest S 2001 Very-high-efficiency double-heterostructure copper phthalocyanine / C60 photovoltaic cells *Appl. Phys. Lett.* **79** 126

[5] Hoppe H and Sariciftci N 2004 Organic solar cells: An overview *J. Mater. Res.* **19**/7 1924-45

[6] Szybowicz M, Runka T, Drozdowski M, Baia W, Wojdyia M, Grodzicki A, Piszczek P and

Bratkowski A 2007 Temperature study of Raman, FTIR and photoluminescence spectra of ZnPc thin layers on Si substrate *J. Molec. Struct.* **830** 14–20

[7] Zhivkov I, Spassova E, Danev G, Andreev S and Ivanov Tz 1998 Vacuum deposited copper phthalocyanine thin films – structure and surface morphology *Vacuum* **51** 2 189-92

Study of the charge transport mechanism in pulsed laser deposited AlN:Si films

I P Minkov[1,5], S Simeonov[2], A Szekeres[2], A. Cziraki[3], G Socol[4], C Ristoscu[4] and I N Mihailescu[4]

[1] Technical University of Sofia, 8, St. Kliment Ohridski Blvd., 1000 Sofia, Bulgaria
[2] Institute of Solid State Physics, Bulgarian Academy of Sciences, 72, Tsarigradsko Chaussee, 1784 Sofia, Bulgaria
[3] Eotvos Lorand University, Department of Materials Physics, 1/A Pazmany Peter setany, 1117 Budapest, Hungary
[4] National Institute for Lasers, Plasma, and Radiation Physics, PO Box MG-54, RO-77125, Magurele, Ilfov, Romania

E-mail: ipmincov@tu-sofia.bg

Abstract. AlN films doped with Si (AlN:Si) were synthesized on *p*-Si(100) substrates by pulsed laser deposition. Al-AlN:Si-Si metal-insulator-silicon (MIS) structures were formed and their current-voltage characteristics measured at 77 K and 290 K were analyzed. The results revealed that the charge transport is carried through the AlN:Si-Si MIS structures by the mechanism of trap space charge limited current.

1. Introduction

Aluminuim nitride has wide application prospects in the area of high-power, high-temperature electronic and UV optoelectronic devices, especially for UV solid-state LED and lasers. For these applications, it is indispensable to prepare AlN films and structures with good and controllable *n*- and *p*-type conduction. AlN is a semiconductor with the wide energy gap of 6.2 eV. Because of this, AlN doping is a complex process. To obtain *n*-type AlN, Si doping is widely used [1,2]. During the Si doping, compensation of Si donors occurs, as it has been observed in MO VPE AlN films [1]. Deep levels introduced in MBE deposited AlN films create a high-resistivity state in the films irrespectively of the Si doping [2]. Therefore,studying the electrical conductivity of Si doped AlN films prepared by different methods is essential for prospective applications of such AlN semiconductor structures.

Pulsed laser deposition (PLD), because of its capability of reproducing the target stoichiometry in the deposited films, is suitable for preparation of AlN:Si films with controllable Si doping level. Establishing the charge-transport mechanism in PLD AlN:Si films is an unavoidable step in achieving reliable and effective Si doping of AlN. Measuring the current-voltage (*I-V*) characteristics at different temperatures, which reveals, in particular, the deep levels role in the charge transport through the deposited films, is a powerful technique for studying the electro-physical properties of semiconductor structures.

[5] To whom any correspondence should be addressed.

In this paper we present results on the study of the charge transport mechanism in pulsed laser deposited AlN films doped with Si. For this purpose, metal-insulator-silicon (MIS) structures with PLD AlN:Si films were prepared and their DC *I-V* characteristics were recorded. The charge transport mechanism was established by analyzing these characteristics.

2. Experimental details

Si-doped AlN films were synthesized on *p*-type Si(100) substrates by pulsed laser deposition. The PLD targets were prepared by homogeneously mixing AlN and Si (10 at. %) powders, followed by pressing and sintering. The distance between the target and substrate was set at 5 cm. After loading the substrate and target, the chamber was evacuated to a residual pressure of 10^{-4} Pa and the substrate was heated up to 800 °C. During ablation, the target was rotated and translated along two orthogonal axes to avoid piercing and to improve the morphology of the films deposited. The deposition was performed in nitrogen at a dynamic pressure of 0.1 Pa or 10 Pa. AlN:Si films with ~200 nm thickness were deposited by applying 20 000 laser pulses generated by an UV KrF* laser source (λ = 248 nm, τ_{FWHM} = 25 ns) operating at 10-Hz repetition rate and incident laser fluence of 10 J/cm^2 on the target.

For the electrical measurements, MIS capacitors with PLD AlN:Si films were formed by vacuum evaporation through a metal mask of Al dots on the AlN surface and a continuous Al film on the Si wafer backside. The *I-V* characteristics were measured at 290 K and 77 K. The measurement sequence began from 0 V toward the negative or positive voltages applied to the Al-dot contact on the film surface followed by a voltage reversal toward zero applied bias voltage. For clarity, we denote below the initial voltage passage from zero to minimal negative or maximal positive voltage as "up", and the voltage passage going back to zero as "down" directions.

Our earlier investigations on PLD AlN films have revealed that the their structure is strongly dependent on the deposition conditions and, in particular, on the nitrogen pressure [3,4]. Therefore, it was expected that the AlN:Si films would also exhibit different structures and, thus, different electrical properties. Taking this into consideration, the AlN:Si films were examined by a large-angle (2Θ = 0–90°) X-ray Philips X'Pert diffractometer with Cu radiation (λ = 0.154056 nm).

3. Results and discussion

Figure 1 presents the recorded XRD patterns of the films studied. The appearance of Bragg peaks related to a crystalline phase and identified using the ASTM data basis [5] is evidence of the presence of crystallites in the mainly amorphous matrix growing in a cubic phase predominantly in the (111) crystallographic direction. The broadening of the peaks observed as compared with the diffraction Bragg peak of a perfect crystal revealed that the randomly oriented crystallites had grown up to sizes in the nanoscale range. The films deposited at 0.1 Pa N_2 pressure had a higher degree of crystallization.

Figure 2 presents the *I-V* characteristics of a MIS capacitor with an AlN:Si film deposited at 10 Pa. As is evident, no hysteresis is observed at electric fields up to 4.5×10^4 V cm^{-1}. Under these electric fields, the current is proportional to the applied voltage and the *I-V* curves recorded with positive (forward direction) or negative (reverse direction) voltages, applied to Al-dot contacts, are with equal or slightly different slopes (inset in figure 2). The slope of the curves corresponds to a specific resistivity of 1.0×10^4 Ω cm of the AlN:Si film deposited at the given conditions.

In the case of MIS capacitors with an AlN:Si film deposited at 0.1 Pa, the *I-V* characteristics are non-ohmic and a hysteresis effect appears (figure 3). In both reverse and forward directions, the current is different in an up-and-down cycle. During the initial voltage passage from zero to minimal negative or maximal positive voltage (up), the current density is smaller than that obtained by going back

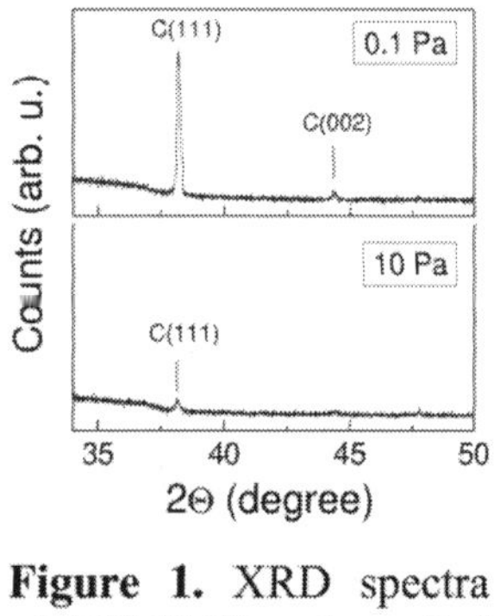

Figure 1. XRD spectra of AlN:Si films deposited at different nitrogen pressures.

to zero voltage (down). This means that during the first stage, the charge carriers are captured by traps in the AlN:Si films. In the forward direction, when the electrical field exceeded 1.2×10^5 V/cm the current became unstable. Keeping constant the voltage at ~2.5 V for several minutes, the current increased approximately twice. Because of this, in the forward direction the hysteresis effect was much stronger.

The specific resistivity in forward direction at the lower electrical fields of 5.4×10^3 V/cm (up) and 4.2×10^3 V/cm (down) (figure 3) was equal to 6.6×10^5 Ω cm and 9.6×10^4 Ω cm, respectively. Correspondingly, in reverse direction at the lower electrical fields of 1.0×10^4 V/cm (up) and 4.2×10^3 V/cm (down) its value was 5.1×10^5 and 2.4×10^5 Ω cm, respectively. At the higher electrical fields of 1.2×10^5 V/cm, the resistivity decreased in both directions and was in the range of $(9\times10^4\div2.4\times10^5)$ Ω cm.

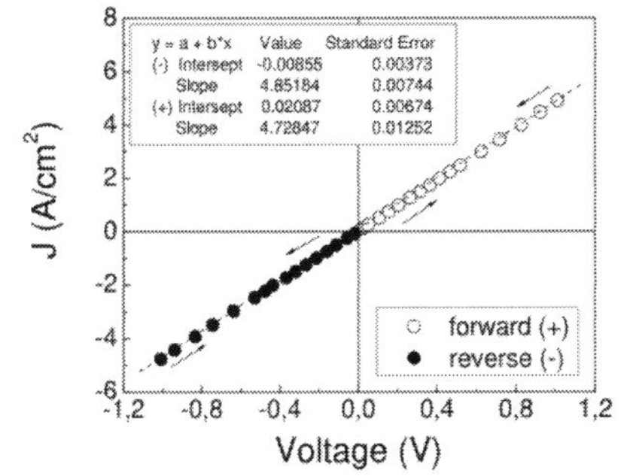

Figure 2. *I-V* curves of a MIS capacitor with AlN:Si film deposited at 10 Pa N_2 pressure.

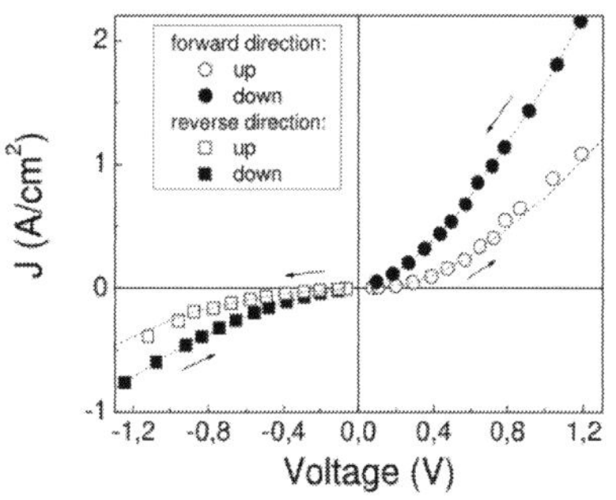

Figure 3. *I-V* curves of a MIS capacitor with AlN:Si film deposited at 0.1 Pa N_2 pressure.

The *I-V* characteristics for the 0.1 Pa case (figure 3) were further analyzed by replotting them as ln*J* versus ln*V* plots (not shown here). The slope of the plots (excluding the first three points near zero applied voltage) indicates that the current density *J* is proportional to V^ξ. In the initial voltage passages (up) in forward direction the value of this ξ parameter is equal to ξ = 1.97, while in reverse direction ξ = 2.02. In the return passages (down), the ξ values are ξ = 1.59 and ξ = 1.60 for reverse and forward bias, respectively. As it is established in [6], a $I(V^\xi)$ dependence with ξ = 2 is evidence that the current in the insulating film is space-charge limited and occurs via single deep levels with equal position in the insulator energy bandgap. In such a case, both the density of trapped charges and their effective velocity are proportional to the applied voltage *V* and, therefore, the current is proportional to V^2. For the AlN:Si MIS structures, it is the case when the current is measured in the initial voltage passages (up). For the return voltage passages (down), the observed deviation of ξ from 2 (ξ < 2) is connected with charge trapping at deep levels having long relaxation times.

In AlN, the electron mobility exceeds largely the hole mobility. For example, at 300 K in a pure AlN film, the electron mobility is 800 $cm^2 V^{-1} sec^{-1}$ [6], while in a Mg doped AlN film, the hole mobility is (10÷26) $cm^2 V^{-1} sec^{-1}$ [7]. In our case, the current through the MIS structures with AlN:Si film deposited at 0.1 Pa is only 2-3 times as high in forward direction as that in reverse direction, while the difference between the forward and reverse currents through the MIS structures with 10 Pa AlN:Si film is even smaller (2–3 %). Therefore, in both cases the current is carried by electrons.

The above considerations allow us to infer that the electrical current though the AlN:Si films investigated is carried by electrons via deep levels with single energy level; in other words, the current through the AlN:Si films is the so-called space charge limited current.

Typical *I-V* curves of AlN:Si MIS structures measured at 77 K and 290 K and in the initial voltage passages are shown in figure 4. It is evident that in both reverse and forward directions the current density at 290 K is slightly higher than that at 77 K. At low electric fields, the *I-V* curves are linear for both temperatures, while for the AlN:Si film deposited at 10 Pa the linear part extends over a larger range of voltages. The specific resistivity ρ of the AlN:Si films is determined from the slope of the linear part of the curves and the results are summarized in table 1.

The temperature behavior observed of the current through the AlN:Si MIS structures is consistent with charge transport by electron hopping from occupied deep levels to non-occupied ones. It is known that in AlN the defects related to Al vacancies are with acceptor properties [8] and, therefore, the Si donors are partly compensated in the AlN:Si film by Al-vacancy acceptors. Such a process considerably reduces the efficiency of AlN doping with Si donors with the purpose of obtaining *n*-type conduction. Further investigations are needed to distinguish between the mechanisms of electron variable-range hopping and electron hopping to nearest non-occupied acceptor in these AlN:Si films.

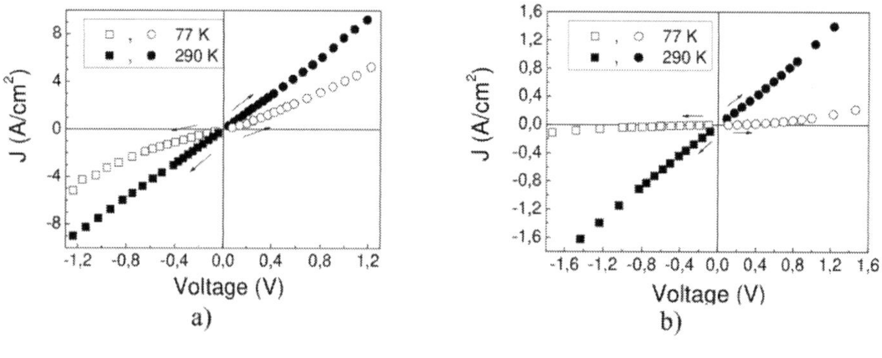

Figure 4. *I-V* characteristics measured at 290 K (full symbols) and 77 K (empty symbols) of MIS capacitors with AlN:Si film deposited at a nitrogen pressure of 0.1 Pa (a) and 10 Pa (b).

Table 1. Specific resistivity ρ of AlN:Si films deposited at different nitrogen pressures.

Nitrogen pressure	**0.1 Pa**		**10 Pa**	
Calculated quantity	ρ (77 K) [Ω cm]	ρ (290 K) [Ω cm]	ρ (77 K) [Ω cm]	ρ (290 K) [Ω cm]
Reverse *I-V*	1.56×10^4	6.22×10^3	1.29×10^6	4.38×10^4
Forward *I-V*	1.18×10^4	6.22×10^3	6.52×10^5	4.60×10^4

Conclusions

We established that the electrical current through PLD AlN:Si films synthesized at the conditions described is carried by electrons via deep levels in the AlN:Si energy bandgap. The trap space charge limited current occurs by electron hopping from occupied deep acceptor levels to non-occupied ones.

Acknowledgements

The authors acknowledge the support of this work under the 2009–2011 Collaboration Agreement between the Bulgarian and Romanian Academies of Sciences.

References

[1] Taniyasu Y, Kasu M and Kobayashi N 2002 *Appl. Phys. Lett.* **81** 1255–7
[2] Boger R, Fiederle M, Kirste L, Mayer M and Wagner J 2006 *J. Phys. D Appl. Phys.* **39**/21 4616–20
[3] Bakalova S, Szekeres A, Cziraki A, Lungu C P, Grigorescu S, Socol G, Axente E and Mihailescu I N *Appl. Surf. Sci.* 2007 **253** 8215–9
[4] Szekeres A, Fogarassy Zs, Petrik P, Vlaikova E, Cziraki A, Socol G, Ristoscu C, Grigorescu S and Mihailescu I N *App. Surf. Sci.* 2011 **257** 5370–4
[5] *Index to the Powder Diffraction File 2000*, (Joint Comm. on Powder Diffractions Standards)
[6] Lampert M and Mark P 1970 *Current injection in solids* (New York, London: Academic press)

[7] Yarar Z, Ozdemir B and Ozdemir M 2007 *J. Electronic Mater.* **36**/10 1303–12
[8] Stampfi C and van de Walle C G 2002 *Phys. Rev. B* **65**/15 155212–21

Properties of thin silicon carbide films prepared by rapid thermal annealing

M Beshkova[1], K Grigorov[1, 2, 4], I Nedkov[1], M Massi[2], B Sismanoglu[2], H Maciel[2] and N B Velchev[3]

[1] Emil Djakov Institute of Electronics, Bulgarian Academy of Sciences, 72, Tsarigradsko Chaussee, 1784 Sofia, Bulgaria
[2] Plasmas and Processes Laboratory, Department of Physics, Technological Institute of Aeronautics, 12228-900, S.J. Campos, Brazil
[3] Paisii Hilendarski University of Plovdiv, 24, Tzar Asen Str., Plovdiv, Bulgaria

E-mail: kgrigoro@netissat.bg

Abstract. We report experimental results related to the structural and electrical properties of thin SiC films. Thin carbon films with thicknesses 50 Å and 300 Å were deposited by R.F. sputtering and processed by rapid thermal annealing (RTA) for 3 min at temperatures of 800 °C and 1400 °C in a vacuum chamber at 2×10^{-5} Torr. The thin films properties were studied by Raman spectroscopy and electrical cross-conductance.

1. Introduction

Silicon carbide combines a wide band gap, high thermal conductivity, high breakdown field and high saturated electron drift velocity, thus proving to be the best choice for a semiconductor for high-power, high-temperature and high-frequency devices, where Si and GaAs devices cannot be used [1]. SiC is also appropriate for devices operating in aggressive environments because of its stability to ionizing radiation and chemical attacks due to the high energy Si–C bond.

SiC layers fabricated by sublimation [2], chemical vapor deposition (CVD) [3, 4], magnetron sputtering [5] and laser ablation [6] have a relatively large thickness, which makes them unattractive for certain device applications.

In this work, we report the preparation of thin SiC nano-layers by rapid thermal annealing (RTA) of carbon films on silicon substrates. The samples were investigated by Raman spectroscopy and electrical conductivity measurements.

2. Materials, experimental procedure and equipment

We used P type (100) silicon substrates with resistivity 40 - 60 Ω cm. The silicon wafers were cleaned by a standard chemical procedure. Carbon films with thicknesses of 50 Å and 300 Å were deposited by RF sputtering at 0.5 Pa pressure in a gas mixture of 30 % H_2 and Ar. The thicknesses of the films were determined ellipsometrically. The target-substrate distance was fixed at 70 mm.

Further, the samples were submitted to RTA in vacuum of 5×10^{-5} Torr (6.7×10^{-3} Pa) for 3 min at temperature of 800 °C and 1400 °C. The heating from room to annealing temperatures was performed

[4] To whom any correspondence should be addressed.

in 2 s, while cooling down to 650 °C required 3 - 5 s. For the Raman measurements, a Raman SPEX spectrometer (RAMLOG) system was used. The electrical conductivity was determined by measuring the cross-conductance of the layers. These measurements were carried out by a mercury probe with a contact area of 7×10^{-2} cm^2 to the layers. Good ohmic contact to the backside of the silicon substrate was ensured by a silver sealing covering the entire backside area of the wafer.

3. Experimental results and discussion

Raman spectra of the layers were been taken before and immediately after the rapid thermal annealing procedure. Figure 1(a) shows the Raman spectrum of a carbon layer with thickness of 300 Å before the RTA process; a large peak can be clearly seen in the range 1500 - 1550 cm^{-1}, which is characteristic for amorphous carbon [7]. The Raman spectrum of the layer after RTA at 1400 °C for 3 min presented in figure 1(b) show the presence of rounded peaks in the ranges 910 - 1050 cm^{-1} and 775 - 850 cm^{-1}, which is indicative of the existence of hexagonal and cubic SiC polytypes [8-10], respectively. The spectrum in figure 1(a) consists of the TO and LO phonon modes for both phases, which are not resolved.

Typically, the cubic phase is considerably less in quantity. Calculated as a percentage with respect to the basic phase (the hexagonal SiC), it is about 14%. We can thus assume that the cubic SiC phase is distributed in the basic matrix of hexagonal silicon carbide.

Figure 2 shows a Raman spectrum of the SiC layer with initial thickness of 50 Å subjected to RTA at 800 °C for 3 minutes. A further increase of the annealing temperature does not change the Raman spectra intensity, indicating that the solid-state reaction has been completed. The most clearly expressed peaks are those of the layer with thickness 300 Å and annealed at 1400 °C, which indicates that both phases increase as the annealing temperature and layer thickness are increased.

The analysis of the electrical cross-conductance of the carbon and SiC layers deposited onto silicon substrates were carried out for samples with thicknesses of 300 Å because of their good homogeneity, as assessed by scanning electron microscopy (SEM) images (not shown here). The current-voltage characteristics are shown in figure 3. The I-V curves are typical for a Schottky type diode with high density of defects at the interface between the metal (a mercury contact with an area of 7×10^{-2} cm^2) and the p-type silicon substrate.

The analysis of the curve displayed in figure 3 for an as-deposited sample shows that it behaves simply as an additional resistance connected in series with the rectifying contact between the carbon layer and the silicon substrate. After the RTA procedure at 1400°C for 3 min, the SiC layer behaves as a semiconductor with a large density of

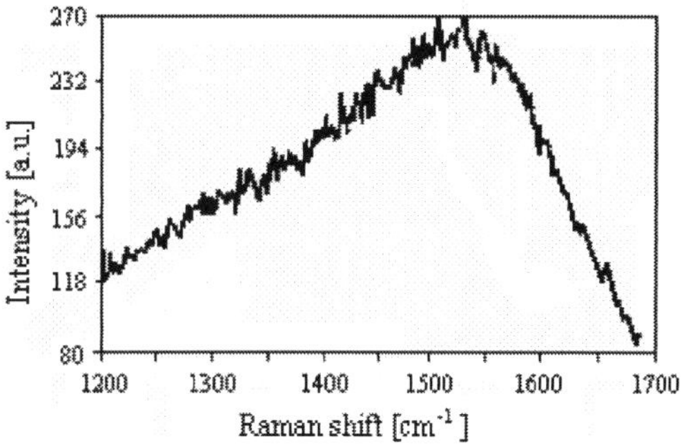

(a)

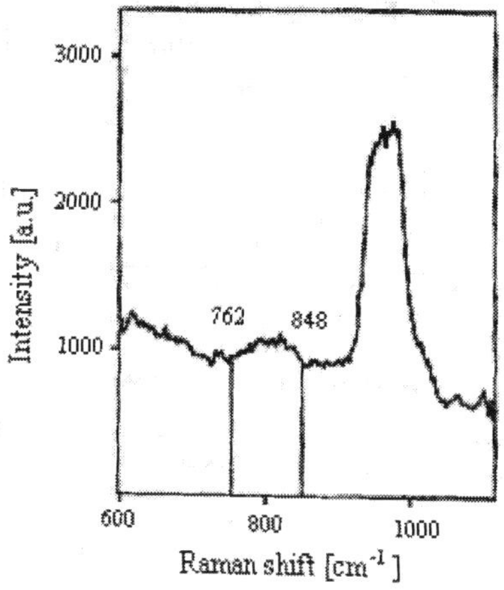

(b)

Figure 1. Raman spectrum of a) as-deposited C layer with thickness 300 Å, b) SiC layer with thickness 300 Å after annealing for 3 min at 1400 °C.

defects. The rectifying properties of the structure are due to the contact between the mercury probe and the SiC layer. The high current observed under reverse biasing is generated by the defects in the SiC layer.

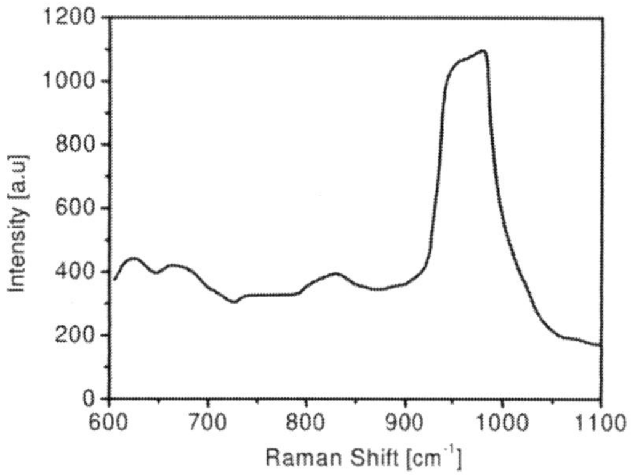

Figure 2. Raman spectrum of SiC layer with thickness of 50 Å after annealing for 3 min at 800 °C.

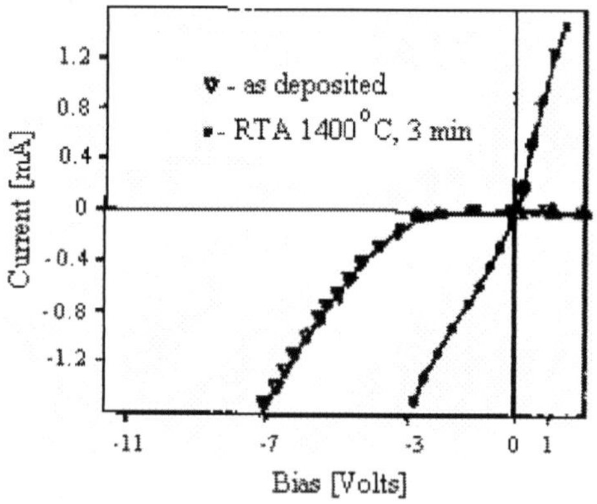

Figure 3. Effect of rapid thermal annealing on I-V characteristic of Hg-C-Si and Hg-SiC-Si structures.

Conclusions

The effect of RTA in vacuum on the properties of carbon and SiC layers were studied using both Raman and electrical conductivity measurements. The experiments show that samples annealed at temperatures as high as 1400 °C exhibit a large current in the forward direction. Therefore, at 1400 °C formation of SiC/Si hetero-junctions by RTA takes place. The changes discussed of the electrical properties of the SiC layers are correlated with the Raman spectra which reveal peaks corresponding to hexagonal and cubic SiC phase after RTA treatment.

Reference

[1] Baliga B J 1996 *Physics of Semiconductor Power Devices* (JWS Publishing)
[2] Lebedev A, Zelenin V, Abramov P, Lebedev S, Nel'son D, Razbirin B, Scheglov M, Tregubova A, Syväjärvy M and Yakimova R 2007 *Mater. Sci. Forum* **556-557** 175-8
[3] Nishino K, Kimoto T and Matsunami H 1997 *Japan J. Appl. Phys.* **36** 5202-7
[4] Soueidan M, Ferri G, Nsouli B, Cauwet F, Dazord J, Younes G and Monteil Y 2006 *Mater. Sci. Eng. B* **130** 66-72
[5] Saito N, Inui Y, Yamaguchi T and Nakaaki I 1998 *J. Appl. Phys* **83** 2067
[6] Ghosh S, Battacharyya P and Bose D N 1996 *Appl. Phys. Lett.* **68** 2979
[7] Huong P V 1991 *Diamond and Related Materials* **1** 35
[8] Burton J C, Sun L, Pophristic M, Lukacs S J, Long F H, Feng Z C and Ferguson I T 1998 *J. Appl. Phys.* **84** *6268*
[9] Zarrabian M, Fourchescoulon N, Turban C, Marhic C and Lancin M 1997 *Appl. Phys. Lett.* **70** 2535
[10] Shroder R E, Nemanich R J and Glass J T 1990 *Phys. Rev. B* **41** 3738

Characterization of electrophoretic suspension for thin polymer film deposition

D Mladenova[1,2,5], M Weiter[1], P Stepanek[3], I Ouzzane[1], M Vala[1], V Sinigersky[4] and I Zhivkov[1,2]

[1] Centre for Materials Research, Faculty of Chemistry, Brno University of Technology, 118 Purkynova, 612 00 Brno, The Czech Republic
[2] Institute of Optical Materials and Technologies, Bulgarian Academy of Sciences, Acad. G. Bonchev Str. Bl. 101, 1113 Sofia, Bulgaria
[3] Department of Supramolecular Polymer Systems, Institute of Macromolecular Chemistry, Academy of Sciences of The Czech Republic, 2 Heyrovskeho nam., 162 06 Praha 6, The Czech Republic
[4] Institute of Polymers, Bulgarian Academy of Sciences, Acad. G. Bonchev Str., Bl. 103A, 1113 Sofia, Bulgaria

E-mail: dnl@clf.bas.bg

Abstract. The optical absorption and fluorescence spectra of poly [2-methoxy-5-(3′,7′-dimethyloctyloxy)-1,4-phenylenevinylene] toluene solutions and 50:50% toluene/acetonitrile suspensions show clearly distinguishable differences (e.g., peak broadening and shifting), which could be used for characterization of suspensions with different acetonitrile content. The dynamic light scattering (DLS) measurement of the suspensions prepared showed a particle size of 90 nm. Thin films with thicknesses of about 400 nm were prepared by electrophoretic deposition (EPD) and spin coating. As the films are very soft, a contactless optical profilometry techique based on chromatic aberration was used to measure their thickness. AFM imaging of spin coated and EPD films revealed film roughness of 20÷40 nm and 40÷80 nm, respectively. The EPD film roughness seems to be less than the suspension particle size obtained by DLS, probably due to the partial film dissolving by the toluene present in the suspension.

1. Introduction

Among the variety of the methods for "wet" polymer thin film deposition [1] (e. g. spin and dip coating, spray and Langmuir-Blodgett deposition, and ink-jet printing), the electrophoretic deposition (EPD) [2] has the advantages of using a diluted suspension for deposition of relatively thick films, ability for covering a large area and the unique property of separating the solidification stage (precipitation of a solid phase in the suspension) from the film formation stage, which takes place on the electrode.

One of the main problems impeding the wide usage of EPD for thin polymer film preparation is the instability – the suspension particle size depends strongly on the precipitation conditions and tends to

[5] To whom any correspondence should be addressed.

grow with the time. This effect creates difficulties in controlling the EPD thin film structure and morphology.

This study aims to establish methods for characterization of critical stages of the EPD process of deposition of thin poly[2-methoxy-5-(3′,7′-dimethyloctyloxy)-1,4-phenylenevinylene] (MDMO-PPV) films.

2. Experimental details

The EPD suspension with MDMO-PPV (Sigma-Aldrich, catalogue number 546461) concentration of 0.0033 g l^{-1} and toluene/acetonitrile ratio of 50:50% (toluene 99.8%, acetonitrile 99.8%) was prepared and used immediately for measurement or deposition to prevent further particle coagulation.

Dynamic light scattering (DLS) measurements of the suspension prepared were performed on a Malvern Zetasizer Nano ZS instrument equipped with a helium-neon laser; the scattering angle was 173°. The data were processed taking into account the viscosity of the toluene/acetonitrile mixture with 50 % acetonitrile content [3].

The EPD thin film was deposited on the positive electrode at a current of about 50÷70 μA.

A solution with concentration of 8.95 g l^{-1} was prepared and spin coating at approx. 2500 rpm for 60 s was carried out on a KW-4A Chemat Technology Inc. spin coater.

The film thickness of the deposited MDMO-PPV films was determined by a MicroProf® FRT optical profilometer based on chromatic aberration. The method has the advantage of performing fast high-resolution contactless and non-destructive measurements, which is of paramount importance in our case of soft polymer films. The films were scratched by a sharpened tungsten wire then a thin (about 100 nm) Al film was deposited in vacuum to equalize the optical reflection from both scratched and unscratched areas.

Surface morphology images of MDMO-PPV films were taken by NTEGRA Prima AFM. The measurements were carried out in semicontacting mode (frequency of 170 kHz, amplitude 80÷100 nm and scanning rate of 0.5 Hz).

3. Results and discussion

3.1. Suspension characterization

3.1.1. Absorption spectra. Optical absorption spectra of a solution and a 50:50 % toluene/acetonitrile suspension with the same MDMO-PPV concentration of 0.0033 g l^{-1} are presented in figure 1. The solution spectrum (curve 1) consists of the characteristic MDMO-PPV absorption peak [4].

In the suspension (curve 2) spectrum, a "red" shoulder appears which leads to a broadening of the peak. This effect could be related to the appearance of a precipitated solid phase during the suspension formation caused by the precipitating polar acetonitrile.

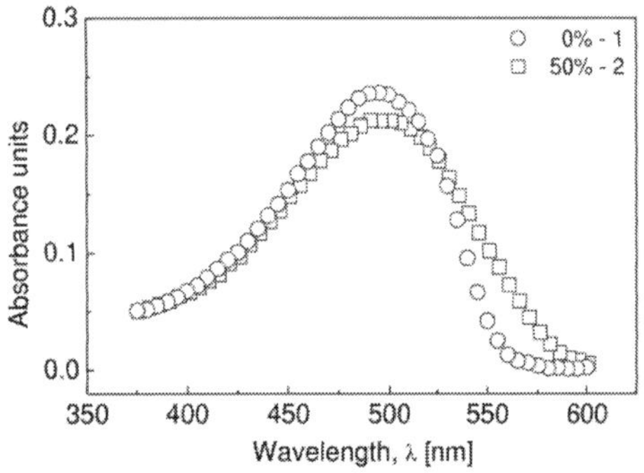

Figure 1. Absorption spectra of a MDMO-PPV solution (curve 1) and a suspension with 50% acetonitrile content (curve 2).

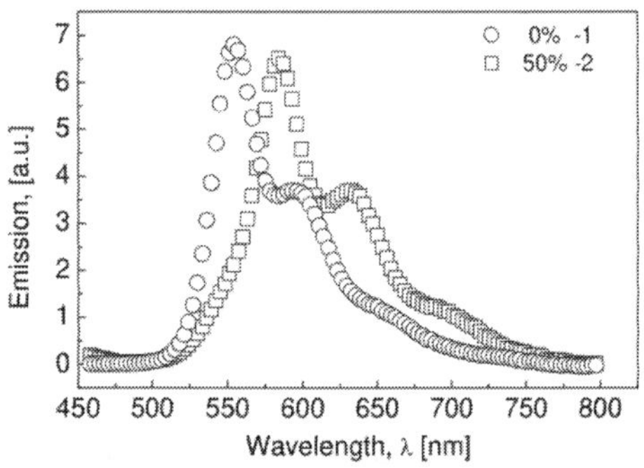

Figure 2. Fluorescence spectra of a MDMO-PPV solution (curve 1) and a suspension with 50% acetonitrile content (curve 2).

3.1.2. Fluorescence spectra. More detailed information about the material under study can be obtained from the fluorescence spectra, as they present information from two processes – absorption and subsequent emission. In figure 2, fluorescence spectra of a MDMO-PPV solution (curve 1) and a suspension with 50 % acetonitrile (curve 2) are plotted. The maximum of the spectrum in the suspension is "red" shifted by about 30 nm. The spectrum shift reflects more precisely the formation of the solid phase.

It could be concluded that effect of the peak broadening in the absorption spectra could be used for a qualitative estimation of the suspension properties, while the peak shifts in the fluorescence spectra yield more detailed information about the formation of a solid phase in the suspension.

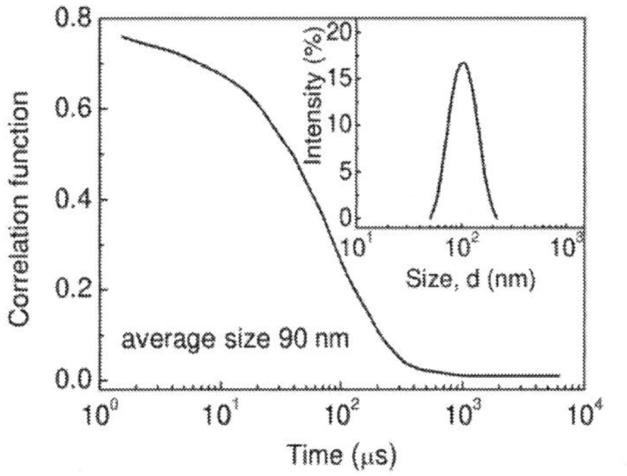

Figure 3. DLS measured curves.

3.1.3. Dynamic light scattering. The size of the particles forming the solid-state phase in the toluene/acetonitrile suspension with 50 % acetonitrile content was estimated by DLS. The typical correlation function obtained (figure 3) allows a straightforward determination of the particle size. The inset in the figure presents the particle size distribution by the intensity obtained using inverse Laplace transformation. The data obtained indicate an average particle diameter of 90 nm.

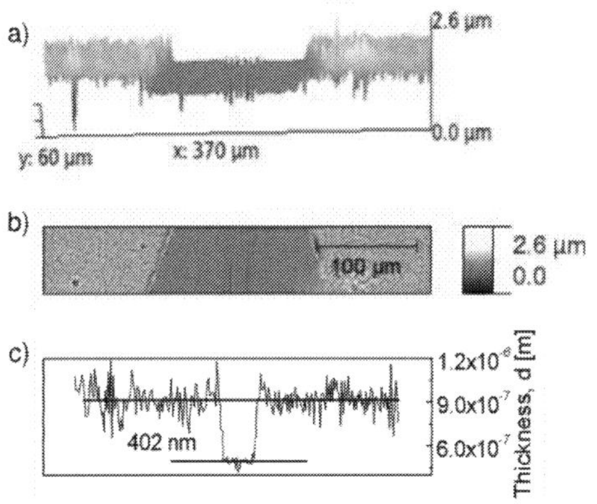

Figure 4. Optical aberration scanning of a scratched MDMO-PPV film: a) – 3D, b) – 2D and c) – 1D images.

3.2. Thin film characterization

3.2.1. Optical profilometer film thickness measurement. Figure 4, a) and b) present scanned optical aberration images of a scratched MDMOPPV. The scratched area is clearly distinguished, which allows a satisfactory film thickness determination. A film thickness of about 400 nm was determined by processing the data from a single profile line (figure 4, c). Optical-aberration 3D and 2D images can be used as a preliminary film-surface morphology estimation.

A film thickness of about 400 nm was measured by the same procedure for the spin-coated films.

3.2.2. AFM imaging. 2D and 3D AFM images of EPD deposited films are shown in figure 5. The surface roughness observed is less than but comparable to the particle size as obtained by DLS. Despite the general assumption that the size of the particle in the suspension should be preserved after EPD of a film [5], partial dissolving of the particles deposited on the substrate by the solvent (50% toluene in the suspension) is possible, which decreases the film roughness. Thus, varying the toluene/acetonitrile ratio gives an opportunity to control the film roughness.

For comparison, the AFM image of a spin-coated film with similar thickness is presented in figure 6. The picture shows a predominant film surface roughness of 20÷40 nm, which is smoother than the surface of the EPD film. The peaks observed with height approx. 50÷70 nm could be connected with the presence of undissolved or aggregated MDMO-PPV particles due to the relatively high solution concentration.

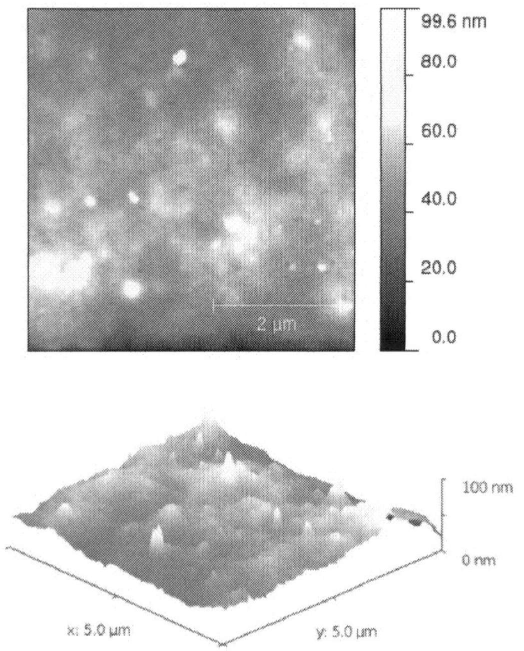

Figure 5. AFM 2D and 3D images of an EPD film measured on a 5×5 µm scanned area.

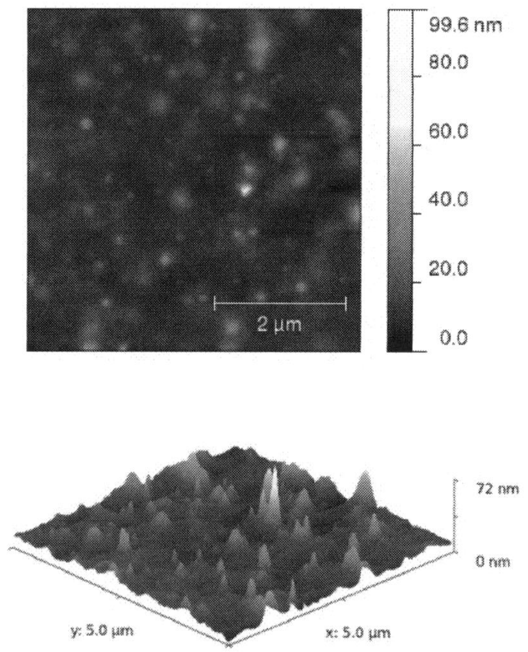

Figure 6. AFM 2D and 3D images of a spin coated film measured on 5×5 µm scanned area.

Conclusions

A combination of experimental methods was applied to the characterization of the different stages of the polymer electrophoretic deposition process. The methods could be used to control the suspension stability and the film structure and morphology, which are critical parameters during the EPD of thin polymer films for solar energy conversion purposes.

Acknowledgments

This work was supported by the South Moravian Region and the 7th Framework Program for Research and Development (Grant SIGA 885), and by the Bulgarian National Science Fund at the Ministry of Education, Youth and Science (Grant DO 02-254).

References

[1] Tada K and Onoda M 2009 *Molecular Crystals and Liquid Crystals* **505** 124-9
[2] Boccaccini A R and Zhitomirsky I 2002 *Current Opinion in Solid State and Materials Science* **6** 251-60
[3] Rltroulls G, Papadopoulos N and Jannakoudakls D 1986 *J. Chem. Eng. Data* **37** 146-8
[4] Quoc T V 2006 *Electrophoretic deposition of semiconducting polymer metal/oxide nanocomposites and characterization of the resulting films* (Dresden Technischen Universität Dresden Germany) p 109
[5] Landfester K, Montenegro R, Scherf U, Gqnter R, Asawapirom C, Patil S, Neher D and Kietzke T 2002 *Adv. Mater.* **14** 651

Journal of Physics: Conference Series **356** (2012) 012041 doi:10.1088/1742-6596/356/1/012041

Structural and dielectric properties of TiO_2 thin films deposited by the sol-gel method on Si substrates

P Vitanov, A Harizanova and T Ivanova [1]

Central Laboratory for Solar Energy and New Energy Sources,
Bulgarian Academy of Sciences, 72, Tsarigradsko Chaussee, 1784 Sofia, Bulgaria

E-mail: tativan@phys.bas.bg

Abstract. Titanium dioxide thin films with good dielectric properties were obtained on Si substrates using the sol-gel method. The metal oxide films were deposited by spin coating from sol solution and additionally annealed at 600 °C and 850 °C. XRD analysis, FTIR and Raman spectroscopy were used for revealing the film structure, the crystal phases and the formation of a polycrystalline material. The electrical measurements on metal-insulator-silicon (MIS) structures showed good dielectric properties. An optimal regime for high temperature annealing (850°C/15 min) was found, for which no effects related to the thermal instability of TiO_2 film on Si were observed. The leakage current density of a 38-nm thick TiO_2 film annealed at 850°C/15 min/O_2 was approximately 1.95×10^{-9} A/cm^2 at 1.6×10^{-6} V/cm.

1. Introduction

Titanium dioxide has been recognized as one of the most important materials for applications in the semiconductor electrochemistry, as a gas sensor [1], as antireflective coatings [2], in electrochromic devices and solar cells [3], as photocatalysts and optical filters [4]. TiO_2 is also used as a dielectric material for electronic devices [5].

The TiO_2-Si system has been intensively studied for high-k applications in memory capacitors and in nanoscale MOSFET [3, 6]. It is attractive as it has high permittivity k = 80-110 depending on the crystal structure and the method of deposition [7]. TiO_2 crystallizes at temperatures above 400°C, therefore it is expected to exhibit a polycrystalline morphology. In many cases, the polycrystalline structure might be problematic due to the grain boundaries acting as high leakage paths. In order to address this issue, lanthanide dopants in TiO_x films are utilized to create and maintain an amorphous phase for capacitor applications [8]. Another approach is to use pseudobinary alloys [9]. It is important to note that TiO_2 is not thermodynamically stable on Si during deposition by CVD and other technological processes [10]. When TiO_2 films are obtained by different methods and undergo various types of annealings, TiO_2 may react with the Si surface to form Ti–O–Si bonds. This leads to a decrease of the maximum capacitance of metal-insulator-silicon (MIS) structures as a result of the decrease of the effective dielectric constant.

Recently, sol-gel technology has been intensively developed to synthesize various functional coating films. This process possesses many advantages, such as the utilization of simple equipment, high homogeneity of the films deposited and the possibility to use a great variety of substrates and

[1] To whom any correspondence should be addressed.

sizes [11]. It is used with a low temperature treatment that has the potential to prevent the TiO_2/Si reaction and produce high-quality MIS structures.

The purpose of this work is to show that the sol-gel method allows one to prepare thin TiO_2 films on Si with good quality. Th TiO_2 thin films were deposited by spin coating. The structural properties as a function of the annealing temperature were studied by XRD and Raman and FTIR spectroscopy. The electric properties of TiO_2 films were investigated using MIS structures.

2. Experimental details

The TiO_2 layers were prepared by the sol-gel technology. Titanium ethoxide (TEOT) (Fluka type) was used as a precursor. The introduction of glacial acetic acid (Fluka type), with molar ratio TEOT:acetic acid 1:1,5 caused acetate modification, resulting in an exothermic reaction. Further, the solution was modified by 60 moles of water. This led to hydrolysis and condensation reactions, resulting into the gel formation. Acetylacetone acts as a peptizing agent and stabilizer and is added in a molar ratio TEOT:acetylacetone 1:1. More details about titanium sol formation can be found in [12].

The films were deposited by spin coating at 8000 rpm for 30 sec. The annealing procedures of the films consisted of low temperature firing at 350 °C for 30 min and two high-temperature treatments at 600 °C for 60 min and at 850 °C for 15 min. The film thickness was measured with a laser ellipsometer at the wavelength of 632.8 nm. The film thickness is 38 nm. The films obtained are dense, homogenous and smooth with good thickness uniformity of ± 1% across the wafer.

The starting material was a polished p-type Si wafer, with orientation <100> and resistivity of 7 Ω.cm. The substrates were carefully cleaned ultrasonically in acetone and ethanol for 5 min. The second step was etching in 10 % HF solution followed by placing the wafers $H_2SO_4:H_2O_2$ = 1:1 solution (80 °C) for 60 min. This treatment leads to the formation of chemical SiO_2 with thickness of 1-2 nm. This initial oxide is used to create an interfacial layer between the Si wafer and the deposited film. Finally, the wafers were cleaned in deionized water in an ultrasonic bath for 10 min (60 °C).

The crystallization behavior of the TiO_2 film was studied by X-Ray diffraction (at grazing incidence angle of 7°). The Raman scattering spectra were measured by a double spectrometer SPEX 403 with a photomultiplier detector. The FTIR measurements were performed by Shimadzu FTIR Spectrophotometer IRPrestige-21.

The electrical properties of the layers were studied using MIS structures. Aluminium was evaporated on the backside of the Si and a Hg probe was used s an upper electrode. The MIS structures were formed using TiO_2 films annealed at different temperatures – 600 °C (60 min) and 850 °C (15 min).

3. Results and discussions

The X-ray diffraction (XRD) pattern recorded at the grazing incidence angle of 7° shows that the metal oxide layer annealed at 850 °C has a polycrystalline structure corresponding to the rutile phase (see figure 1). The line at $2\theta = 33.6°$ is due to the Si substrate. The strongest peak at 27.5° corresponds to the main XRD line of a rutile phase (110) (JCPDS card 21-1276). The rutile phase is manifested also by additional less intense peaks. Only one weak XRD maximum at $2\theta = 25.3°$ can be assigned to an anatase modification (JCPDS card 21-1272).

The percentage of the rutile phase in the film structure can be estimated from the respective integrated XRD peak intensities using the equation

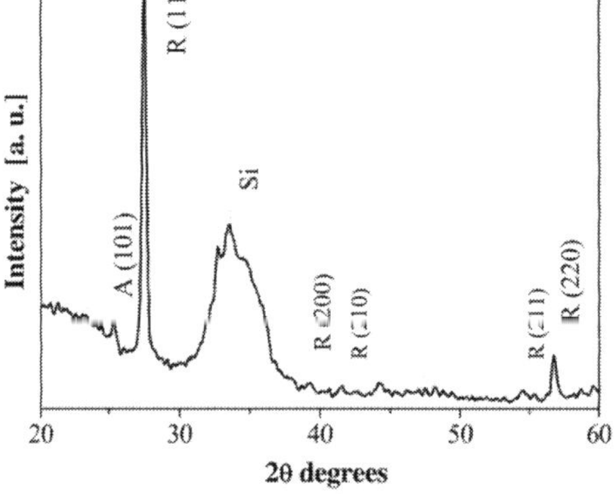

Figure 1. XRD spectrum of a TiO_2 film annealed at 850 °C in oxygen for 15 min.

given in ref [13], considering the anatase peak (101) and the strongest rutile (110) peak. The value obtained is 82 % of rutile phase and, respectively, 18 % anatase titanium dioxide. Based on the XRD data, the lattice parameters are determined as a = 4.583 Å and c = 2.989 Å. They are very close to the standard ones (a = 4.5933 Å, c = 2.9592 Å, JCPDS card 21-1276). The short high-temperature annealing at 850 °C causes the film to crystallize into a thermodynamically stable rutile phase, despite the metastable anatase structure. Such thermal behavior has been observed by other authors [14].

The structural transformation of TiO_2 films deposited on Si was studied by two other characterization techniques as a function of the annealing temperatures. Both FTIR and Raman spectroscopy can reveal the crystalline phases existing in the thin film structure.

FTIR spectra of the films are shown in figure 2. The TiO_2 film annealed at the lower temperature (600 °C) exhibits two absorption bands. The strong line at 437.8 cm^{-1} is assigned to Ti-O bonding in crystalline anatase TiO_2, very close to the value (435 cm^{-1}) of anatase crystal [15]. The band at 653.7 cm^{-1} is attributed to Ti-O stretching vibrations.

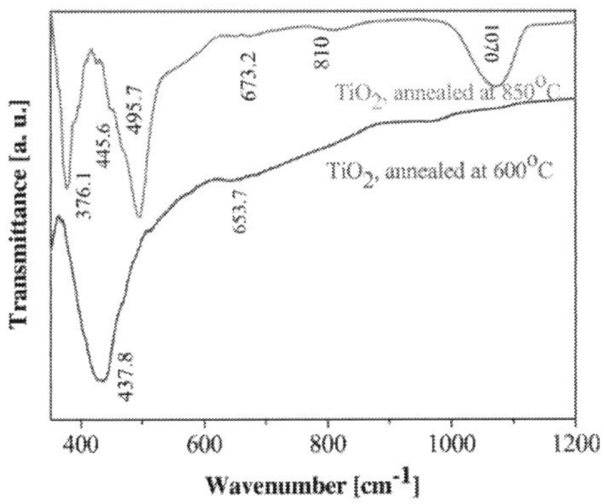

Figure 2. FTIR spectra of sol-gel TiO_2 films annealed at 600 °C (60 min) and at 850 °C (15 min).

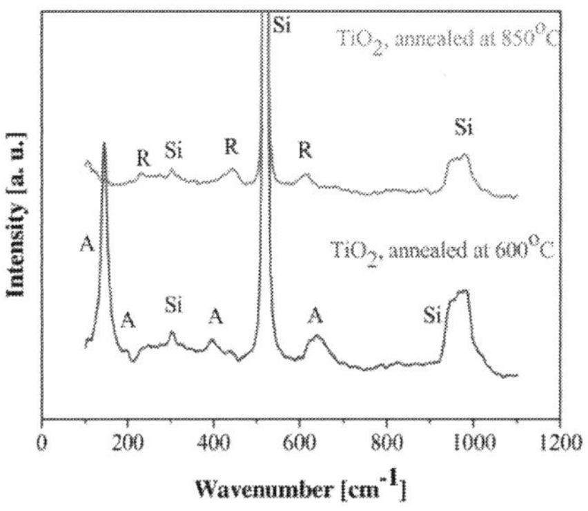

Figure 3. Raman spectra of TiO_2 films annealed at 600 °C (60 minutes) and at 850 °C (15 min).

The thermal treatment at 850 °C leads to the appearance of more IR lines. The absorption lines located at 376.1, 445.6, 495.7 cm^{-1} and at 810 cm^{-1} are referred to a rutile phase [16]. The 673.2 cm^{-1} band is due to the Ti-O stretching mode. The existence of the absorption band at 1070 cm^{-1} is a sign of Ti-O-Si bonding and penetration of Si into the film [17]. The FTIR analysis confirms the XRD results, namely, that the short treatment at 850 °C causes crystallization in a rutile phase and the TiO_2 film annealed at 600 °C shows the presence of an anatase phase. These results were confirmed by the Raman study as well.

The effect of the two annealing temperatures was also studied by Raman spectroscopy (Fig. 3). The common Raman lines appearing in the two spectra are due to the Si substrate. The strong peak at 520 cm^{-1} is related to the phonon vibrations of Si, while the other Si peaks are located at 305 cm^{-1} and 960 cm^{-1}. The TiO_2 films annealed at the lower temperature, 600 °C, revealed Raman peaks at 144, 197, 398 and 635 cm^{-1}. The well-defined line at 144 cm^{-1} is attributed to the anatase phase. This peak is reported to be a very sensitive feature for anatase [18]. The small shoulder at 197 cm^{-1} and the lines at 398 and 635 cm^{-1} are due to anatase as they have been found in the Raman spectrum of anatase single crystal [18]. Meanwhile, increasing the annealing temperature leads to the disappearance of the strong peak at 144 cm^{-1}, while new lines at 230, 444 and 612 cm^{-1} are detected. They are assigned only to rutile TiO_2; their positions coincide with the literature data [19]. This suggests that the annealing at 600 °C/60 min results in an anatase crystalline structure, while annealing at 850 °C/15 min induces the film crystallization in a rutile phase. This conclusion is in good agreement with the XRD and FTIR studies.

Journal of Physics: Conference Series **356** (2012) 012041 doi:10.1088/1742-6596/356/1/012041

We also measured the high-frequency (1 MHz) and the quasistatic-pulsed C-V characteristics of MIS structures.

As a comparison, table 1 presents the results for the dielectric properties of TiO_2 films annealed at 600 °C (60 min) and 850 °C (15 min). The dielectric constant of TiO_2 films as determined from the quasistatic CV measurements slightly decreases as the annealing temperature is increased. The leakage current density also decreases when the annealing temperature is raised.

Table 1. The results of measurements of MIS structures with TiO_2 thin film.

Annealing conditions	Dielectric constant at 1 MHz	Dielectric constant from quasi-static C-V tests	Density of fixed charge N_f [cm^{-2}]	Lowest surface state density D_{it} [eV^{-1}cm^{-2}]
600 °C/60 min/air	20.0	24.0	-8.26×10^{10}	1.1×10^{12}
850 °C/15 min/air	20.4	20.4	-6.55×10^{11}	2.4×10^{11}

Conclusions

The sol-gel approach proposed and the spinning procedure result in the formation of thin films (below 100 nm) of TiO_2 on Si substrates with good dielectric properties. The main properties were investigated of MIS structures fabricated by this method. An optimal regime for high temperature annealing (850 °C/15 min) was found, for which no effects related to the thermal instability of TiO_2 film on Si were observed. The results obtained show that this deposition method can find applications in the semiconductor technology.

References

[1] Zakrzewska K, Radeska M and Rekas M 1997 *Thin Solid Films* **310** 161

[2] Tachibana Y, Akiyama H Y and Kuwabata S 2007 *Solar Energy Mater. Solar Cells* **91** 201

[3] Ma Y, Ono Y and Hsu S T 1999 *Mater. Res. Soc. Symp. Proc.* **567** 355

[4] Jiang K, Zakutayev A, Stowers J, Anderson M D, Tate J, McIntyre D H, Johnson D C and Keszler D A 2009 *Solid State Sci.* **11** 1692

[5] Jung H, Oh S W, Ce J, Jayaprakash N and Sun Y 2009 *Electrochem. Comm.* **11** 756

[6] Khan M Z R, Hasko D G, Saifullah M S M and Welland M E 2009 *J. Phys.: Condens Matter.* **21** 215902

[7] Wilk G D, Wallace R M and Anthony J M 2001 *J. Appl. Phys.* **89** 5243

[8] van Dover R B 1999 *Appl. Phys. Lett.* **74** 3041

[9] Vitanov P, Harizanova A, Ivanova T, Alexieva Z and Agostinelli G 2006 *Japan J. Appl. Phys.* **45** 5894

[10] Taylor C J, Gilmer D C, Colombo D, Wilk G D, Campbell S A, Roberts J and Gladfelter W L 1999 *J. Am. Chem. Soc.* **121** 5220

[11] Kumar M, Kumar M and Kumar D 2010 *Microelctr. Engin.* **87** 447

[12] Harizanov O and Harizanova A 2000 *Solar Energy Mater. Solar Cells* **63** 185

[13] Zhang Q, Gao L and Guo J 2000 *Appl. Catal. B* **26** 207

[14] Ge L and Xu M 2007 *J. Sol-Gel Sci. Technol.* **43** 1

[15] Fallet M, Permpoon S, Deschanvres J L and Langlet M 2006 *J. Mater. Sci.* **41** 2915

[16] Rao A R and Dutta V 2007 *Solar Energy Mat. Solar Cells* **91** 1075

[17] Warrier K G K, Kumar S R, Sibu C P and Werner G 2001 *J. Porous Mater.* **8** 311

[18] Ma W, Lu Z and Zhang M 1998 *Appl. Phys. A* **66** 621

[19] Fukuda H, Namioka S, Miura M, Ishikawa Y, Yoshimo M and Nomura Sh 1999 *Japan J. Appl. Phys.* **38** 6034

Investigation of magnetic properties of thin films using computer simulation

N A Balakirev[1,3], V A Zhikharev[1] and G G Gumarov[2]

[1] Kazan State Technological University, Kazan, Russia
[2] Zavoisky Physical-Technical Institute RAS, Kazan, Russia

E-mail: balakirevnikita2009@gmail.com

Abstract. A two-dimensional dendrite is generated within the diffusion limited aggregation (DLA) model in the presence of an external magnetic field. The magnetic interaction between a grown dendrite and diffusing atoms results in the elongation of the dendrite in the direction of the magnetic field. The dependence is studied of the dendrite elongation on the grid occupation. The energy of the magnetic anisotropy is calculated for an elongated dendrite. The FMR spectra are calculated in geometries when the static magnetic field is either perpendicular or parallel to dendrite plane. It is shown that the FMR signals in latter case depend on the static magnetic field orientation with respect to the elongation direction.

1. Introduction

The fabrication of thin magnetic film with pronounced in-plane magnetic anisotropy on the surface of a semiconductor is a current problem in spintronic technologies [1]. Such a film is considered to serve as an injector of spin polarized electrons into the semiconductor. One method to solve this problem is the deposition of films of ferromagnetic metals or their compounds on a semiconductor. There have been many investigations performed to obtain in-plane magnetic anisotropy in thin metal films by ion beam methods (see for example [2–4]). However, the efficiency of spin injection from metal into semiconductor is drastically reduced due to the large difference between their conductivities [5,6]. It is suggested that the problem can be solved by using a ferromagnetic semiconductor as an injector. In silicon spintronics, the ferromagnetic silicide Fe_3Si is promising in this respect [7]. Recently, thin films of magnetic iron silicide Fe_3Si with uniaxial anisotropy were synthesized by high-dose Fe^+ ion implantation (ion-beam synthesis) into silicon in an external magnetic field [8]. In that work it was suggested that the magnetic anisotropy of the films synthesized is determined by the shape anisotropy of ferromagnetic silicide dendrites created due to the anisotropy of diffusion of implanted atoms. During ion implantation, clusters of a new phase are synthesized within a thin layer near the target surface. Due to radiation damages, the in-plane diffusion of implants in the layer is more intensive than that out of the plane. In a model approximation, ion implantation can be considered as a two-dimensional DLA problem. In [9], a modified multicenter DLA model was developed to take into account the magnetic dipole-dipole interaction between diffusing atoms and created dendrites. The importance of such interactions were emphasized in investigations of DLA problems [10,11], the diffusion limited deposition processes [12], and the transformation of the shape of magnetic precipitate

[3] To whom any correspondence should be addressed.

during Ostwald ripening processes [13]. It was shown that this interaction results in the elongation of grown clusters in the direction of the external magnetic field. In the present work, the magnetic anisotropy and the ferromagnetic resonance (FMR) of a model system - a single elongated 2D-dendrite - were numerically investigated.

2. DLA model in the case of ion-beam synthesis

The formation of a new phase during ion-beam synthesis is considered as a two dimensional DLA problem. The computer model is based on a kinetic Monte Carlo simulation on a square lattice. The simulation is performed at a square 400×400 grid. The implanted atom is simulated as a particle generated at some random site of the grid. Then it undergoes random walking until it attaches to the growing dendrite. If the generated particle walks out of the grid, it is discarded and a new particle is introduced into a random point of the grid, etc. The motion of i-th adatom is governed by the probability of a jump between adjacent sites of the grid. To jump from point $(x_i;y_i)$ to point $(x_i+1;y_i)$ the random walking i-th adatom must overcome the energy barrier $\Delta E_b(x_i,y_i\,|\,x_i+1,y_i)$, whose maximum is located at point $(x_i+\lambda;y_i)$, where $0 \le \lambda \le 1$. Here x_i and y_i are dimensionless coordinates in unit of grid spacing a. The probability of the jump is:

$$P(x_i,y_i\,|\,x_i\pm1,y_i)=\frac{\exp\left\{-\dfrac{\Delta E_b(x_i,y_i\,|\,x_i\pm1,y_i)}{k_B T}\right\}}{W(x_i,y_i)},$$

$$W(x_i,y_i)=\exp\left\{-\frac{\Delta E_b(x_i,y_i\,|\,x_i+1,y_i)}{k_B T}\right\}+\exp\left\{-\frac{\Delta E_b(x_i,y_i\,|\,x_i-1,y_i)}{k_B T}\right\}+ \qquad (1)$$

$$+\exp\left\{-\frac{\Delta E_b(x_i,y_i\,|\,x_i,y_i+1)}{k_B T}\right\}+\exp\left\{-\frac{\Delta E_b(x_i,y_i\,|\,x_i,y_i-1)}{k_B T}\right\},$$

where W(x$_i$; y$_i$) is a normalizing coefficient, T is the temperature, and k_B, the Boltzman constant. Usually, in two-dimensional DLA models statistically isotropic random walking is assumed, that is the energy barriers around point $(x_i;y_i)$ are supposed to be equal, or $\Delta E_b(x_i,y_i\,|\,x_i\pm1,y_i\pm1)=\Delta E_b^0(x_i:y_i)$, providing probability $P_0(x_i;y_i)$ = 1/4 for all four available jump directions in a square grid.

In the case of Fe ion implantation, the growing silicide aggregate is assumed to be ferromagnetic (in the experiments reported in [8], the presence of aferromagnetic phase in implanted layer was confirmed by Mossbauer measurements). The interaction of the adatom's magnetic moment with the magnetic field $\vec{B}_{dd}$ created by the ferromagnetic dendrite changes the energy barrier:

$$\Delta E_b(x_i,y_i\,|\,x_i\pm1,y_i)=\Delta E_b^0(x_i,y_i)\pm\frac{\partial E_{dd}}{\partial x}\lambda\,, \qquad (2)$$

where

$$E_{dd}=-\vec{\mu}_i\cdot\vec{B}_{dd}=\vec{\mu}_i\cdot\frac{\mu_0}{4\pi a^3}\sum_k\left[\frac{\vec{\mu}_k}{r_{ik}^3}-\frac{3\vec{r}_{ik}(\vec{\mu}_k\vec{r}_{ik})}{r_{ik}^5}\right]$$

is the energy of dipole-dipole interaction between magnetic moment $\vec{\mu}_i$ of i-th adatom and the moment of k-th dendrite atom.

If the model of isotropic random walking without magnetic interaction is taken into account, then the jump probability (1) becomes:

$$P(x_i, y_i \mid x_i+1, y_i) = \frac{\exp\left\{-\dfrac{\partial E_{dd}(x_i, y_i)}{\partial x}\dfrac{1}{2k_B T}\right\}}{2\left[ch\left(\dfrac{\partial E_{dd}(x_i, y_i)}{\partial x}\dfrac{1}{2k_B T}\right) + ch\left(\dfrac{\partial E_{dd}(x_i, y_i)}{\partial y}\dfrac{1}{2k_B T}\right)\right]}. \tag{3}$$

The DLA model described is used below to generate ferromagnetic dendrites on a two dimensional square grid.

3. Results

3.1 Dendrite simulation

The examples of simulated isolated dendrites created after the addition of 20000 particles on the 400×400 grid in the presence of an external magnetic field $\vec{B}_c$ and without it are presented in figure 1 and figure 2, correspondingly. The simulation assums that field $\vec{B}_c$ is sufficient to saturate the ferromagnetic dendrite magnetization. The magnetic moments of both the moving adatom and the dendrite are supposed to be parallel to $\vec{B}_c$. It is evident that the dendrite created in a magnetic field is elongated by $\varepsilon=\langle R_x\rangle/\langle R_y\rangle$, where $\langle R_x\rangle$ and $\langle R_y\rangle$ are the average sizes of the dendrite along the external field direction (OX axis) and in the perpendicular one (OY axis). The $\langle R_x\rangle$ and $\langle R_y\rangle$ were calculated relatively to the center of mass of the dendrite. The elongation ε as function of the grid occupation is presented in figure 3. The function exhibits a maximum and then decreases. Such a dependence is due to boundary effects.

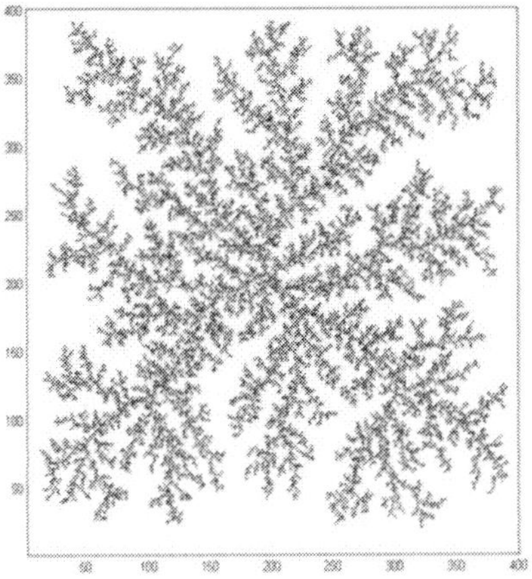

Figure 1. The dendrite grown at the grid 400x400 without magnetic field, 20000 particles are added.

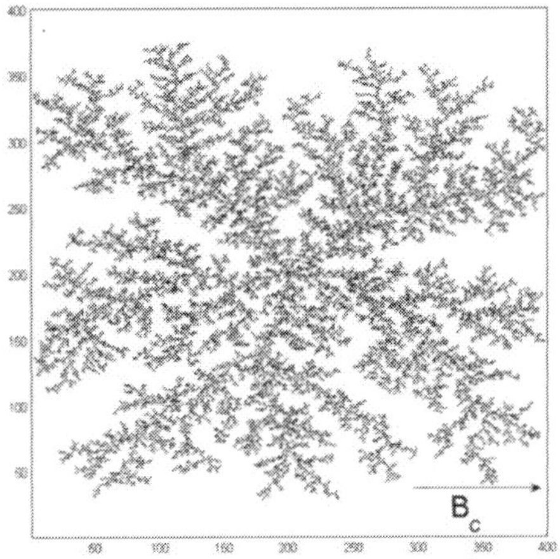

Figure 2. The dendrite grown at the grid 400x400 in presence of magnetic field, 20000 particles are added.

3.2. In-plane magnetic anisotropy

The ferromagnetic dendrite elongation results in a magnetic in-plane anisotropy. The energy of the anisotropy E_{anis} can be numerically calculated as the sum of the dipole-dipole energies of all atoms of the dendrite. This energy depends on the angle φ between the direction of elongation (OX axis) and the magnetic moments of the dendrite atoms. An example of the dependence is presented in figure 4. The elongation for the given dendrite is ε = 1,55. Other dendrites within this simulation have elongations in the range [1.4 - 1.65] and exhibit an angular dependence of E_{anis} similar to that in figure 4:

$$E = K * \sin^2 \varphi + const, \tag{4}$$

where K is the anisotropy constant $K = \alpha \frac{\mu_0 \mu^2}{4\pi a^3} * \rho_{Fe}$. The coefficient α strongly depends on the dendrite elongation and for the given dendrite (ε = 1,55) is equal to 1.4, ρ_{Fe} is the density of Fe atoms in the dendrite. If ρ_{Fe} corresponds to the silicide Fe_3Si then K is in the order of $10^2 - 10^3 J/m^3$.

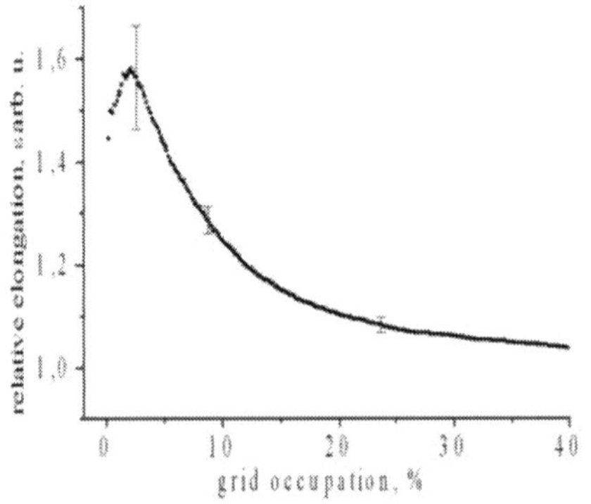

Figure 3. Relative elongation ε versus grid occupation.

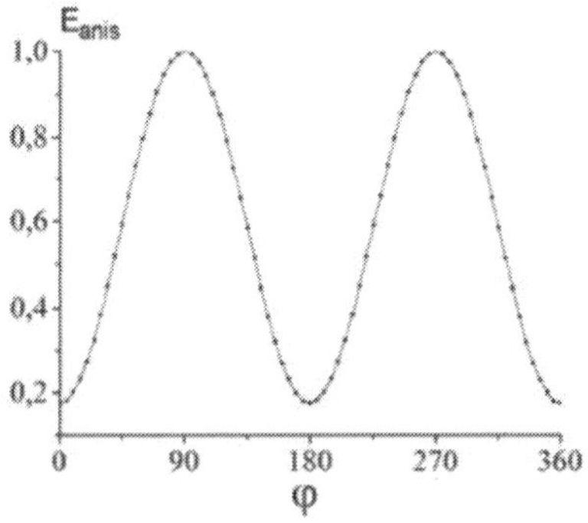

Figure 4. The dependence of E_{anis} on angle φ between dendrite magnetization and X axis.

3.3. Ferromagnetic resonance (FMR)

The FMR signal $P(H_0)$ is calculated as the sum of the resonance absorption $P_i(H_0)$ of i-th magnetic moment μ_i involved in the dendrite.

$$P(H_0) = \frac{1}{N} \sum_i P_i(H_0) \tag{5}$$

According to Kittel theory resonance absorption function $P_i(H_0)$ is obtained from the solution of Bloch equation for magnetic moment μ_i [14]. The magnetic moment μ_i experiences the static magnetic field H_0, microwave magnetic field $\vec{h}_1$ and dipole magnetic field H_d created by all other magnetic moments of the dendrite.

Below two geometries of the magnetic fields arrangement are considered: case A (figure 5) and case B (figure 6). In both cases resonance absorption function $P_i(H_0)$ is described by formula:

$$P_i(H_0) = \frac{2\pi\omega_{0i}\omega\mu_i h^2 \gamma(\cos^2(\varphi_i) + \sin^2(\varphi_i)\cos^2(\theta))}{T[\{\omega^2 - \omega_{0i}^2 + \frac{1}{T^2}\}^2 + \frac{4\omega_{0i}^2}{T^2}]}; \tag{6}$$

$$\omega_{0i} = \gamma\sqrt{\left(H_0 + H_{d\parallel}\right)^2 + H_{d\perp}^2}, \tag{7}$$

where $\vec{h}_1$ and ω are the amplitude and frequency of microwave field, T is the relaxation time. The fields $H_{d\parallel}$, $H_{d\perp}$ and the angles φ and θ are shown in the figures below.

Case A: H_0 is perpendicular to the dendrite plane (OZ axis), the microwave field $\vec{h}_1$ lies in the dendrite plane (parallel to the OX axis).

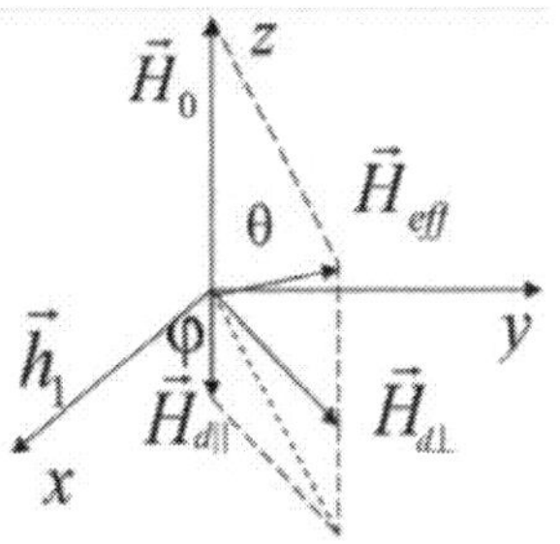

Figure 5. A: H_0 is directed perpendicular to the dendrite plane.

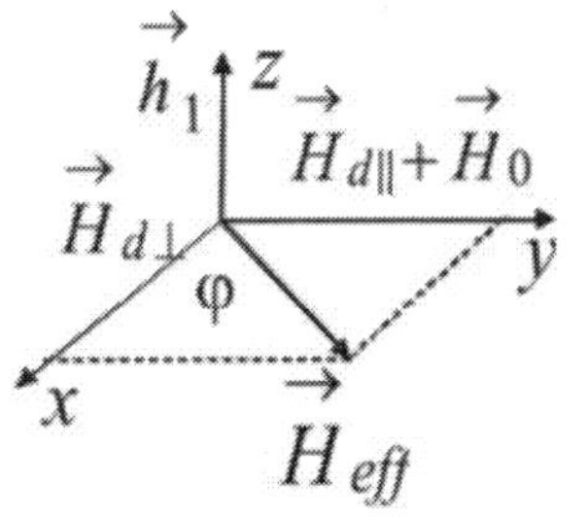

Figure 6. B: H_0 is parallel to the dendrite plane.

The components of the dipole field H_d strongly depend on the H_0 value. Without external magnetic field, the dipole field H_d lies in the plane of the dendrite, or it has only x- and y-components. When H_0 increases, the magnetic moments rotate out of the plane and have an angle θ with respect to H_0. Consequently, in (7) the components of the dipole field are $H_{d\|} = H_{dz}(\theta)$, $H_{d\perp} = \sqrt{H_{dy}^2 + H_{dx}^2}$. At the critical value $H_0 = H_r$ all magnetic moments become perpendicular to the dendrite plane (θ=0) and H_d has only a z-component, or $H_{d\|} = H_{dz}(0)$, $H_{d\perp} = 0$. The value of H_r is determined by grid spacing a and dendrite size:

$$\mathrm{H_r} = \frac{3\dfrac{\mu_0\mu}{8\pi a^3}\sum_{i=0}^{N}\sum_{j=0}^{N}\dfrac{\left(r_{ij}^x\right)^2}{r_{ij}^3}}{N}, \qquad (8)$$

where N is the number of particles within the dendrite. The results of the computer calculation of the resonance absorption $P(H_0)$ in this case are presented in figure 7 as a function of the grid occupation.

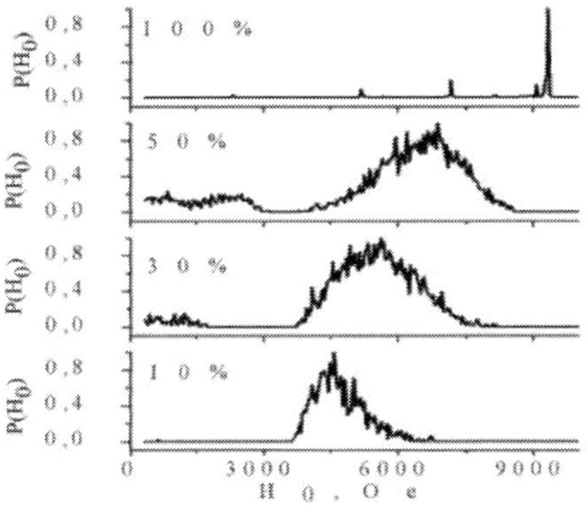

Figure 7. The FMR signal in **A** geometry at different degree of grid occupation 10, 30, 50 and 100 %.

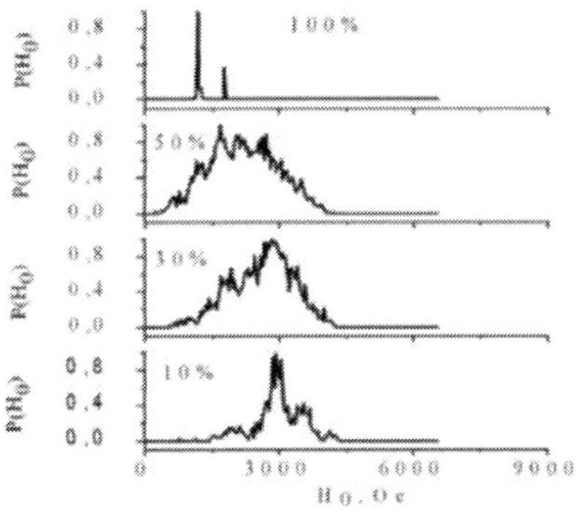

Figure 8. The FMR signal in **B** geometry at different degree of grid occupation 10, 30, 50 and 100 %.

Case B: H_0 lies in the dendrite plane, the microwave field $\vec{h}_1$ is perpendicular to the dendrite plane (|| OZ axis). If H_0 is perpendicular to the OX axis (the direction of elongation of the dendrite), the absorption signal is described by formula (6) with $\cos(\theta)=0$ and $\omega_{0i}=\gamma\sqrt{\left(H_0+H_{dxi}\right)^2+H_{dyi}^2}$. The absorption signals are presented in figure 8.

If H_0 is parallel to the OX axis, then for grid occupation greater then 10 % one has similar absorption spectra. But for low grid occupation ($\le 5\%$) the absorptions for the case $H_0 \parallel OX$ axis and $H_0 \perp OX$ axis are substantially different, as presented in figure 9.

4. Discussion

The resonance absorption spectra presented in figure 7 and figure 8 reveal the main feature of the FMR - strong dependence of the position of the resonance absorption line on the sample shape due to the influence of a demagnetization field. In ferromagnetic spherical samples, resonance absorption occurs at $H_{res} = H_{0res}$ (H_{0res}=3300 Oe at ν = 9,33 GHz). For thin continuous ferromagnetic films when a static magnetic field H_0 is perpendicular to the film plane, the resonance signal is shifted to $H_{res} = H_{0res} + 4\pi M$, where M is the ferromagnet magnetization (so called "high field" signal) [14]. When H_0 lies in the film plane, the resonance takes place at H_{res} lower then H_{0res}. It was shown in [15] that a similar orientational dependence of the resonance signal is pertains to a thin granular film consisting of spherical particles. The resonance spectra of the isolated dendrites simulated in the present work also exhibit a shift of the resonance line position at different orientation of H_0 with respect to the dendrite plane. The case of 100% grid occupation is well described by Kittel formulas and corresponds to a continuous ferromagnetic film. When occupation is lower than 50%, one has low density clusters and the resonance signals become broader and assymetric. In this case the simple Kittel formulae for resonance absorption cannot be used and only numerical calculation of the absorption signal is possible. For a very low degree of grid occupation (less then 5%) a noticeable dependence of the absorption line on the static magnetic field orientation relatively to the direction of dendrite elongation is seen. Dendrites with the highest degree of elongation are typical for such a low grid occupation (see figure 3). If H_0 is perpendicular to the long dendrite axis, satellite signals on the high field side are observed in addition to the main absorption signal at H_0 =3300 Oe. If H_0 is parallel to the long dendrite axis, then the additional signals appear on the low field side. Therefore, monitoring the FMR absorption can be used as a method for studying the induced uniaxial anisotropy of a film consisting of dendrites grown in the presence of an external magnetic field. When the grid occupation increases, the dendrite elongation decreases, which is connected with the boundary conditions. The dendrite becomes much more "symmetric". In the case of an isolated dendrite, the boundary conditions are determined by the grid edges. For multicenter DLA problems, the reduction in the dendrite elongation is caused by the influence of other dendrites.

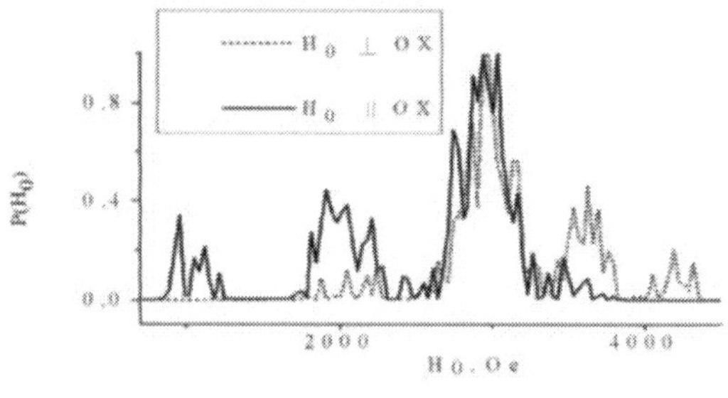

Figure 9. The FMR signals for dendrite simulated on 400x400 grid at 5% degree of grid occupation.

Conclusions

Two-dimensional isolated dendrites are generated within the DLA model in the presence of an external magnetic field. It is shown that the magnetic interaction between the grown dendrite and a diffusing atom leads to elongation of the dendrite in the direction of the magnetic field. The dependence is demonstrated of the dendrite elongation on the grid occupation. The energy of the in-plane magnetic anisotropy is numerically estimated for an elongated dendrite. The FMR spectra are calculated for the cases when the static magnetic field is perpendicular or parallel to dendrite plane. It is shown that the

FMR signals in the latter case depend on the static magnetic field orientation with respect to the direction of dendrite elongation.

Acknowledgments

The authors are grateful to the Joint Supercomputer Center of the Russian Academy of Sciences (Kazan branch) for the computational resources provided.

References

[1] Fabian J, Matos-Abiague A, Ertler C, Stano P, Zutic I 2007 *Acta Phys. Slovacia* **57**/4 565
[2] Chang G S, Moewes A, Kim S H, Lee J, Jeong K, Whang C N, Kim D H and Shin S-C 2006 *Appl. Phys. Lett.* **88** 092504
[3] Zhang K, Uhrmacher M, Hofsass H and Krauser J 2008 *J. Appl. Phys.* **103** 103083507
[4] Gupta R, Lieb K P, Muller G A, Schaaf P and Zhang K 2005 *Hyperfine Interaction* **160** 107
[5] Schmidt G, Ferrand D, Molenkamp L W, Filip A T and van Wees B J 2000 *Phys. Rev. B* **62** R4790
[6] Rashba E I 2000 *Phys. Rev. B* **62** R16 267
[7] Ionescu A, Vaz C A F, Trypiniotis T, Gurtler C M, Garcнa-Miquel H, Bland J A C, Vickers M E, Dalgliesh R M, Langridge S, Bugoslavsky Y, Miyoshi Y, Cohen L F and Ziebeck K R A 2005 *Phys. Rev. B* **71** 094401
[8] Gumarov G G, Petukhov V Yu, Bukharaev A A, Biziaev D A, Nuzhdin V I 2009 *Nucl. Instr. Meth. Phys. Res. B* **267** 1307
[9] Balakirev N A, Gumarov G G, Zhikharev V Aand Petukhov V Yu 2011 *Comp. Mater. Sci.* **50** 2925-9
[10] Meakin P 1983 *Phys. Rev. A* **27** 2616
[11] Mizuseki H, Tanaka K, Kikuchi K, Ohno K and Kawazoe Y 1998 *Com. Mat. Sci.* **10** 46-50
[12] Tasinkevich M, Tavares J M and de los Santos F 2006 *J. Chem. Phys.* **124** 064706
[13] Beaugnon E and Arras E 2006 *J. Phys.:Conf. Series* **51** 439-45
[14] Kittel C 1947 *Phys. Rev.* **71** 270
[15] Azarov I A and Zhikharev V A 1995 *Appl. Mag. Res.* **9**/2 165

Monte Carlo simulation of electron kinetics in a hollow cathode discharge

S Karatodorov[1,3], D Mihailova[2], J van Dijk[2], J van der Mullen[2] and M Grozeva[1]

[1] Georgi Nadjakov Institute of Solid State Physics, Bulgarian Academy of Sciences, 72, Tzarigradsko Chaussee, 1784 Sofia, Bulgaria
[2] Department of Applied Physics, Eindhoven University of Technology, Den Dolech, 25612 AZ Eindhoven, The Netherlands

E-mail: skarat@issp.bas.bg

Abstract. A kinetic model is reported computing the electron behavior in a hollow cathode discharge based on the Monte Carlo technique. It is a part of the PLASIMO modelling toolkit. The model allows the electrons to be closely followed while they travel and undergo collisions in the discharge. The Monte Carlo modulewas applied to the case of a HCD used as an excitation medium of atoms obtained by laser ablation. Results are obtained on the electron energy distribution function and the mean electron energy under typical discharge conditions. The output data and future development of the model and its applications are analyzed and discussed.

1. Introduction

The elemental composition of different materials can be investigated using a hybrid technique employing laser ablation (LA) and a hollow cathode discharge (HCD). The HCD in this so-called LA-HCD technique [1], is employed for additional excitation of the ablated material.

The focus of this work is modelling the processes in the plasma of a HCD. For this purpose we use the PLASIMO modelling platform [2], on which the HCD was already simulated by the MD2D model [3] - a time dependent two-dimensional multi uid model. The uid model calculates the spatial density distribution of various discharge species, the mean electron energy and the electrostatic potential. The results of the MD2D model are found to agree well with experimental and analytical observation [3,4,5].

In the HCD, there are regions, for instance in the cathode fall (CF), where the particles do not experience many collisions, or particles exist, such as pendulum electrons, whose free path exceeds the plasma size. This is why the kinetic approach is applied to simulate them. As the electrons are the main energy carriers from the field to the hollow cathode (HC) plasma, their behaviour is first simulated.

The aim of the present work is to develop a kinetic model for the electrons in a HCD and describe their behavior in terms of kinetic quantities, such as the electron energy distribution function (EEDF), the mean electron energy and the distribution of ionization events. The model is used for the optimization of the discharge; it provides guideline information for the fluid MD2D model.

[3] To whom any correspondence should be addressed.

2. Operation parameters of the kinetic model

Figure 1 presents the geometry - a hollow cathode cylinder with length of 20 mm and inner radius of 4 mm. The spatial distribution of the electric field inside the cylinder is obtained from the fluid model and is used as input for the kinetic model. The operating gas is helium (He) at pressure of 1 Torr. In this environment, the simulated particles are the electrons. Their collision processes with the He ground state atoms are listed in table 1, while the cross-sections [6] used for these processes are taken from [7].

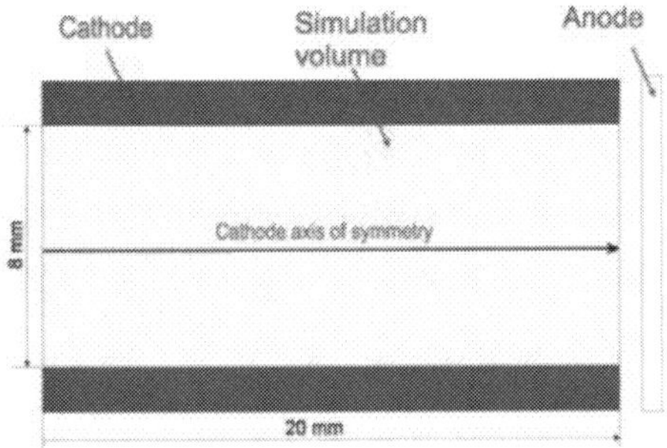

Figure 1. Scheme of the experimental HC with the simulation volume.

Table 1. Reactions used in the kinetic MC model for electrons.

Type	Reaction	Reaction threshold energy [eV]
Elastic collision	He + e → He + e	0
Excitation	He + e → He* + e	19.8
Ionization	He + e → He$^+$ + 2e	24.57

3. MC modelling results

3.1. MC simulations of wall secondary electron emission

The electrons emitted from the cathode wall are accelerated in the CF region and dissipate their energy in exciting and ionizing the plasma. However, some of the electrons can be accelerated to such an extent that their mean-free-path exceeds the plasma size. As a result, they can make multiple passes through the plasma with high velocity and are rejected by the CF on the opposite side. These electrostatically trapped electrons are known as pendulum electrons.

In order to observe the time evolution of the emitted electrons from the wall and the presence of pendulum electrons in the discharge, we inject 10^6 electrons at time t = 0 s from random points lying on the inner cathode wall. Since their transport is mainly radial, we impose only the radial electric field component.

Figure 2 shows the radial distribution of the ionization events involving the electrons. At the beginning of the simulation (figure 2a) the profiles of the ionization events have two pronounced peaks close to the walls, where the electrons have gained enough energies for ionization. At a later stage, these two peaks move towards the center of the HC (figure 2b) where they overlap (figure 2c). Finally, the ionization profile turns into a smooth curve with a sharp peak in the center (figure 2d).

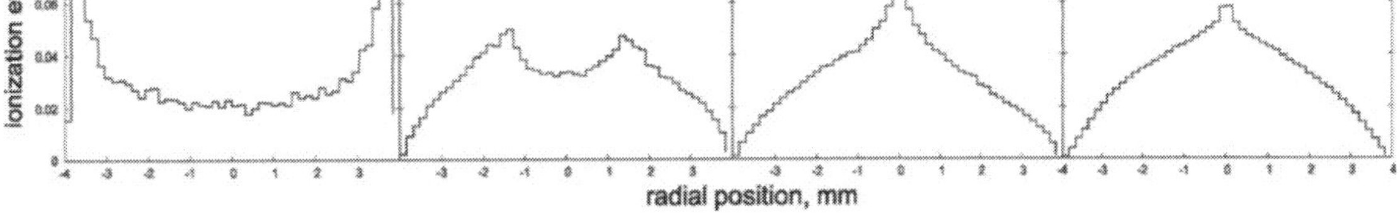

Figure 2. Time evolution of the ionization events obtained in the simulation of secondary emitted electrons.

This focusing effect of ionization events in the center of the discharge is attributed to the presence of pendulum electrons. As for them, it is more probable that they will pass through the center of the HC and, accordingly, it is more probable that they ionize there, which results in the sharp peak. Similar results have been reported earlier [8, 9] indicating the presence of pendulum electrons in HCDs.

3.2. MC simulations of HCD

For the simulation of the HCD setup, we impose the radial and axial electric field components obtained from the uid model. The axial electric field component has an important role for the electron transport to the anode [5]. The secondary electron emission from the wall is simulated by a continuous injection of electrons from the cathode wall at a rate of 1×10^{14} electrons/s. This value is also taken from the uid modelling results.

The output from the simulation are the EEDF and the spatially resolved EEDF. Figure 3 shows the EEDF obtained. The EEDF can be viewed as consisting of three parts. The first part comprises the low energy electrons with energies in the range $0 \div 40$ eV. These electrons are localized in the bulk of the discharge and are undergoing predominantly elastic collisions as their energy is not suficient for more than one inelastic collision. Thus, the Maxwellization of these electrons is mainly due to elastic collisions. The presence of Maxwellization is indicated by the smooth line formed by the lower energy electrons in the EEDF. The second part consists of electrons in the range $40 \div 250$ eV . This group of high energy electrons is characteristic for the HCD. Electrons with such energies can excite not only atomic but also ionic states of the He atoms. The last part includes the electrons which have energies up to the full CF potential energy - 350 eV . This is the group of secondary emitted electrons and pendulum electrons. The secondary emitted electrons have just been ejected from the cathode and still carry the energy absorbed in the CF. The pendulum electrons tend to have low number of collisions and oscillate between the opposite CFs.

Figure 4 presents the spatially resolved EEDF. It is obtained by dividing the whole volume of the HC to a number of coaxial sub-volumes in which the EEDF is determined. The middle part of the EEDF in sub-volume 4 is higher than what is found for other regions. This is due to the fact that sub-volume 4 overlaps with the CF region, where a high radial electric field is present. The radial electric field is negligible in the other three sub-volumes, which results in similar shapes of the EEDF. However, some differences of sub-volumes 1 to 3 exist. The electrons in sub-volume 3, coming with high energy from the CF, start spending their energy in

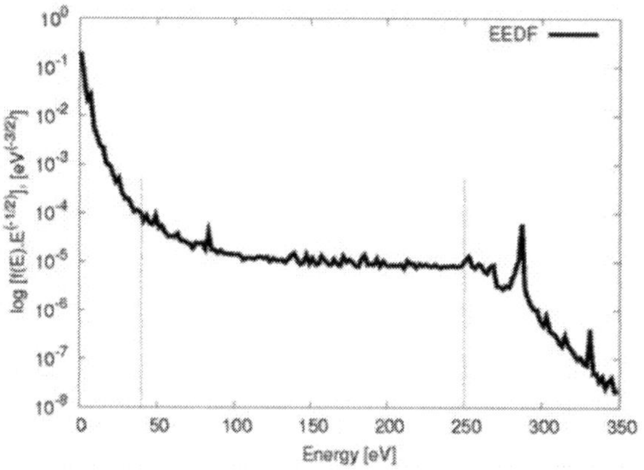

Figure 3. EEDF, obtained by MC simulations of all electrons in the HCD.

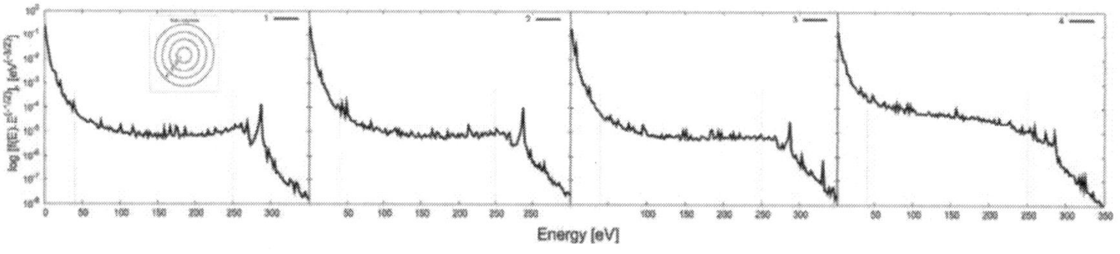

Figure 4. Spatially resolved EEDFs obtained in the MC simulation. The scheme of the fourth sub-volumes is shown in the graph.

effective impact ionization with the He atoms, so the EEDF has a lower number of highly energetic electrons. The EEDFs of the two inner sub-volumes are a little higher than in sub-volume 3. This is ascribed to the influence of the axial electric field. The electrons which have reached the inner part of the HC start to absorb energy from the axial electric field, which is accelerating them in the direction of the anode. The influence of the axial field is discussed in [5].

The EEDFs show high concentration of highly energetic electrons that are more abundant in the outer regions of the discharge and, due to the presence of strong radial field, enter in the central part of the discharge.

Conclusions

A kinetic model is developed simulating the behavior of electrons in the HCD. It is demonstrated that the ionization distribution has a radial profile with a sharp peak in the center of the HC. The steepness of the peak is ascribed to the pendulum electrons which predominantly pass through the axis of symmetry. A high concentration of energetic electrons in the whole volume of the HCD is found, thus the inelastic processes (excitation and ionization) are very probable. This shows that the HCD is indeed an appropriate medium for the excitation and ionization of ablated atoms.

In the future the kinetic model can be joined together with the uid MD2D in a hybrid model for the HCD.

Acknowledgments

The work is supported by the Bulgarian Science Fund under project DO 02 – 274/2008.

References

[1] Naeem T, Matsuta H and Wagatsuma K 2004 *Anal. Sci.* **20** 1717
[2] van Dijk J, Peerenboom K, Jimenez M, Mihailova D and van der Mullen J 2009 *J. Phys. D* **42** 19402
[3] Mihailova D, Grozeva M, Hagelaar G, van Dijk J, Brok W and van der Mullen J 2008 *J. Phys. D* **41** 24502
[4] Mihailova D, van Dijk J, Grozeva M, Hagelaar G and van der Mullen J 2010 *J. Phys. D: Appl. Phys.* **43** 145203
[5] Hagelaar G, Mihailova D and van Dijk J 2010 *J. Phys. D: Appl. Phys.* **43** 465204
[6] Petrović Z, Dujko S, Marić D, Malović G, Nikitović Š, Šašić O, Jovanović J, Stojanović V and Radmilović-Radenović M 2009 *J. Phys. D: Appl. Phys.* **42** 194002
[7] Phelps A 1955 *Phys. Rev.* **99** 1307
[8] Kolobov V and Tsendin L 1995 *Plasma Sources Sci. Technol.* **4** 551
[9] Den Hartog E, Doughty D and Lawler J 1988 *Phys. Rev. A* **38** 2471

Journal of Physics: Conference Series **356** (2012) 012044 doi:10.1088/1742-6596/356/1/012044

Simulation tools for computer-aided design and numerical investigations of high-power gyrotrons

M Damyanova[1,4], E Balabanova[1], S Kern[2], S Illy[2], S Sabchevski[1], M Thumm[2], E Vasileva[1] and I Zhelyazkov[3]

[1] Emil Djakov Institute of Electronics, Bulgarian Academy of Sciences, Association EURATOM-INRNE, 72, Tsarigradsko Chaussee, 1784 Sofia, Bulgaria
[2] Karlsruhe Institute of Technology, Association EURATOM-KIT, Institute for Pulsed Power and Microwave Technology, Hermann-von-Helmholtz-Platz 1, D-76344 Eggenstein-Leopoldshafen, Karlsruhe, Germany
[3] Faculty of Physics, St. Kliment Ohridski University of Sofia, Association EURATOM-INRNE, 5, James Bourchier blvd., 1164 Sofia, Bulgaria

E-mail: milena@ie.bas.bg

Abstract. Modelling and simulation are essential tools for computer-aided design (CAD), analysis and optimization of high-power gyrotrons used as radiation sources for electron cyclotron resonance heating (ECRH) and current drive (ECCD) of magnetically confined plasmas in the thermonuclear reactor ITER. In this communication, we present the current status of our simulation tools and discuss their further development.

1. Introduction

Gyrotrons are among the most powerful sources of coherent microwaves in the millimeter wavelength region operating in a continuous wave (CW) mode and find diverse use ranging from fundamental research to industrial applications [1]. A typical example is fusion research, where gyrotrons are used as high-power sources of radiation for the start-up (ignition) of magnetically confined plasmas in various reactors for controlled thermonuclear fusion, for electron cyclotron resonance heating (ECRH), electron cyclotron current drive (ECCD)p as well as for stabilization and diagnostics of plasmas. In recent years, the gyrotrons for fusion developed in Europe, Japan, Russia and USA (generating at frequencies between 140 and 170 GHz) have reached the megawatt level of output power. Following this progress, the European Gyrotron Consortium (EGYC) consisting of several research institutions lead by the KIT-IHM, Germany and CRPP-EPFL, Switzerland, is currently developing a coaxial gyrotron with an output power of 2 MW and higher, delivered through a synthetic diamond output window. Despite this progress, however, there still remain certain serious problems and challenges to be addressed. Some of them (e.g., the parasitic oscillations in the beam tunnel; stray magnetic fields; Penning discharges in the technical part of the gun) have already been addressed and partially resolved. At the same time, other problems (for instance, various sources of beam instabilities, non-uniformity of the emission from the cathode [2, 3], trapped particles, space charge compensation, etc.) still pose serious theoretical and technical problems. Some of them are due

[4] To whom any correspondence should be addressed.

to physical factors and phenomena that are three-dimensional and non-stationary (3D) by nature [2, 3]. Therefore, they cannot be adequately described using the currently available physical models and computer codes, which are predominantly two-dimensional (2D) and quasi-static. All this underlines the importance of both improving the available simulation tools as well as developing novel, more efficient, models and programs that can be used for the analysis, computer aided design (CAD) and optimization of high-power gyrotrons with improved performance. A concept for the further development of these simulation tools is formulated in [4]. The first steps towards its realization are presented in [2, 3, 5–7]. In this paper, we review the current status of the simulation tools used and illustrate the functionality of their latest versions.

2. Current status and functionality of the simulation tools for analysis and CAD of gyrotrons

The structure of the simulation software used for analysis, optimization and CAD of gyrotrons is illustrated in figure 1, together with some of the computational platforms onto which different packages are operational. Although some of them (e.g., ***DAPHNE***, ***ESRAY*** [8], ***CAVITY-IHM*** [9], and various components of ***GYROSIM*** [10]) are well validated, benchmarked and debugged, they are undergoing constant adaptation and upgrade to the ever-changing computational environments (hardware, operating systems, novel versions of the compilers and numerical libraries). Alongside with the maintenance of these codes and their use in numerical experiments, we are working on the further development of the ***GYROSIM*** [10] and ***GYREOSS*** packages.

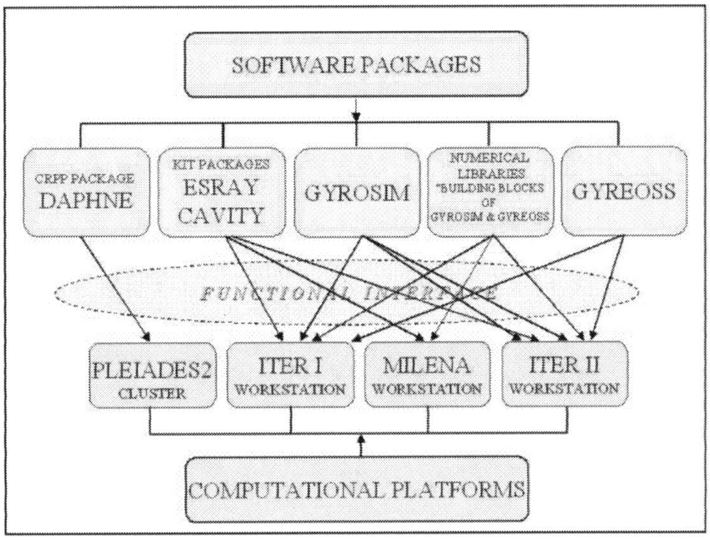

Figure 1. Structure of the software packages for simulation of gyrotrons and computational platforms on which they are operational used by the Bulgarian research team.

DAPHNE and ESRAY are ray-tracing and PIC (Particle in Cell) codes for trajectory analysis of the electron-optical system (EOS) of the tube. The same functionality has GUN-MIG/CUSP of the GYROSIM package. All they are based on adequate fully-relativistic self-consistent 2.5D physical models, but differ in the algorithms and methods employed for the solution of the field problem (boundary value problem for the electrostatic potential), the integration of the equations of motion (particle pushing), as well as in the program implementation. The most advanced amongst them is ESRAY. Advantageous and distinguishing features of this package are: (i) the use of a boundary fitted mesh to discretize the computational region with great precision; (ii) object-oriented program implementation and efficient numerical algorithms that speed-up the computation; (iii) advanced postprocessor for visualization of the results of numerical experiments. The latter is illustrated by a collage of screenshots in figure 2(a).

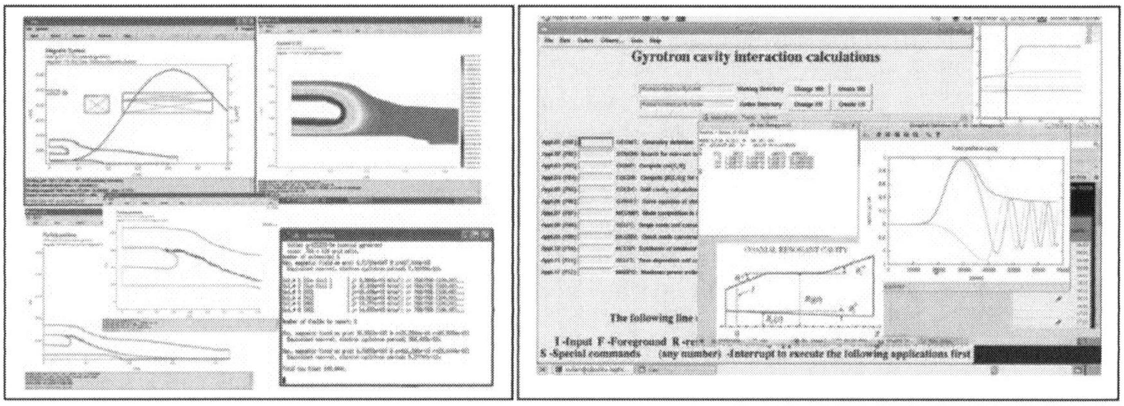

a) b)

Figure 2. Illustrations of the visualization capabilities of the IHM-KIT packages: a) ESRAY; b) CAVITY.

Both packages, CAVITY (IHN-KIT) and GYROSIM [10], contain a hierarchy of computer codes ranging from simple cold-cavity codes for preliminary analysis to more sophisticated time-dependent and multimode codes. In its current version, the GYROSIM package is specialized only to regular (cylindrical) cavities, while CAVITY (IHM-KIT) can also simulate coaxial gyrotron resonators. The most advanced component of CAVITY is the SELFT code, in which a self-consistent, time dependent, multimode physical model is realized [9]. Different components can be invoked through a GUI (front-end), the navigation screen of which is shown in figure 2(b), together with some illustrative visualization windows.

The main goal of the GYREOSS (which stands for GYRotron EOS Simulation) is the realization of a physical model which is able to take into account the physical factors that are three dimensional in nature. Since the simulation in 3D requires enormous computational resources (as a rule available only on parallel systems, e.g., clusters, grids, clouds etc.) its development involves utilization of advanced programming techniques, numerical libraries and integrated development environments (IDE). Currently, the work on GYREOSS is focused on the development of an efficient 3D field solver based on the finite element method (FEM) on a tetrahedral mesh. The first version (prototype) of GYREOSS was built around the GetDP solver and some of the essential, pre- and post-processing modules were integrated using gmsh. After studying, benchmarking and comparing different numerical libraries, FreeFEM++ has been selected as an appropriate building block to replace GetDP. It will be used for the realization of a novel more efficient solver as well as for programming of some of the other components of the PIC code. In the novel solver, the boundary value problem for the Poisson equation is solved in a mixed finite element formulation. A significant advantage of such an approach is the possibility to approximate the electric field with accuracy higher than when applying the conventional formulation and calculating it through numerical differentiation of the potential. This is especially important for the PIC algorithm since it is the electric field rather than its potential that is required for the calculation of the Lorentz force in view of tracing particle motion. A screenshot of the GUI of GYREOSS is shown in figure 3.

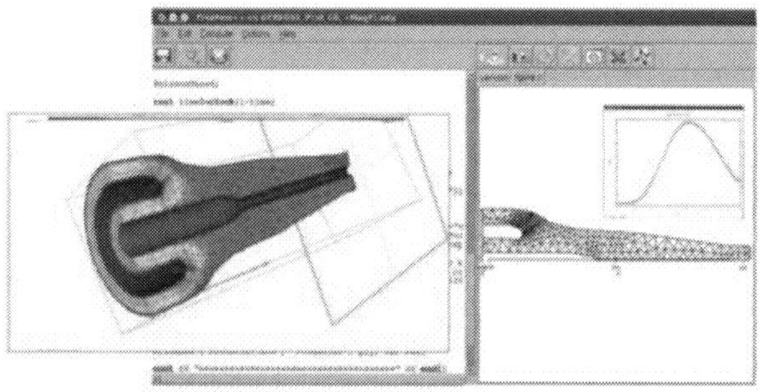

Figure 3. Illustrative screenshots of the GYREOSS code.

Journal of Physics: Conference Series **356** (2012) 012044 doi:10.1088/1742-6596/356/1/012044

Conclusions

The codes outlined above have been used in a series of numerical experiments carried out to study the designs of high-power gyrotrons that are under consideration and/or development at present. The simulations performed provide a deeper physical insight into the operation of high-performance gyrotrons of the megawatt class and demonstrate the improved capabilities and functionality of the upgraded codes. Moreover, these results suggest some further experiments for more detailed studies of the correlation between the beam-quality parameters and the efficiency, on the one hand, and the particular design (configuration of the electrodes and the magnetic field, etc.), on the other. It is expected that the upgraded versions of the simulation packages will contribute to the development of the next generation of improved-performance high-power gyrotrons for fusion.

Acknowledgements

This work was carried out in the framework of Task 2.1.2 of the EURATOM-INRNE Association scientific program.

References

[1] Thumm M 2011 State-of-the-art of high power gyro-devices and free electron masers (Update 2010) *KIT Sci. Reports* **7575** (KIT Scientific Publishing 2010) 1–131

[2] Sabchevski S, Illy S, Piosczyk B, Borie E and Zhelyazkov I 2008 Towards the formulation of a realistic 3D model for simulation of magnetron injection guns for gyrotrons (a preliminary study) *Wissenschaftliche Berichte FZKA* **7409** 1–34

[3] Malygin A, Pagonakis I, Illy S, Avramides K, Piosczyk B, Kern S and Thumm M 2011 Three-dimensional numerical studies of non-uniform emission in magnetron injection guns *Proc. 8th Int. Workshop Strong Microwaves and Terahertz Waves: Sources and Applications* (Nizhny Novgorod Russia July 9–162011) 99–100

[4] Sabchevski S, Zhelyazkov I, Thumm M, Piosczyk B, Tran T M, Hogge J-Ph and Pagonakis J Gr 2007 Recent evolution of the simulation tools for computer aided design of electron-optical systems for powerful gyrotrons *Computer Model. Engin. & Sci.* **20** 203–220

[5] Sabchevski S and Zhelyazkov I 2007 Modelling and simulation of new generation powerful gyrotrons for the fusion research *J. Phys.: Conf. Series* **63** 012003

[6] Damyanova M, Sabchevski S and Zhelyazkov I 2010 Pre- and post-processing of data for simulation of gyrotrons by the GYREOSS software package *J. Phys.: Conf. Series* **207** 012032

[7] Damyanova M, Kern S, Illy S, Thumm M, Sabchevski S, Zhelyazkov I and Vasileva E 2010 Modelling and simulation of gyrotrons for ITER *Proc. 4th Int. Workshop & Summer School Plasma Phys.* (Kiten Bulgaria July 5–10 2010)

[8] Illy S 1997 Untersuchungen von strahlinstabilitaten in der kompressionzone von gyrotron-oszillatoren mit Hilfe der kinetischen theorie und zeitabhangiger particle-in-cell-simulationen *Wissenschaftliche Berichte FZKA* **6037** 1–145

[9] Kern S 1996 Numerische simulation der gyrotron-wechselwirkung in koaxialen resonatoren, *Wissenschaftliche Berichte FZKA* **5837** 1–245

[10] Sabchevski S, Idehara T, Saito T, Ogawa I, Mitsudo S and Tatematsu Y 2010 Physical models and computer codes of the GYROSIM (GYROtron SIMulation) software package *FIR Center Report FIR FU-99* (http://fir.u-fukui.ac.jp/FIR_FU99S.pdf)

The role of plasma radius as a condition for sustaining a coaxial discharge at various wave modes

K Ivanov[1], T Bogdanov [1] and E Benova[2,3]

[1] Faculty of Physics, St. Kliment Ohridski University of Sofia,
5, James Bourchier Blvd., 1164 Sofia, Bulgaria
[2] Department for Language Teaching and International Students,
St. Kliment Ohridski University of Sofia
27, Kosta Loulchev Str., 1111 Sofia, Bulgaria

E-mail: benova_phys@deo.uni-sofia.bg

Abstract. A gas discharge can be produced and sustained by travelling electromagnetic waves in various geometries: planar, spherical, cylindrical and coaxial. An electromagnetic wave travelling along a dielectric tube can produce plasma outside the tube when a metal rod is placed along the tube axis, which is the typical arrangement of a coaxial surface-wave-sustained discharge (CSWD). The CSWD has been studied intensively both theoretically and experimentally since 1998.

In the case of a SWD in cylindrical geometry, plasma is mainly produced and sustained by the azimuthally symmetric waves. In coaxial geometry, there are both experimental and theoretical indications showing that higher wave modes may also produce and sustain plasma under certain conditions. In order to find out these conditions theoretically, we developed a one-dimensional fluid model.

The purpose of this work is to investigate theoretically the behavior of wave phase diagrams under various discharge conditions and to find the discharge conditions under which plasma can be produced, as well as those conditions when this is not possible.

1. Introduction

The coaxial structure is a relatively new type of plasma source, which was proposed recently [1,2] and has since been intensively studied [3,4]. A typical configuration of this type of gas discharges is presented in figure 1. This plasma source produces large-volume clean electrodeless plasma and is very attractive because of the widening range of possible technological applications as in UV lamps [5], in surface cleaning, material treatments, plasma sterilization and thin film deposition [6]. At the same time, fundamental research provides deeper knowledge and better understanding of the physics of the discharge. The plasma is both radially and axially inhomogeneous. It has been found both experimentally and theoretically that the cylindrical plasma column is produced by one wave mode only. Usually this is the azimuthally symmetric one ($m = 0$, m being the azimuthal wave number). A travelling dipolar wave ($m = 1$) can sustain plasma only under some special conditions. The purpose of this work is to investigate the characteristics of the various wave modes, the ability of a given mode to sustain plasma in a coaxial structure, and the role of the dielectric tube. We studied theoretically four

[3] To whom any correspondence should be addressed.

configurations: (i) vacuum–plasma; (ii) metal–vacuum–plasma; (iii) dielectric–plasma; (iv) metal–dielectric–plasma, varying the azimuthal wave number m, the plasma parameter σ and the discharge conditions.

2. Basic assumptions and equations

At low pressures, the electron–neutral collisions frequency ν is smaller than the wave angular frequency ω and the plasma can be considered a weakly dissipative medium. The ratio ν/ω can than be neglected in the plasma permittivity expression, i.e., we can use the simple form $\varepsilon_{\mathrm{p}} = 1-\omega_{\mathrm{p}}^2/\omega^2$, ($\omega_{\mathrm{p}} = \left(4\pi e^2 n/m\right)^{1/2}$ being the plasma frequency) [7]. The coaxial structure is investigated on the basis of a one-dimensional fluid model. The plasma is produced and sustained by a surface electromagnetic wave (f = 2.45 GHz), which propagates along the interface between the plasma and the dielectric tube or vacuum.

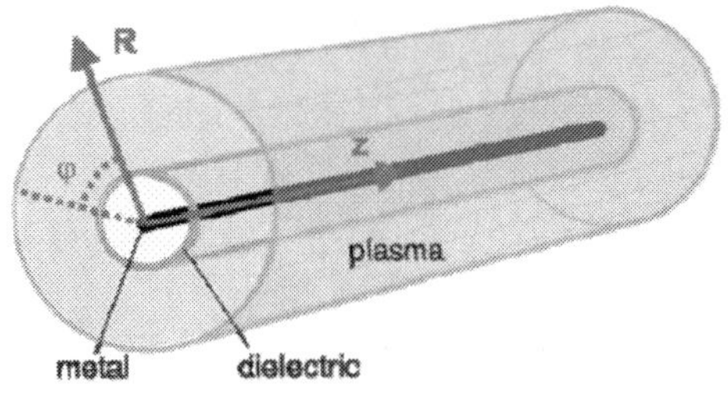

Figure 1. Coaxial discharge configuration.

Our model is based on Maxwell's equations in cylindrical coordinates from which we obtain the wave equation. The solutions of the wave equation yields the wave-field components amplitude as combinations of Bessel or modified Bessel functions:

$$\begin{aligned} F_z^{\mathrm{v}}(a_{\mathrm{v}}\rho) &= C_1 I_0(a_{\mathrm{v}}\rho) + C_2 K_0(a_{\mathrm{v}}\rho) \\ F_z^{\mathrm{d}}(a_{\mathrm{d}}\rho) &= \begin{cases} C_3 I_0(a_{\mathrm{d}}\rho) + C_4 K_0(a_{\mathrm{d}}\rho) & x > \sigma\sqrt{\varepsilon_{\mathrm{d}}} \\ C_3' J_0(a_{\mathrm{d}}\rho) + C_4' N_0(a_{\mathrm{d}}\rho) & x < \sigma\sqrt{\varepsilon_{\mathrm{d}}} \end{cases}, \\ F_z^{\mathrm{p}}(a_{\mathrm{p}}\rho) &= C_5 I_0(a_{\mathrm{p}}\rho) \end{aligned} \tag{1}$$

where

$$a_{\mathrm{p}} = (x^2-\sigma^2\varepsilon_{\mathrm{p}})^{1/2}, \quad a_{\mathrm{d}} = (x^2-\sigma^2\varepsilon_{\mathrm{d}})^{1/2}, \quad a_{\mathrm{v}} = (x^2-\sigma^2)^{1/2}$$

$$\rho = \frac{r}{R}, \quad x = kR, \quad \sigma = \frac{\omega R}{c}$$

The boundary conditions are the conditions for continuity of the electromagnetic field's tangential components at the medium boundaries and the condition for annulment of the E_z–component at the metal cylinders. By solving this system of equations, we obtain all wave-field components, as well as the local dispersion relation. It is difficult to present the dispersion equation in a compact form; we give it below in a more general form:

$$\mathrm{D}(m, \sigma, \eta, \gamma, k, \omega, \omega_{\mathrm{p}}) = 0. \tag{2}$$

In our case, the dispersion relation (2) is local because of the axial inhomogeneity of the plasma. Since the wave frequency ω is constant and the plasma frequency ω_{p} changes along the column, the dispersion equation gives the dependence of the normalized plasma density $N = (\omega_{\mathrm{p}}/\omega)^2$ on the dimensionless wave number $x = kR$ (usually presented as $\omega/\omega_{\mathrm{p}}$ vs. x, or the so-called phase diagrams) at different wave modes. The geometry of the discharge is accounted for by the dimensionless parameters presented in table 1.

Table 1. Dimensionless parameters of the discharge geometry.

Geometric factors	Notations	Parameters
Plasma radius	R (outer radius of the dielectric tube)	$\sigma = \omega R/c$ (c–speed of light)
Dielectric tube radius	R_d (inner radius of the dielectric tube)	$\gamma = R_d/R$
Tube thickness	$d = R - R_d$	$\gamma = 1 - d/R$
Radius of the metal rod on the axis	R_m	$\eta = R_m/R$

3. Results and discussion

By solving the local dispersion relation at fixed $\omega = 2\pi f$ ($f = 2.45$ GHz) and varying the wave modes, we obtain the phase diagrams for the vacuum–plasma configuration (figure 2). The dotted line in figure 2 (and in the other figures) corresponds to the resonance plasma density $N_{res} = 1 + \varepsilon_v = 2$ (or resonance frequency $(\omega/\omega_p)_{res} = 1/\sqrt{2}$). All phase diagrams approach this value as the dimensionless wave number (k_zR) increases. For small values of k_zR, the phase diagrams are divided in two groups: (i) at higher plasma radii (parameter σ), the phase diagrams follow the standard behavior of ω/ω_p increasing with k_zR, which corresponds to the plasma density decreasing from the wave launcher to the column end (dashed lines). (ii) for smaller values of the plasma parameter σ and for $m = 0$, only backward wave propagation is observed in the phase diagrams (ω/ω_p decreasing of with k_zR, black line in figure 2), while for dipolar and quadrupolar azimuthal wave modes, a small region of forward wave propagation is also observed. For small σ, all phase diagrams are in the region of very low plasma density ($N < N_{res}$). We, therefore, assume that the azimuthally symmetric wave mode cannot sustain the discharge at small σ. There exists a value of σ (the so-called critical value σ_{cr}) which divides the two groups of phase diagrams [3]. For the vacuum–plasma configuration, the "critical" value of σ corresponds to $(fR)_{cr} \approx 7.45$ GHz cm meaning that one can expect to produce plasma in such configuration by using sufficiently high enough wave frequency and plasma

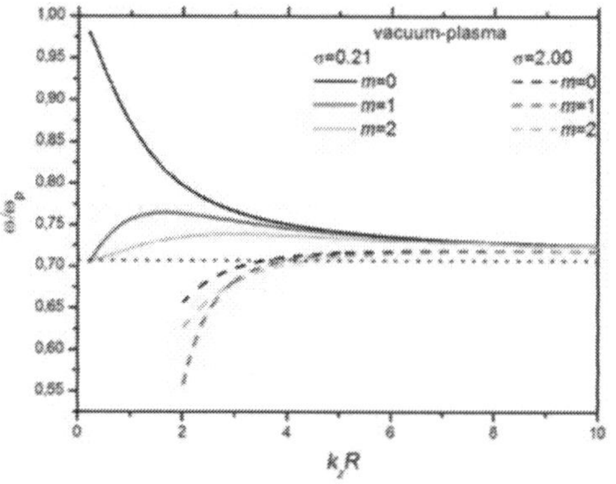

Figure 2. Phase diagrams for the vacuum-plasma configuration in various wave modes and plasma parameter values.

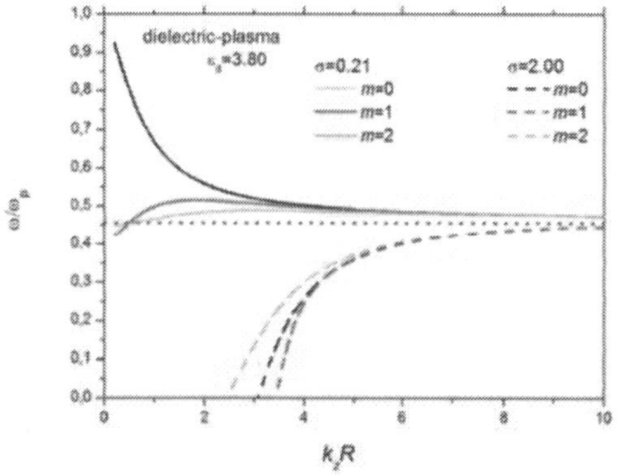

Figure 3a. Phase diagrams for the dielectric–plasma configuration in various wave modes and plasma parameter values.

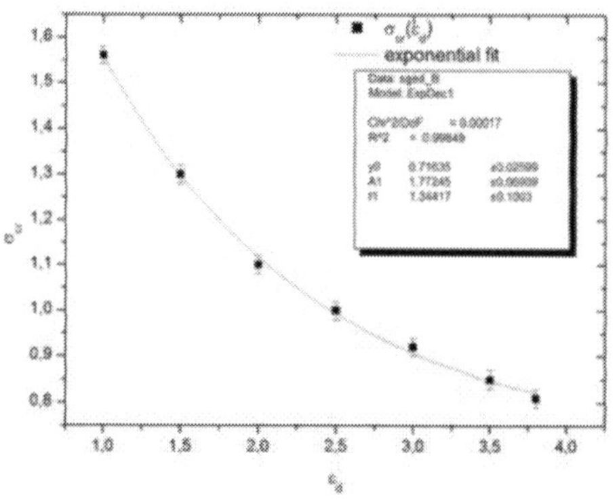

Figure 3b. Dependence of the critical plasma parameter on the dielectric permittivity.

radius. The situation is similar for the dielectric–plasma configuration, as can be seen in figure 3a. In this case, the "critical" value of σ is smaller than that for the vacuum–plasma configuration and depends on the dielectric permittivity of the tube material. For a dielectric with $\varepsilon_d = 3.8$ (quartz), it corresponds to $(fR)_{cr} \approx 3.86$ GHz.cm. Our results show that the value of σ_{cr} decreases exponentially as the permittivity of the medium increases (figure 3b).

The next step in the study is changing the configuration by adding a metal rod on the tube axis of the previous configurations. As one can see in figures 4a and 4b, this leads to a significant change in the phase curves at $m = 0$. Even a thin metal rod (a fraction of 10^{-5} of the plasma radius) is sufficient to change the phase diagrams, so that a region of forward wave propagation appears at small σ. This means that now the azimuthally symmetric wave is also able to create and sustain the discharge for $\sigma < \sigma_{cr}$, while for $\sigma > \sigma_{cr}$ the plasma density is higher.

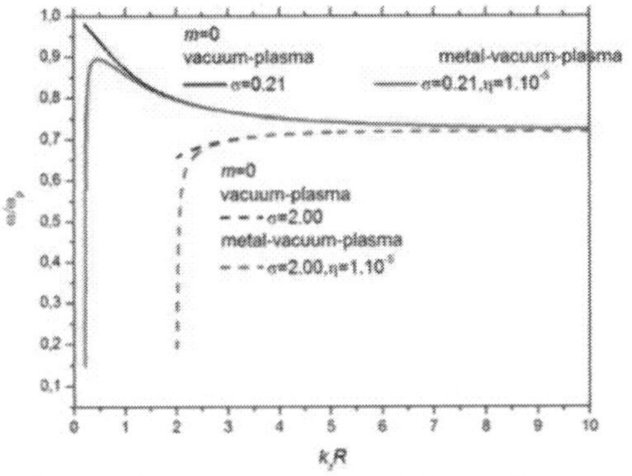

Figure 4a. Comparison of phase diagrams for the vacuum–plasma and metal–vacuum–plasma configurations for $m = 0$.

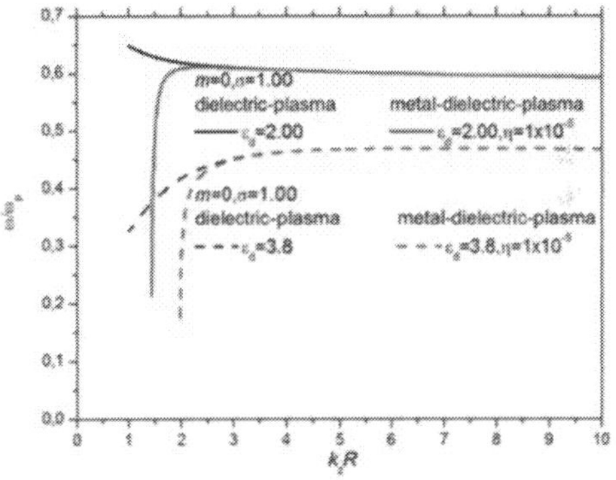

Figure 4b. Comparison of phase diagrams for the dielectric–plasma and metal–dielectric–plasma configurations for $m = 0$.

The role of the metal rod thickness is the same for both configurations, as seen in figure 5. Increasing the normalized metal rod radius η leads to an increase in the plasma density. The maximum of the phase diagrams divides them into two regions, namely, a region of forward wave propagation (from small values of the dimensionless wave number to the maximum value) and a region of backward wave propagation. We assume that only the forward wave can produce plasma and the plasma column ends upon reaching the maximum of the phase diagram. As η increases, the phase diagrams approach zero and the region of backward wave propagation disappears.

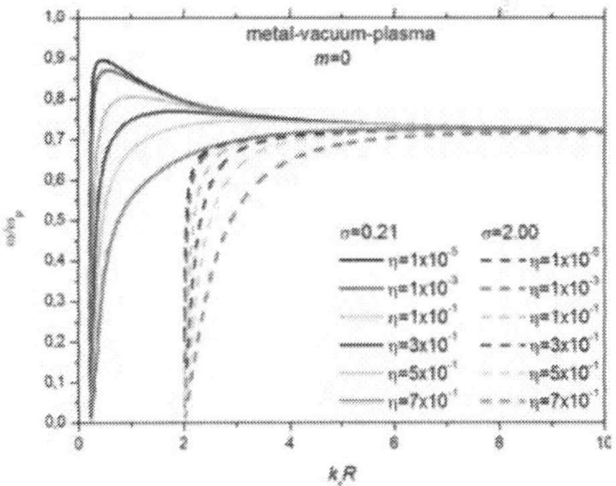

Figure 5. The role of metal rod thickness.

Conclusions

The investigation presented shows that the plasma radius is a key discharge condition for sustaining th plasma. At small values of the plasma radius (small values of the plasma parameter σ), the azimuthally symmetric wave cannot sustain the plasma, because in the phase diagrams of both vacuum–plasma and dielectric–plasma configurations there is only a backward wave propagation region. The higher wave mode phase diagrams are above the resonance frequency, which corresponds to a plasma density below resonance. We assume that the plasma cannot be sustained under these conditions. At high values of the plasma radius (high σ value), all phase diagrams are below the resonance line at small

k_zR, i.e., all wave modes can sustain plasma with high density in both vacuum–plasma and dielectric–plasma configurations.

There exists a critical value of plasma parameter σ which divide the phase diagrams in two parts: at $\sigma < \sigma_{cr}$, the wave cannot sustain plasma. The value of σ_{cr} decreases exponentially with the increase of the dielectric permittivity. The maximum value of σ_{cr} (= 1.56) occurs in the plasma–vacuum configuration.

Adding even a very thin metal rod on the tube axis significantly improves the conditions for sustaining plasma.

Acknowledgments

This work was supported by the Fund for Scientific Research of the University of Sofia under Grant 197/2011.

References

[1] Gritsinin S, Kossyi I, Malykh N, Misakyan M, Temchin S and Bark Y 1999 *Preprint No 1* (Russian Academy of Science, General Physics Institute, Moscow)

[2] Benova E and Neichev Z 2002 Axial model of a coaxial discharge sustained by travelling UHF electromagnetic wave *Czechoslovak J. Phys.* **52** D659–65

[3] Bogdanov T and Benova E 2009 Theoretical study of electromagnetic wave modes sustaining the the coaxial discharge *Proc. 36th EPS Conf. Plasma Phys.* **33E** P-1.087 (29 June - 3 July 2009 Sofia Bulgaria)

[4] Benova E and Bogdanov T 2009 Modelling of a coaxial discharge sustained by travelling electromagnetic wave at various wave mode *Microwave Discharges: Fundamentals and Applications* eds M Kando and M Nagatsu (CURREAC Hamamatsu) pp 21–29

[5] Barkhudarov E M, Misakyan M A, Kossyi I A, Kiselev V A, Kozlov Y N, Taktakishvili I M and Tulupov A V 2006 *Microwave Discharges: Fundamentals and Applications* (YANUS-K Moscow) pp 279–284

[6] Shullz A, Walker M, Feichtinger J, Räuchler E and Schumacher U 2003 *Microwave Discharges: Fundamentals and Applications* (INP, Greifswald) pp 231–241

[7] Benova E 2006 Self-consistent modelling of surface-wave-sustained discharge from low to atmospheric pressure *Microwave Discharges: Fundamental and Applications* ed Lebedev Yu (YANUS-K Moscow) pp 9-18

Nanoscale biomaterial interface modification for advanced tissue engineering applications

V Safonov[1,6], A Zykova[1], J Smolik[2], R Rogovska[2], N Donkov[3], A Goltsev[4], T Dubrava[4], I Rassokha[4] and V Georgieva[5]

[1] Kharkov Institute of Physics and Technology NSC, Kharkov, Ukraine
[2] Institute for Sustainable Technologies, NRC, Radom, Poland
[3] Emil Djakov Institute of Electronics, Bulgarian Academy of Sciences, 1784 Sofia, Bulgaria
[4] Institute for Problems of Cryobiology and Cryomedicine, NASU, Kharkov, Ukraine
[5] Georgi Nadjakov Insitute of Solid State Physics, Bulgarian Academy of Sciences, 1784 Sofia, Bulgaria

E-mail: ais@kharkov.ua

Abstract. Recently, various stem cells, including mesenchymal stem cells (MSCs), have been found to have considerable potential for application in tissue engineering and future advanced therapies due to their biological capability to differentiate into specific lineages. Modified surface properties, such as composition, nano-roughness and wettability, affect the most important processes at the biomaterial interface. The aim of the present is work is to study the stem cells' (MSCs) adhesive potential, morphology, phenotypical characteristics in *in vitro* tests, and to distinguish betwen the different factors influencing the cell/biomaterial interaction, such as nano-topography, surface chemistry and surface free energy.

1. Introduction

The modification of an interface between the surface of biomaterials and the biological environment is a very important task for the modern tissue and cell engineering methods. The most intensively developed technologies are methods combining biomaterial engineering with biological substances, such as various growth factors, proteins and collagens [1-3]. Functional coatings deposition is an effective way of surface modification that provides direct control of the stoichiometry, impurity elements, functional groups and surface charges [4,5]. As the characteristic size-scale of cellular receptors interaction is a few 100 nm, the surface relief modifications should be also controlled at a nanometer-submicrone level [6,7]. Studying the stem cells' (MSCs) adhesive potential, morphology, phenotypical characteristics and distinguishing between the different factors influencing the cell/biomaterial interaction in *in vitro* tests is thus of great interest.

2. Materials and methods

In the present study, the samples were glass substrates (Petri dishes), uncoated and oxide-coated (Al_2O_3, ZrO_2 (MS) magnetron-sputtered and Ta_2O_5 (EB) e-beam-evaporated) with different roughness

[6] To whom any correspondence should be addressed.

parameters – 20, 200 and 400 nm. The Al_2O_3 (MS) coatings deposition was performed in a high-vacuum pumping system with base pressure of about 10^{-3} Pa. The main process parameters were: magnetron discharge power 1 – 8 kW, oxygen source power 1 kW, deposition rate 8 μm/hour [8]. The Ta_2O_5 films evaporation process was carried out at initial vacuum of 7×10^{-4} Pa, operational mode vacuum of 3×10^{-3} Pa, anode current of 50 mA and calculated evaporation power of 350 W [9]. The deposition rate under these conditions was 28 nm/min.

The surface roughness was measured by means of a Hommel T-2000 profilometer. The surface topography was estimated by AFM (Quesant Instrument Corporation, USA). The coatings' structure was investigated by means of XPS and XRD. The X-ray photoelectron spectroscopy measurements were carried out using an ESCALAB MkII (VG Scientific) electron spectrometer at a base pressure in the analysis chamber of 5×10^{-8} Pa (1×10^{-6} Pa during the measurement), using an Al_α X-ray source (excitation energy hν=1486.6 eV).

Other parameters, such as surface free energy (SFE), its polar and dispersion components and fractional polarity were determined by means of the Wu and Owens-Wendt-Rabel-Kaelble methods. The contact angles were measured on a Kruss K12 Tensiometer at a temperature of 20 °C [9].

Adhesion is an integral index of the cells' structural and functional state reflecting their functional activity. Therefore, we started the *in vitro* experimental studies by investigating the adhesive potential of bone marrow cells (BMs) cultured on glass uncoated and oxide-coated substrates with different roughness parameters. The MSCs phenotypical characteristics were determined by a FACS Calibur cytofluorimeter using fluorochromal monoclonal antibodies to CD73, CD106, CD44 structures.

3. Results and discussion

The dependences between the conditions of deposition and treatment of the coatings and the BMs adhesive potential were statistically and significantly different depending on the substrate materials. The best results were obtained on the glass and glass/Al_2O_3 (MS) surfaces. A statistical difference was observed between the cells' adhesive and survival parameters depending on the surface roughness in the range 20 – 400 nm. In the same range, the cell adhesion decreased with the increase of the surface roughness parameters. Figure 1 shows the dependence between the cell adhesion (percents) and the roughness parameters for various substrate materials.

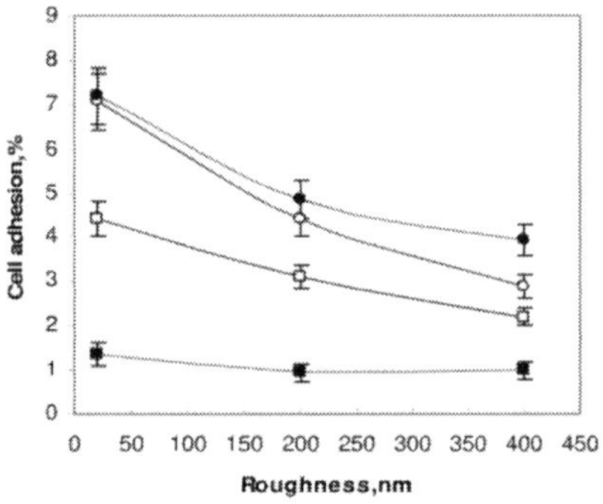

Figure 1. The cell adhesive potential (percents) on the various substrate materials with roughness parameters in the range 20-400 nm after 2 days of culturing: (◦) glass, (•) glass/Al_2O_3 (MS), (▫) glass/ZrO_2 (MS), (▪) glass/Ta_2O_5 (EB).

The structure of the Al_2O_3 (MS), ZrO_2 (MS) and Ta_2O_5 (EB) thin films was investigated by means of XPS and XRD. The X-ray diffraction profiles of as-deposited Al_2O_3 (MS), ZrO_2 (MS), and Ta_2O_5 (EB) coatings demonstrated an amorphous nature (no peaks were observed). It is always difficult to adjust the various parameters involved in thin film deposition to achieve tailored surface chemistry, stoichiometry and surface states of the coatings, these being important for the cell/biomaterial response. We also analysed the structure of the Al_2O_3 (MS) and ZrO_2 (MS) oxide coatings by XPS.

Using these spectra, we estimated the O/Al and O/Zr stoichiometric ratios. The photoelectron spectra of Al_2O_3 (MS) oxides exhibited the binding energy peaks E(Al2p) – 74.4 eV and E(O1s) – 531.3 eV (figure 2). For ZrO_2 (MS) coatings, the binding energy peaks E(Zr3d) – 182.4 eV and E(O1s) – 530.2 eV were observed. Two peaks exist in the Zr3d XPS spectrum, showing an energy shift of 2,43 eV of the 3d level due to spin orbit coupling (figure 3).

The XPS spectra of the Ta_2O_5 (EB) films were also obtained. All spectra consisted of well-defined XPS lines of Ta 4f, 4d, 4p and 4s, O 1s. The Ta 4f doublets are typical for e-beam evaporated Ta_2O_5 and exhibit two peaks: Ta 4f7/2 at ~26.3-26.6 eV and Ta 4f5/2 with binding energy higher by 1.9 eV. The Ta 4f lines of the deposited films agree well with the Ta 4f doublet representative of the Ta-0 bond in Ta_2O_5. The O/Ta ratio estimated from the spectra is ~3 for all samples. Replacing the surface bonds with oxygen by either thermal or plasma treatment results in a shear in the more chemically stable hydrophilic surface region. However, the annealing process results in some additional impurities which were observed in the XPS spectra of e-beam evaporated Ta_2O_5 (C1s, Si2p, Na1s). The results show that the surface properties are strongly influenced by the preliminary treatment. The deposition and treatment conditions affect the properties of the coatings and the positive cell response [9].

The surface free energy (SFE), its polar and dispersion parts and fractional polarity estimations were made by the Wu method for two-liquids system and by Owens-Wendt-Rabel-Kaeble' methods for the following liquid system: α-bromonaphthalene-formamide-ethylene glycol-diidomethane-glycerol-water. The SFE varies practically independently from the surface roughness parameters in the range 2 – 400 nm. The SFE values were in the range 52 – 53 mN/m for Al_2O_3 (MS), 48 – 51 mN/m for ZrO_2 (MS) and 41 – 44 mN/m for Ta_2O_5 (EB) coatings. These results make it possible to distinguish between the influence of roughness and surface free energy effects on the cell/nanomaterial interactions.

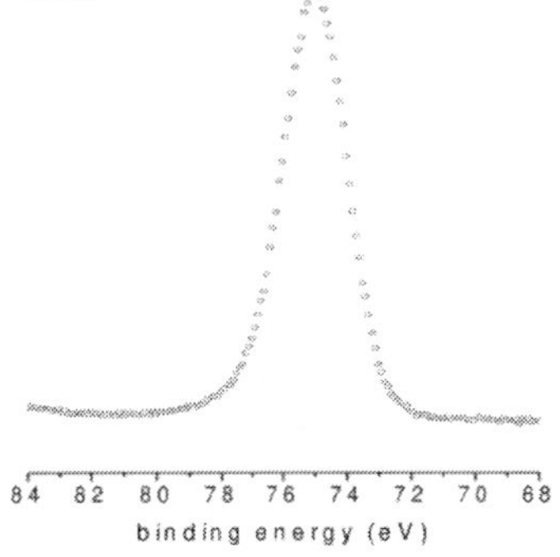

Figure 2. Al2p XPS spectra of Al_2O_3 (MS) coatings.

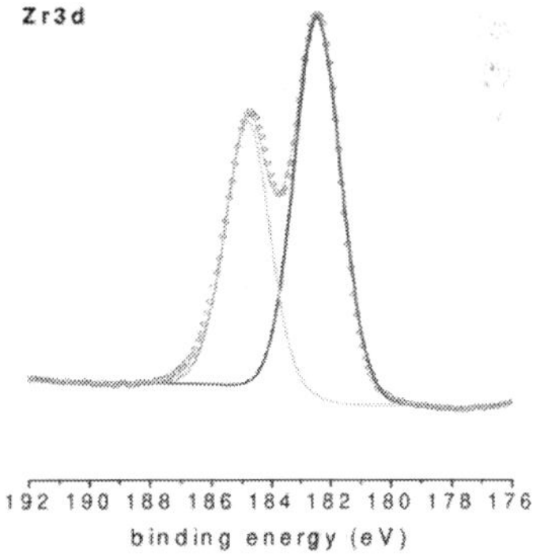

Figure 3. Zr3d XPS spectra of ZrO_2 (MS) coatings.

We further studied the MSCs genes expression. An important component of the MSCs immune modulating activity is the enzyme indoleamine 2,3-dioxygenase (IDO) produced by these cells.

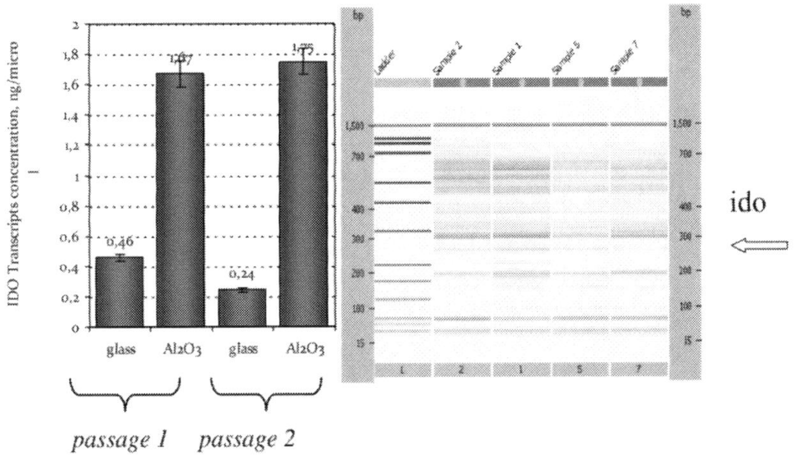

Figure 4. Expression rate of the IDO gene in cultured BMs on various substrates. Transcription of the IDO gene in the zone of 200-359 n.p. is shown by an arrow.

Figure 4 demonstrates the rise in the expression rate of the IDO gene in BMs after culturing on a glass/Al_2O_3 (MS) substrate in comparison with BMs cultured on a glass substrate after the 1st and the 2nd passage. Due to the increase of the cell content with MSCs markers during culturing on oxide nanocoating, a rise in the expression rate of IDO gene is to be expected, since in the culture of BM cells, namely MSCs, are IDO producers.

Conclusions

Our data demonstrate that the cells' adhesive potential and phenotypical characteristics are different for the different oxide coatings deposited by magnetron sputtering and e-beam evaporation. The best results were obtained in the case of magnetron-sputtered oxide coatings with minimal roughness in the range 20 – 400 nm, intermediate values of the surface free energy as calculated by means of the Wu, Owens-Wendt-Rabel-Kaelble methods in the range 50 – 60 mN/m and the greater part of SFE polar components and fractional polarity. Changes were observed in the molecular-genetic apparatus of MSCs (IDO gene expression degree), together with an increase in the MSCs marker number on the oxides nano-structured surface. The results show the effect of the surface parameters modification on of the nano-materials interaction with mesenchymal stem cells and open prospects for a direct control of such parameters as adhesion, proliferation and differentiation of MSCs during their culturing.

Acknowledgment

The study was supported by grant 4974 Targeted Research and Development Initiatives Program of the Science and Technology Center in Ukraine (STCU) and by the Program for Cooperation between the Bulgarian Academy of Science and the National Ukrainian Academy of Science.

References

[1] Anselme K 2000 *Biomaterials* **21** 667
[2] Zinger O, Zhao G, Schwartz Z, Simpson J, Wieland M, Landolt D and Boyan B 2005 *Biomaterials* **26** 1837
[3] Goltsev A, Gordiyenkko Ye, Rassokha I, Dubrava T, Zykova A, and Safonov V 2011 *J. Numerical Analysis Industrial and Applied Mathematics* (*JNAIAM*) in press
[4] Dhayal M, Cho S I, Moon J Y, Cho S J and Zykova A 2008 *J. Appl. Surf. Sci.* **254** 3331
[5] Xavier S P, Carvalho P S P, Beloti M M and Rosa A L 2003 *J. Dent.* **31** 173
[6] Li D, Ferguson S J, Beutler T, Cochran D L, Sittig C, Hirt H P and Buser D 2002 *J. Biomed. Mater. Res.* **60** 325
[7] Dalby M J, Riehle M O, Sutherlandb D S, Agheli H and Curtis A S 2004 *Biomaterials* **25** 5415
[8] Zykova A, Safonov V, Luk'yanchenko V, Walkowicz J and Rogovska R 2007 *Problems Atomic Sci. and Technol. Series: Plasma Phys.* **13** 200
[9] Safonov V, Zykova A, Smolik J, Rogovska R, Donkov N and Georgieva V 2010 *J. Phys.: Conf. Series* **253** 012068

IR studies of impurities in chalcogenide glasses and thin films of the Ge-Sb-S-Te system

V Pamukchieva[1,3], K Todorova[1], O C Mocioiu[2], M Zaharescu[2], A Szekeres[1] and M Gartner[2]

[1] Georgi Nadjakov Institute of Solid State Physics, Bulgarian Academy of Sciences, 72, Tsarigradsko Chaussee, 1784 Sofia, Bulgaria
[2] Institute of Physical Chemistry, Romanian Academy of Sciences, Spl. Independentei 202, 060021 Bucharest, Romania

E-mail: vesela@issp.bas.bg

Abstract. Chalcogenide glasses with composition $Ge_xSb_{40-x}S_{50}Te_{10}$ (x = 10, 20 and 27 at.%) were synthesized and thin films from these parent materials were evaporated onto glass substrates. The vibrational properties of the impurities in the powdered glasses and the corresponding films were studied by FTIR spectroscopy. Vibrational modes attributed to O-H hydroxyl groups, molecular H_2O and carbon impurity atoms were detected in the IR spectra of powdered glasses, but were absent in the IR spectra of the films. Vibrational modes of oxygen atoms bonded to basic elements, such as Ge-O and Te-O bonds, were present in the IR spectra of both kinds of samples.

1. Introduction

In general, chalcogenide glasses transmit longer IR wavelengths than silica and fluoride glasses and thus find applications in civil, medical and military areas [1,2,3]. The transparency range of these glasses depends on their composition. For example, for S-based glasses the transparency is located within the 0.5 – 7 μm spectral range; for Se-based glasses it is within 0.8 – 12 μm; and for telluride glasses this range is 1.2 – 16 μm. The applications of chalcogenide glasses in the IR optics are mainly restricted by vibrational absorption bands of impurities introduced during and/or after the glass synthesis [4]. Therefore, choosing proper glass compositions for which the ability for chemical interaction between the absorbed impurities and the covalent-bonded glass-forming network is minimal is of great importance. Useful information about the impurity atoms properties and their chemical bonding can be obtained by IR spectroscopy.

Recently, we started studying the physico-chemical properties of a new quaternary chalcogenide glassy system, namely Ge-Sb-S-Te [5,6]. We report here the results of an IR spectroscopic study of synthesized $Ge_xSb_{40-x}S_{50}Te_{10}$ (x = 10, 20 and 27 at. %) glasses and of thin films evaporated from these parent materials, focusing on the impurities present in the samples. The information obtained on the IR absorption observed for various bonded impurity atoms is discussed in terms of the samples composition.

[3] To whom any correspondence should be addressed.

2. Experimental details

The glassy alloys of $Ge_xSb_{40-x}S_{50}Te_{10}$ (x = 10, 20, 27) were synthesized from 5N purity elements by the conventional melt-quenching method. The synthesis was performed in a rotary furnace, the glass components of a proper composition being placed in a quartz ampoule evacuated down to 10^{-3} Pa. The ampoules were loaded in the furnace and heated up to 950°C. They were kept at this temperature for 24 h rotating the furnace for homogeneous melting and, upon ending the process, the ampoules were pulled out and quenched in ice water.

Part of the bulk glasses thus synthesized was powdered. For IR measurements, the powdered glasses (~1 mg each) were mixed with spectroscopic grade KBr (200 mg each) and the mixtures were pressed into pellets and dried at 100°C for one hour. The other part served as a parent material for thermal vacuum evaporation of corresponding thin films onto ordinary glass substrates. The thickness of the films (~1 μm) was controlled in situ by a MIKI FFV quartz sensor device.

The pellets were measured on a Nicolet 6700 Fourier Transform Infrared (FTIR) Spectrometer, while the thin films, on a SHIMADZU FTIR Prestije-2 operating with a high-energy ceramic light source (350 – 7800 cm^{-1}) and a high signal-to-noise ratio of 40000:1. For both apparatuses, the FTIR measurements were performed at room temperature in the spectral range of 4000 – 400 cm^{-1} and 4000 – 350 cm^{-1}, respectively, with a resolution of 4 cm^{-1}. The spectral range of these apparatuses allows one to detect the infrared absorption bands for various bonded impurity atoms.

3. Results and discussion

Generally, mainly oxide and hydride impurities may be present in chalcogenide glasses. Most of the absorption bands observed in the IR spectra could be assigned to vibrational bands of these impurity bonds. The IR spectra of powdered glasses andthe corresponding thin films are presented in figures 1 and 2, respectively, while the position of the detected absorption bands are summarized in tables 1 and 2. The bands were identified on the basis of experimental observations reported in refs [9-17].

3.1. IR transmittance spectra of powdered glasses

The glasses showed good transparency in the entire spectral range studied. The transmittance values showed a weak tendency to decrease with decreasing the Ge content. The IR transmittance spectra of all three glassy compositions were similar; thus, only the IR spectra of $Ge_{10}Sb_{30}S_{50}Te_{10}$ and $Ge_{27}Sb_{13}S_{50}Te_{10}$ compositions, i.e. the glasses with the lowest and highest Ge amount, are illustrated in figure 1. Various absorption bands are seen, which are related to chemical bonds of different extrinsic impurities present in the glassy material. The position of these impurity bands are summarized in table 1. It should be noted that the vibrational bands related to the basic chemical bonds in the glassy compositions are below the measurement limit of 400 cm^{-1} and, therefore, cannot be detected.

For all three compositions, the most intensive and broadest band appeared within the range 3300–3700 cm^{-1} with a minimum centered at around 3446 cm^{-1}. This absorption band can be associated with the vibrational modes of O-H hydroxyl groups bonded with external hydrogen and/or atoms of the chalcogenide structural network [9-11]. The other, much weaker band, associated with adsorbed molecular H_2O [10] appeared around 1640 cm^{-1}. The reason for the appearance of these two absorption bands most probably is moisture absorbed by the sample from the environment during the samples preparation (powdering, pressing etc) and during the IR measurements. The

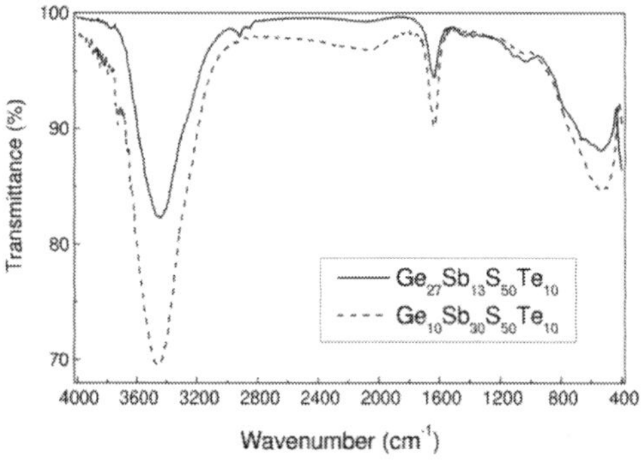

Figure 1. IR transmittance spectra of Ge-Sb-S-Te powdered glasses with compositions, inserted.

Table 1. Position of impurity bands in the IR spectra of powdered Ge-Sb-S-Te glasses.

Vibratioal modes	Vibrational band position (cm^{-1})		
	$Ge_{27}Sb_{13}S_{50}Te_{10}$	$Ge_{20}Sb_{20}S_{50}Te_{10}$	$Ge_{10}Sb_{30}S_{50}Te_{10}$
$\nu_{O\text{-}H}$	3448	3431	3444
$\nu_{C\text{-}H}$	2920 2850	2920	
$\nu_{C\text{-}O\text{-}S}$	2050		2051
$\nu_{O\text{-}H}$	1639	1637	1640
$\nu_{C\text{-}H}$	1430 1383	1440	
$\nu_{C\text{-}O}$; $\nu_{C\text{-}H}$	1160		1106
$\nu_{Ge\text{-}O}$	1032	1048	1047
$\nu_{Te\text{-}O}$	668 537	663	529

intensity of water and O-H related vibrational bands depends on the glassy composition. S-based chalcogenides have larger affinity to absorbing water-related particles and the absorbed impurities bind structurally with S atoms [18]. In our case, the S amount is constant and, therefore, the variation observed of the intensity of these two bands can be explained by the difference in Sb atomic concentration [18]. By increasing the Sb content, the metallic character of chemical bonds increases and, thus, the probability increases of Sb atoms interacting with these impurities [18].

The oxides are among the impurities that are strongly dependent on the chemical composition of chalcogenide glasses. In all IR spectra, absorption bands are present related to vibration of oxygen atoms bonded to basic elements or to carbon impurity atoms. A band has been reported in the spectral range 700 – 800 cm^{-1} assigned to phonon absorption of Ge-O bonds within mixed chalcogenide germanium tetrahedra [14]. Therefore, the shoulder observed around 800 – 730 cm^{-1} can be due to the presence of Ge-O bonds in the powders. In addition, the weak band around 1040 cm^{-1} can also be related to the Ge-O vibration mode [16]. The broad and asymmetric band centered between 668 – 529 cm^{-1} can be, dependent on the composition, attributed to vibrational mode of Te-O bonds [13].

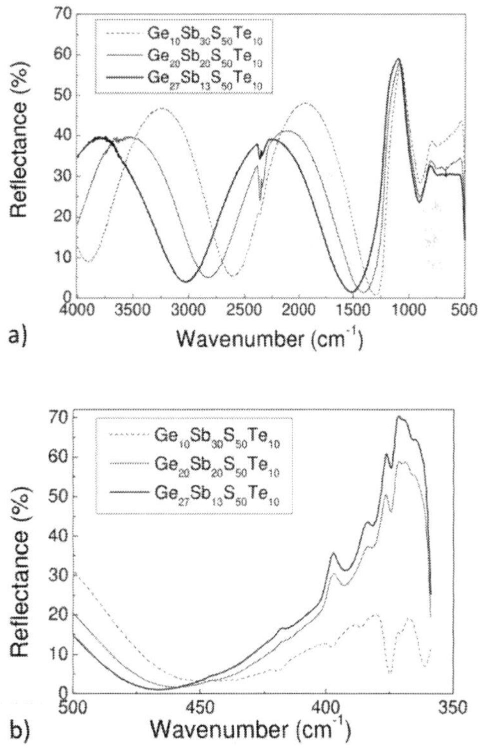

Figure 2. IR reflectance spectra within 4000-500 cm^{-1} (a) and 500-350 cm^{-1} (b) spectral range of films evaporated from the corresponding parent glasses.

All other absorption bands detected in the IR spectra are very weak and do not dependent on the glass composition. They can be assigned to vibrations of carbon impurities in different configurations. The absorption bands in the ranges 2860 – 2922 cm^{-1} and 1430 – 1383 cm^{-1} are assigned to stretching and bending vibration of C-H bonds, respectively, in CH_2 and/or CH_3 impurity groups [10, 19]. Absorption bands in the range 1160-1106 cm^{-1} can be assigned to vibrations of the C-O ester groups and/or the CH_2 group [20]. The broad and very weak band centered at 2050 cm^{-1} has been assigned to C-O-S impurity groups due to contamination of the raw materials [10,12].

In conclusions, it is shown that the influence of O- and C- containing complexes (–C–O–, ≡C–O–S– and –C–H–) on the IR transmittance spectra is insignificant in comparison with that of water related impurities. Since all these impurities are present in powdered chalcogenides, one would expect that they will transfered into the growing film during evaporation of powdered glasses.

3.2. IR reflectance spectra of evaporated films

The IR reflectance spectra of the films are presented in figures 2a and 2b. The positions of the absorption bands of the corresponding chemical bonds are summarized in table 2.

In the spectral range 1200 cm^{-1} – 4000 cm^{-1} (figure 2a), interference fringes appeared, with the vibrational bands of the proper chemical bonds being superimposed on them. Absorption bands of water-related impurities in the IR spectra can be observed neither at ~3600 cm^{-1} nor at ~1640 cm^{-1}, which is evidence that the films are free from these impurities.

The doublet observed in the spectral range 2340 – 2365 cm^{-1} (figure 2) is due to the presence of H_2S molecules in the film surface region [10]. The intensity of this band has a tendency to increase with increasing the Ge content as a result of compositional variations and structural compactness [5]. Such a vibrational band is not observed in the IR spectra of the corresponding parent glasses (figure 1) because the parent materials are with better stoichiometry and higher compactness.

Concerning the oxygen related impurities, some weak vibrational bands are present in the IR spectra of the evaporated films as well. According to literature data [21], the vibrational bands in the spectral ranges 400 – 500 cm^{-1} (figure 2b) and 720 – 780 cm^{-1} (figure 2a) are attributed to intermolecular vibrations of Te-O chemical bonds. In our case, weak bands were detected at 419, 668 and 746 cm^{-1} which could be related to Te-O bond vibrations. The band at 900-912 cm^{-1} was more

Table 2. The position of vibrational bands in the IR spectra of Ge-Sb-S-Te thin films.

Vibrational modes	Vibrational band position (cm^{-1})		
	$Ge_{27} Sb_{13} S_{50} Te_{10}$	$Ge_{20} Sb_{20} S_{50} Te_{10}$	$Ge_{10} Sb_{30} S_{50} Te_{10}$
$\nu_{H\text{-}S\text{-}H}$	2364 2344	2364 2344	2364 2344
$\nu_{Ge\text{-}O\text{-}Ge}$	912	906	900
$\nu_{Te\text{-}O}$	746 668 419	746 668 419	746 668 419
$\nu_{Ge\text{-}S}$	466	460	444
$\nu_{GeS4/2}$	392 381 374 370 366	392 381 373 370 366	397 386 375 370 361

intense and could be assigned to the stretching mode of oxygen in Ge-O-Ge bonds [15]. The presence of oxygen atoms binding to basic elements is quite realistic since the films have been exposed to air after sample loading from the vacuum chamber, as well as during storage and IR measurements. Our earlier XPS studies of chalcogenide films with other compositions [22] revealed some degree of oxidation of the surface layer.

For all three film compositions, a strong and broad band with several superimposed small minima in the spectral range 350 – 500 cm^{-1} was detected (figure 2b). The broad band centered around 460 – 446 cm^{-1} is related to the vibrational mode of Ge-S chemical bonds [12,17]. The small peaks, superimposed on this band, centered at different positions (table 2) can be assigned to the stretching mode of the basic $GeS_{4/2}$ tetrahedral units with possible different configurations [8, 23].

The vibrational modes of the other basic chemical bonds are beyond the IR measurement limit (350 cm^{-1}) and, thus, cannot be detected.

Conclusions

Extrinsic impurities in the powdered glassy alloys of $Ge_xSb_{40-x}S_{50}Te_{10}$ (x = 10, 20, 27) and of films evaporated in vacuum from these parent materials were studied by FTIR spectroscopy. The FTIR transmittance spectra of powdered glasses revealed water related impurities, which are dominant over the impurities of O- and C- containing complexes. The evaporated films are free from water impurities but absorption bands related to Ge-O and Te-O chemical bonds are detected, most probably due to oxidation of the surface region.

References

[1] Keirsse J, Boussard-Pledel C, Loreal O, Sire O, Bureau B, Leroyer, P, Turlin B and Lucas J 2003 *Vibrational Spectrosc.* **32**

[2] Busse L, Moon J, Sanghera J and Aggarwal I 1996 *Laser Focus World* **32** 143

[3] Sanghera J, Shaw L, Busse L, Nguyen V, Pureza P, Cole B, Harbison B, Aggarwal I, Mossadegh R, Kung F, Talley D, Roselle D and Miklosh R 2000 R *Fiber Integr. Opt.* **19** 251

[4] Seddon A 1995 *J. Non-Cryst. Solids* **184** 44

[5] Pamukchieva V, Szekeres A, Todorova K, Fabian M, Svab E, Revay Zs and Szentmiklosi L 2009 *J. Non-Cryst. Solids* **355** 2485

[6] Pamukchieva V, Szekeres A, Todorova K, Svab E and Fabian M 2009 *Opt. Mater.* **32** 45

[7] Kamitsos E I, Kapartsis J A, Culeas I P and Iovu M S 1997 *J. Phys. Chem. B* **101** 11061

[8] Mei Q, Saienga J, Schrooten J, Meyer B and Martin S W 2003 *J. Non-Cryst. Solids* **324** 264

[9] Mironov V A and Yankovskii S A 1985 *Spectroscopy in Organic Chemistry* (Moscow: Khimiya)

[10] Kavetskyy T S, Kovalskiy A P, Pamukchieva V D and Shpotyuk O I 2000 *Infrared Phys. Tech.* **41** 41

[11] Efimov M and Pogareva V G 2000 *J. Non-Cryst. Solids* **275** 189

[12] Smith C, Jackson J, Petit L, Rivero-Baleine C and Richardson K 2010 *J. Solid St. Chem.* **183** 1891

[13] Munoz-Martín D, Villegas M A, Gonzalo J and Fernandez-Navarro J M 2009 *J. Eur. Ceram. Soc.* **29** 2903

[14] Tikhomirov V K, Furniss D, Seddon A B, Savage J A, Mason P D, Orchard D A and Lewis K L 2004 *Infrared Phys. Technol.* **45** 115

[15] Rath S, Kabiraj D, Avasthi D K, Tripathi A, Jain K P, Kumar M, Mavi H S and Shukla A K 2007 *Nucl.Instr. Meth. Phys. Res.* B **263** 419

[16] Ito F, Nakanaga T, Takeo H, Essig K and Jones H 1995 *J. Molecular Spectr.* **174** 417

[17] Nakamoto K 2009 *Infrared and Raman Spectra of Inorganic and Coordination Compounds Part A* (Wiley, New York)

[18] Kavetskyy T, Golovchak R, Shpotyuk O, Filipecki J and Swiatek J 2004 *J. Optoelectron. Adv. Mater.* **6** 1141

[19] Bellamy L J 1958 *The Infra-Red Spectra of Complex Molecules* (London: Methuen)

[20] Trif M, Ansorge-Schumacher M and Socaciu C 2007 *Proc. XVth Int. Workshop Bioencapsulation* (6-8 Sept. Vienna Austria)

http://impascience.eu/bioencapsulation/340_contribution_texts/2007-09-06_P3-07.p

[21] Frumarova B, Oswald J, Krecmer P, Frumar M, Cerny V and Smrcka V 1996 *Opt. Mater.* **6** 217

[22] Pamukchieva V, Skordeva E, Arsova D, Guimon M F and Gonbeau D 2005 *J. Optoelectron. Adv. Mater.* **7** 1265

[23] Julien C, Barnier S, Massot M, Chbani N, Cai X, Loireau-Lozac'h A and Guittard M 1994 *Mater. Sci. Eng. B* **22** 191

Magnetic properties of nanosized $MgFe_2O_4$ powders prepared by auto-combustion

Ch Ghelev[1], T Koutzarova[1,7], S Kolev[1], I Nedkov[1], K Krezhov[2], D Kovacheva[3], B Blagoev[1,6], B Vertruyen[4], C Henrist[4], R Cloots[4], A Zaleski[5] and V Nizhankovskii[6]

[1] Emi Djakov Institute of Electronics, Bulgarian Academy of Sciences, 72, Tsarigradsko Chaussee, 1784 Sofia, Bulgaria
[2] Institute for Nuclear Research and Nuclear Energy, Bulgarian Academy of Sciences, 72, Tsarigradsko Chaussee, 1784 Sofia, Bulgaria
[3] Institute of General and Inorganic Chemistry, Bulgarian Academy of Sciences, Acad. Georgi Bonchev Str., Bl.11, 1113 Sofia, Bulgaria
[4] LSIC, Chemistry Department B6, University of Liege, Sart Tilman, B-4000 Liege, Belgium
[5] Institute of Low Temperature and Structure Research, Polish Academy of Sciences, 2 Okólna Str., 50-422 Wroclaw, Poland
[6] International Laboratory of High Magnetic Fields and Low Temperatures, 95 Gajowicka Str., 53-421 Wroclaw, Poland

E-mail: tanya@ie.bas.bg

Abstract. Targets were prepared to be used for magnetron sputtering and laser ablation and their microstructural and magnetic properties were investigated. The base material was nanosized $MgFe_2O_4$ powder produced by citrate auto-combustion synthesis. The auto-combusted powders were annealed at temperatures in the range 600 - 1000°C in air to study the effect of temperature on thofe formation $MgFe_2O_4$. The saturation magnetization M_s was 24.30 emu/g at room temperature.

1. Introduction

In this paper we describe the synthesis of $MgFe_2O_4$ nanoparticles by auto-combustion and present their structural and magnetic characteristics. Among the different ferrites, magnesium ferrite enjoys special attention because of its various applications in high-density recording media, heterogeneous catalysis, adsorption, sensors and magnetic technologies [1, 2]. The $MgFe_2O_4$ powder is used to prepare targets for thin films deposition by magnetron sputtering and laser ablation. This is one of the reasons why the structural and magnetic properties of this material are being studied, since from technological and theoretical points of view it is important to know if a transfer takes place of the structural and physical properties between as synthesized nanoparticles powder, pressed targets and deposited films. The second most important reason why $MgFe_2O_4$ is studied is that it is present as a second phase during synthesis of multiferroic Y-type hexaferrites, such as $Ba_2Mg_2Fe_{12}O_{22}$ [3]. The presence of $MgFe_2O_4$ in

[7] To whom any correspondence should be addressed.

these hexaferrites may affect their magneto-electric properties, which makes the study of magnesium ferrite magnetic properties important. It is well known that the properties of $MgFe_2O_4$ depend on the method of synthesis. The preparation methods that have usually been used are a solid-state reaction [4], co-precipitation [5, 6], microemulsion [7], sol-gel and auto-combustion [8-10]. In our study, the $MgFe_2O_4$ powder was obtained by citrate auto-combustion synthesis.

2. Experiment

$Fe(NO_3)_3$. $9H_2O$, $Mg(NO_3)_2.6H_2O$ and citric acid all with analytical grade purity were used as starting materials. The molar ratio of metal nitrates to citric acid was fixed at 1:3. The metal nitrates were dissolved together in a minimum amount of de-ionized water to obtain a clear solution. A citric acid solution was slowly added to the mixed solution of nitrates as a chelating agent. The solution was slowly evaporated to form a gel. This gel was dehydrated at 120°C to obtain the barium-magnesium-iron citrate precursor. During the dehydration process the gel turned into a fluffy mass and was burnt in a self-propagating combustion manner. The as prepared auto-combusted powders were annealed at temperatures in the range 600 – 1000 °C in air to study the effect of the temperature on the $MgFe_2O_4$ formation.

The powders were characterized using X-ray diffraction analysis on a Bruker D8 Advance diffractometer with CuKa radiation and LynxEye detector within the 15 - 120 deg. 2θ range with a constant step of 0.02 deg. 2θ. The evaluation of the X-ray powder diffraction pattern of $MgFe_2O_4$ sample heated at 1100 °C was performed using the FullProf software. The magnetization was measured by a homemade low-frequency (3.6 Hz) vibrating sample magnetometer (VSM) [11] at 4, 50, 78, 150, 220 and 300 K. The AC magnetization (ACM) measurement was performed on a Physical Properties Measurement System (PPMS) in an *ac* magnetic field with amplitude 10 Oe and frequency 1 kHz. The sample was first cooled down to 4 K without magnetic field and the ACM was measured as the temperature was increased up to 300 K.

3. Results and discussion

The XRD spectra of the powders after auto-combustion and heat treatment at different temperatures are presented in figure 1, a). The material is predominantly amorphous after the auto-combustion process. Small and broad peaks of $MgFe_2O_4$ and Fe_2O_3 (hematite) could be seen on this diffraction pattern indicating the poor crystalline state of the phases formed at this stage of the synthesis. The spectrum of the samples annealed at 600°C shows that the powder is already crystalline but is not single phase. Besides the $MgFe_2O_4$ peaks, there are lines of a second phase of hematite (Fe_2O_3). The XRD patterns show that the peaks become sharper and narrower as the sintering temperature is increased, indicating that the crystallinity of the samples is improving with the sintering temperature. The spectrum of the sample obtained at 850°C shows that the main phase in the material is $MgFe_2O_4$. The second phase of hematite continues to be present, but the peaks intensity is lower in comparison with those of the sample obtained at 600 °C. The XRD spectrum of the sample synthesized at 1100°C shows the presence of $MgFe_2O_4$ only.

The Rietveld refinement procedure was performed on the $MgFe_2O_4$ sample synthesized at 1100°C. The results of the refinement are summarized in table 1. The Rietveld plot is shown in figure1, b). The tetrahedral and octahedral distances obtained are 1.933(2) Å and 2.031(2) Å, respectively. The mean crystallite size was evaluated at about 270 nm.

Table 1. Structural parameters of the $MgFe_2O_4$ sample. The refined unit cell parameter a = 8.3852(1) Å. Reliability factors: R_{exp}=7.07, R_{wp}=9.95, GOF=1.85, R_{Bragg}=4.14.

Site	Position	X	y	z	Ion	Occupancy	B_{izo}
Tetrahedral	8a	1/8	1/8	1/8	Fe	0.863(4)	0.62(3)
					Mg	0.137(4)	0.62(3)
Octahedral	16d	½	1/2	1/2	Mg	0.863(4)	0.64(3)
					Fe	1.137(4)	0.64(3)
Oxygen	32e	0.2581(2)	0.2581(2)	0.2581(2)	O	4	1.31(7)

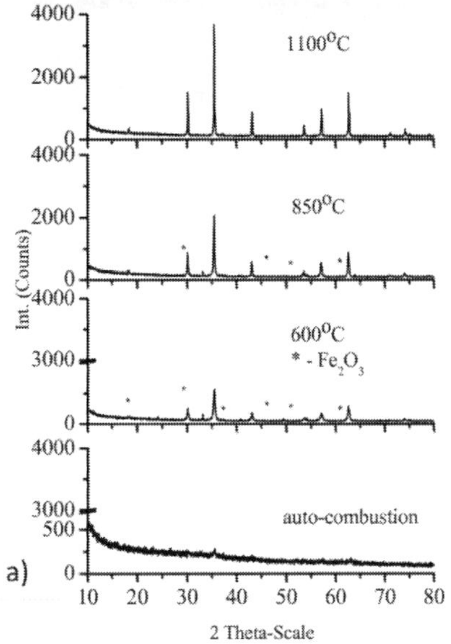

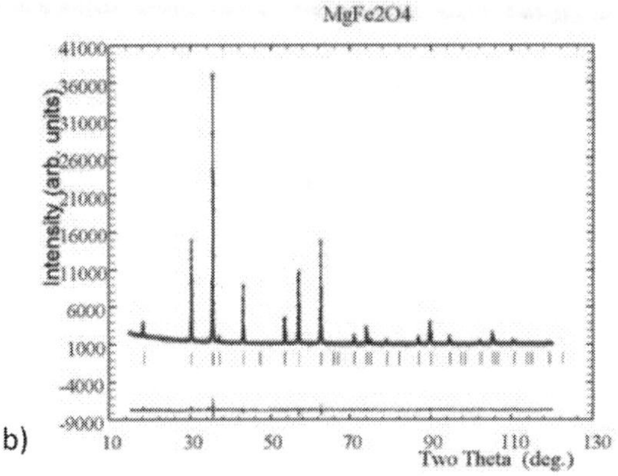

Figure 1. (a) XRD spectrum of samples obtained after auto combustion, and sintered at 600°C, 850°C, 1100°C (*-hematite), (b) Rietveld plot of the structure of the $MgFe_2O_4$ spinel.

Figure 2 shows a SEM image of the sample immediately after the completion of the auto-combustion process. A typical porous structure is observed formed during the burn process, with the pores on the surface having a smaller size than those within the core. The $MgFe_2O_4$ sample heated at 1100°C is homogeneous with respect to the particles size and shape. It is also seen that the particles have an approximately spherical shape.

Figure 2. SEM images of samples obtained after auto-combustion (a) and at 1100°C (b).

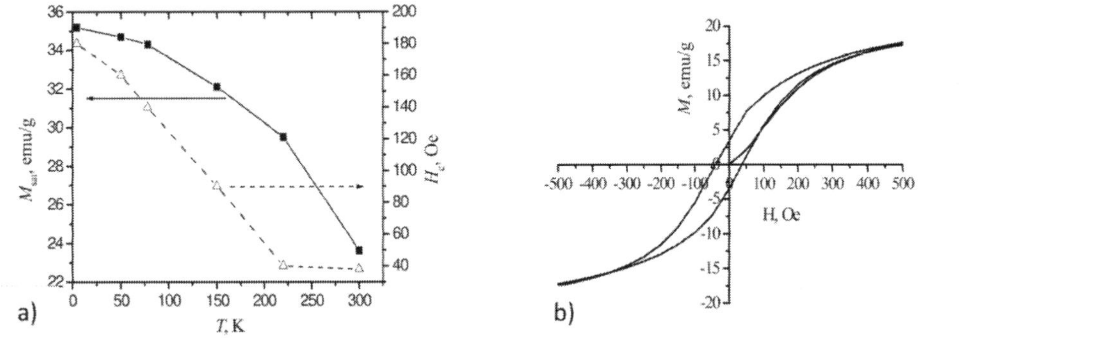

Figure 3. Saturation magnetization at *dc* magnetic field of 1.5 T (filled squares) and coercivity (blank triangles) versus temperature – a); and hysteresis of $MgFe_2O_4$ at 300K – b).

The saturation magnetization M_{sat} and coercivity H_c versus the temperature of the $MgFe_2O_4$ sample synthesized at 1100°C are shown in figure 3, a). The magnetization value at 1.5 T is assumed to be the saturation magnetization value. As the temperature is raised, the saturation magnetization M_{sat} and coercivity H_c decrease quickly. At room temperature (figure 3, b), the coercivity H_c is only 38 Oe, the saturation magnetization M_s is 24.30 emu/g and the remanence M_R is 3.5 emu/g. The coercivity is too small for such relatively large particles. This could be related to fact that the particles probably consist of only a few domains.

Figure 4 presents the real M' (in phase) and imaginary M'' (out of phase) part of the *ac* magnetization. The in-phase component does not exhibit any transitions, while in the out of phase component one can see a few peaks related to energy losses as the energy of the *ac* magnetic field is absorbed by the sample. The spectrum is typical for polydomain-structured soft magnetic materials and demonstrates a motiont and rotation of the domain walls.

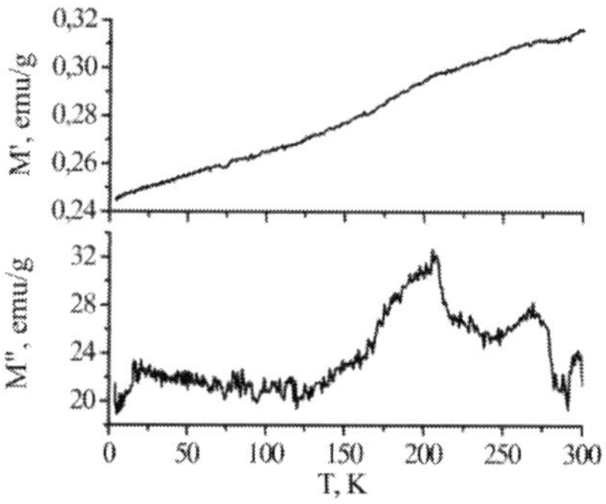

Figure 4. *ac* magnetization (in-phase component of complex magnetization–M' and out of phase component – M'') at *ac* field with amplitude 10 Oe and frequency 1 kHz.

As the temperature is raised, the resonance behavior becomes stronger reaching a maximum at a critical temperature of about 210 K. In this case, the individual particles can be regarded as spherical objects with one or a few domain walls depending of the particles' diameters. Under certain assumptions, the particles could be considered as harmonic oscillators with resonance frequency depending on their diameter. This model gives opportunity to seek a connection between the magnetic properties of the material and the diameter of the particles in the target in the thin film deposition process.

On the other hand, the frequency of maximum dispersion is related to the boundary frequency of the domain wall and is controlled by the attenuation caused by the environment in which the domain wall is moving so that the maximum frequency depends on the anisotropy constant. Further research is needed to study the magnetic dynamics of the particles.

Conclusions

Single phase $MgFe_2O_4$ nanoparticles were synthesized by autocombustion. The mean crystallite size was about 270 nm. The microstructural studies demonstrated that the powders thus produced can be used to form targets for magnetron sputtering and laser ablation. The targets' morphology and magnetic properties were studied systematically, which revealed a polydomain structure with $M''(T)$ value typical for magnetically soft materials. Experiments are being performed on films deposition using these targets, the results of which will be the subject of a future publication.

Acknowledgments

The work was supported in part by the Bulgarian Science Fund under grant DO 02-224 and DHTC 01/4, by the research agreements between WBI, Belgium, and the Bulgarian Academy of Sciences, and by a joint research project between the Polish Academy of Sciences and the Bulgarian Academy of Sciences.

References

[1] Mansour S F and Elkestawy M A 2011 *Ceramics Int.* **37** 1175
[2] Ichiyanagi Y, Kubota M, Moritake S, Kanazawa Y, Yamada T and Uehashi T 2007 *J. Magn. Magn. Mater.* **310** 2378
[3] Koutzarova T, Kolev S, Nedkov I, Krezhov K, Kovacheva D, Blagoev B, Ghelev Ch, Henrist C,

Cloots R and Zaleski A 2011 *J. Supercond. Novel Magnetism* 10.1007/s10948-011-1232-3
[4] Turkin A I and Drebushchak V A 2004 *J. Crys. Grow.* **265** 165
[5] Anis-ur-Rehman M, Malik M A, Akram M, Khan K and Maqsood A 2010 *Phys. Scr.* **83** 015602
[6] Chen Q, Rondinone A, Chakoumakos B and Zhang Z 1999 *J. Magn. Magn. Mat.* **194** 1
[7] Holec P, Plocek J, Niznansky D and Poltierova J 2009 *J. Sol-Gel Sci. Technol.* **51** 301
[8] Pradeep A, Priyadharsini P and Chandrasekaran G 2008 *J. Mag. Mag. Mat.* **320** 2774
[9] Huang Y, Tang Y, Wang J and Chen Q 2006 *Mat. Chem. Phys.* **97** 394
[10] Da Dalt S, Takimi A S, Volkmer T M, Sousa V C and Bergmann C P 2011 *Powder Technol.* **210** 103
[11] Nizhankovskii V I and Lugansky L B 2007 *Meas. Sci. Technol.* **18** 1533

Investigation of optical properties of thin polymer films

N G Sultanova[1,3], S N Kasarova[1] and I D Nikolov[2]

[1] Department of Mathematics and Physics, Assen Zlatarov University of Burgas,
1, Prof. Yakimov Str., 8010 Burgas, Bulgaria
[2] Department of Optics and Spectroscopy, Faculty of Physics,
St. Kliment Ohridski University of Sofia, 5, J. Bourchier Blvd., 1164 Sofia, Bulgaria

E-mail: sultanova@btu.bg

Abstract. Several types of optical polymer materials produced by Eastman Chemical Company were studied. Thin films deposited on glass substrates from solutions were prepared and studied to reveal their refractometric, dispersive and transmission properties in the visible and infrared spectral regions. The refractive indices were measured by means of two modifications of a laser microrefractometer.

1. Introduction

Plastic materials are commonly used in the design and fabrication of optical elements and devices. They have several key advantages over optical glasses, including low cost and weight, high impact resistance and ability to integrate proper mechanical and optical parameters [1]. Polymer optical elements and systems appear in a wide spectrum of industrial, medical, scientific, military and consumer applications. Investigation of thin polymer films (TPFs) is also of great interest for production of optical sensors, displays, fibers and waveguides, etc. [2, 3]. However, the refractometric and dispersive properties of the films differ in comparison with bulk plastics, as established in our previous reports [4, 5]. These differences depend on the material studied, as well as on the film parameters.

In this paper, we report studies on the optical properties of several types of optical polymer materials produced by Eastman Chemical Company. TPFs were deposited on glass substrates from solutions. A laser microrefractometer (LMR), based on critical angle determination, was assembled to measure refractive indices at three wavelengths [6]. A modification of this LMR, as proposed in [7], was also assembled to obtain experimental data at four different laser wavelengths. In both cases, the measurements were conducted based on the principle of diffraction pattern disappearance. The refractometric and dispersive characteristics of the polymer materials were measured for various film thicknesses and types of plastic material. A comparison with the properties of bulk plastic samples is also provided.

2. Experimental

2.1. Sample preparation

Pellets of polyester, polyarylate, polyacrylate and cellulose were dissolved in chloroform, while two

[3] To whom any correspondence should be addressed.

types of copolyesters were dissolved in 1,1,2,2-tetrachloroethane. Solutions of the first four materials were prepared with concentration of 10 wt%, while the concentration of the copolyesters' solutions was 1 wt%. TPFs with different thickness d were obtained by casting a certain amount of the solution on a glass substrate. All of the plastics studied, except for the cellulose layers, were deposited on a heavy TF4 glass plate. A TK21 glass was used for the low refractive cellulose material. The samples were dried at a temperature of 20 °C for 48 hours and then heated up to 60 °C during 6 hours to evaporate the remaining solvent. The TPFs prepared were preserved in a desiccator. A digital micrometer (Mitutoyo Corporation) was used for measuring d with an accuracy of ± 1 μm.

2.2. *Transmission spectra*

The transmittance of the TPFs was measured in the spectral range from 400 to 2500 nm using a Varian Carry 5E UV-VIS-NIR spectrophotometer. The materials studied transmit well in the visible (VIS) and near-infrared (NIR) regions of the spectrum (figure 1). Transmission spectra of polyarylate films with varying thickness d are presented in figure 1-a. For d = 20 μm, transmittance exceeding 80% is observed up to 2260 nm. The thicker layers exhibit greater absorption, especially in the VIS range. The polyester film (figure 1-b) shows transmission of about 90 % in the VIS and NIR regions up to 1700 nm. In both cases, the spectra presented show weak absorption bands between 1660 and 1700 nm due to the first overtone of the C–H group. A considerable transmission decrease is observed at wavelengths greater then 2200 nm where absorption by other C–H groups occurs.

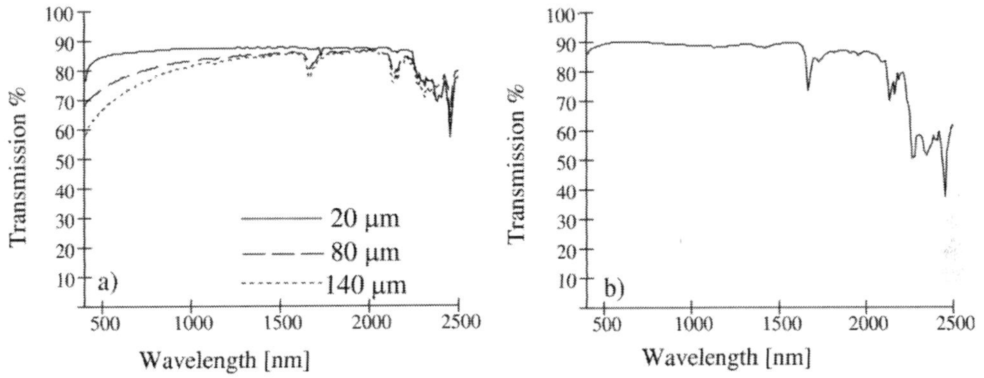

Figure 1. Transmission spectra of: a) polyarylate films with different thicknesses; b) 20 μm thick polyester film.

2.3. *Measuring the refractive indices of optical polymer films*

Two modifications of a LMR, based on the same measuring principle, were used (figure 2). The experimental schemes in both cases differ in the illumination part. As shown in figure 2-a, a He-Ne laser, 1, and two laser diodes, 2, emitting at 632.8 nm, 532 nm and 790 nm, respectively, were used as light sources. The three beams were collected by the two splitters, 3, to form a common beam. In the case of the four-wavelength scheme of the LMR (figure 2-b), illumination is achieved by means of laser diodes, 4, with a power supply, 5, emitting at 406, 656, 910 and 1320 nm. The diodes are placed at four fixed positions and switching between the separate laser sources is realized mechanically. Each laser beam is carefully adjusted in the vertical and horizontal plane to ensure passing of the beam through the two diaphragms, 6, which are separated at a distance of 250 mm and have an equal diameter of 2 mm.

The laser beam in both cases illuminates the internal surface of the prism unit, 7, which is made of heavy flint optical glass TF4 (figure 2-c). Unit 7 is mounted on a rotary goniometric stage, 8, with resolution of 1 arcmin. The sample examined, 9, is placed between the prism and a chromium diffraction grating, 10, with a 40-μm period. The optical contact between the sample measured, the prism and the grating is realized using methylene iodide. At angles less than the critical angle φ_c of the sample film, the incident laser beam passes through the polymer layer and diffracts at the metal grating. The reflected diffraction pattern is observed on a screen, 11.

When the angle of incidence reaches the critical angle of the material, the diffraction orders disappear. An infrared beam-finder card, model IRC32R (Electrophysics), with peak emission at 655 nm and spectral range 800÷1700 nm, was used to visualize the diffraction image in infrared light. The refractive index of the sample is calculated by the expression:

$$n = N \sin\left[A \pm \arcsin\left(\frac{\sin\varphi_c}{N}\right)\right]. \quad (1)$$

The prism's vertex angle is $A = 64.76^{\circ}$, while N is its refractive index at the illumination wavelength λ, which is calculated by fitting the $N(\lambda)$ data published in Lomo Glass Catalogue [8]. The uncertainty of the measurement results was estimated in [6] and a maximal deviation of 2×10^{-4} was obtained for polymer solutions. In case of polymer films, the accuracy established is 2×10^{-3} [5].

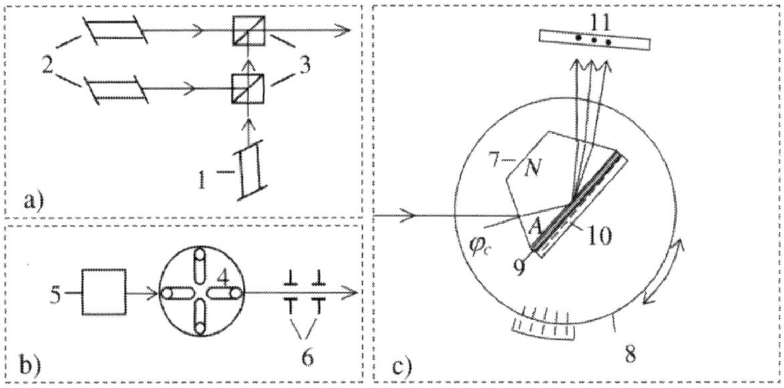

Figure 2. Experimental set-up.

3. Results and discussion

The transmission spectra and the refractive indices measured of the TPFs examined reveal normal dispersion of the materials in the spectral region considered. We used the Cauchy-Schott's and Sellmeier's approximations [4, 5] to calculate the dispersion coefficients, curves and Abbe numbers.

The refractive indices measured of some of the examined films, as well as the calculated dispersion curves, are presented in figure 3. The polymer materials in figure 3-a are measured by means of the four-wavelength LMR, while figures 3-b,-c,-d present results obtained by the three-wavelength LMR.

Dispersion curves of different types of TPFs studied are illustrated in figure 3-a. It is obvious that the dispersion is greater in the VIS spectral region and smaller in the NIR light. The Abbe numbers ν_d for mean dispersion and ν_{1010} for partial dispersion are calculated by the expressions:

$$\nu_d = \frac{n_d - 1}{n_F - n_C} \quad \text{and} \quad \nu_{1010} = \frac{n_{1010} - 1}{n_{700} - n_{1320}}, \quad (2)$$

where n_d, n_F and n_C are the calculated indices at spectral lines 587.6, 486.1 and 656.3 nm. The partial dispersion in the NIR is estimated by the values obtained of the refractive indices n_{700}, n_{1010}, n_{1320}. Our results show that the polyester and polyacrylate films exhibit higher Abbe numbers in comparison with the cellulose samples and, therefore, possess lower dispersion. In case of the three-wavelength LMR results, the Abbe numbers ν_{633} of the films investigated were calculated in a similar manner using the refractive indices measured n_{633}, n_{532}, n_{790}.

Figures 3-b,-c,-d present the dispersion curves of selected polymer materials, namely polyarylate, polyacrylate and copolyester B. The film thickness in microns, as well as the corresponding Abbe number ν_{633}, are quoted in each chart. In the case of the polyarylate films (figure 3-b), the refractive indices values and the Abbe numbers decrease as the film thickness is increased, which could be explained by the increase in the polymer packing near the film surface due to surface tension forces. We established a similar dependence for the copolyester B and cellulose films.

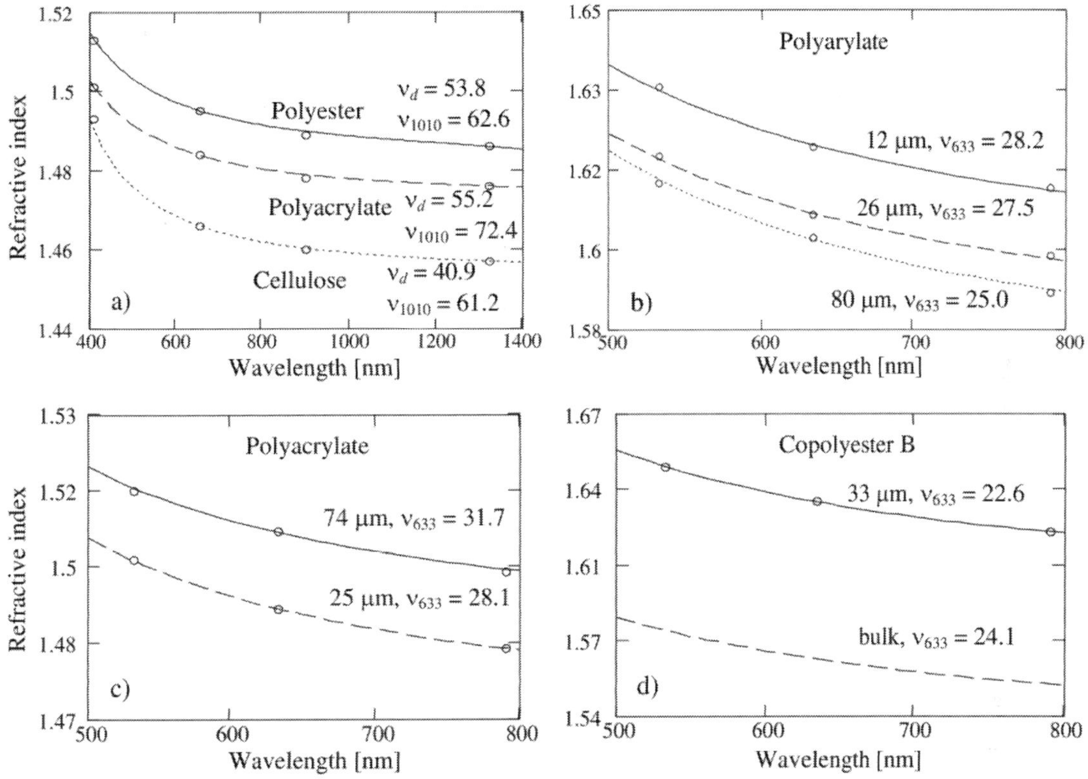

Figure 3. Dispersion curves and Abbe numbers of OPs.

We observed an opposite dependence for the polyacrylate and copolyester A films. The refractive indices and Abbe numbers of the polyacrylate samples decrease for smaller film thickness, while their dispersion grows (figure 3-c). These results show that the dispersion properties of the TPFs depend on the type of the material and its structure. The local refractive index measured is influenced by the film quality and uniformity and is related to the solution viscosity and film adhesion to the glass substrate.

A comparison between the refractive and dispersive properties of a TPF and the bulk material is illustrated in figure 3-d on the example of the copolyester B plastic. In this case, the film has significantly higher refractive indices compared to the bulk sample. However, the slopes of the dispersion curves and the Abbe numbers are similar. An opposite tendency was observed for the polyacrylate film and bulk material. We should note, however, that an average volume refractive index is measured in the case of bulk optical samples, while for the TPFs a local surface value is obtained. The latter depends on the film thickness at the place of laser beam incidence, as well as on the film properties.

Conclusions

The TPFs refractometric and transmission data measured reveal their normal dispersion in the spectral range from 400 nm to 1065 nm. Cauchy-Schott's and Sellmeier's approximations were used to calculate the dispersion coefficients, curves and Abbe numbers. All TPFs studied have greater dispersion in the visible spectrum and lower one in NIR. The refractive and dispersive properties depend on the type of the material and on the film thickness. The results are also influenced by the polymer packing near the surface and by the quality and uniformity of the film, these being related to the solution viscosity and film adhesion to the glass substrate. The refractive and dispersive characteristics of the TPFs differ in comparison to the respective bulk polymer samples, where the average bulk refractive index is measured. In the case of TPFs, local values are obtained, which depend on the type of the material and the film properties.

References

[1] Tolley P 2003 *Phot. Spectra* **10** 76-9
[2] Terui Y and Ando S 2005 *J. Photopolym. Sci. Technol.* **18** 337-40
[3] Nihei E and Shimizu S 2007 *Opt. Commun.* **275** 14-21
[4] Kasarova S, Sultanova N, Ivanov C and Nikolov I 2007 *Opt. Mater.* **29** 1481-90
[5] Kasarova S, Sultanova N, Petrova T, Dragostinova V and Nikolov I 2009 *Proc. SPIE* **7501** 75010P
[6] Kasarova S, Sultanova N, Petrova T, Dragostinova V and Nikolov I 2008 *Proc. SPIE* **7027** 70271O
[7] Vlaeva I, Yovcheva T, Zdavkov K, Minchev G and Stoykova E 2008 *Proc. SPIE* **7027** 70270S
[8] Optical Glass – *Russian standard 13659-68* (LOMO Optical Catalogue)

AUTHOR INDEX

Aleksandrova, M 71
Alexandrova, S 130
Alipieva, E 126
Anastasescu, M 79, 83, 130
Angelov, C 23
Angelov, O 106, 110
Assa, J 156
Atanasov, P A 9, 150
Atanasova, G 150
Avdeev, G 122
Babeva, T 98
Balabanova, E 188
Balakirev, N A 177
Bankova, A 37
Barau, A 79
Benacka, Š 87, 91
Benova, E 32, 192
Beshkova, M 166
Blagoev, B 87, 91, 95, 207
Bodurov, G 71, 75
Boev, V 118
Bogdanov, T 32, 192
Borissov, K 134
Boshkov, N 75
Chernogorova, T P 54
Chromik, Š 87, 91
Cloots, R 207
Conceição, D A S 142
Cziraki, A 13, 161
Damyanova, M 188
Danev, G 102, 156
David, J P R 134
Dejarnac, R 32, 37
Dikova, J 98
Dikovska, A O 9, 150
Dimitrova, M 32, 37
Dimov, D 156
Dimova-Malinovska, D 106, 110
Djermanova, N 37
Dmitriev, S N 23
Donchev, V 134
Donkov, N 58, 114, 197
Dos Santos, A L R 27
Dubrava, T 197
Dulgheru (Nedelcu), N 83
Gadjanova, V 114
Gartner, M 79, 83, 130, 201
Gaži, Š 87, 91
Georgiev, A 156
Georgieva, B 138
Georgieva, V 58, 114, 197
Germanova, K 134
Gesheva, K A 71, 75
Ghelev, C 207
Goltsev, A 197
Grechnikov, A 114
Grigorov, K 166
Grozeva, M 184
Gumarov, G G 177
Gushterov, A 23
Halova, E 130
Harizanova, A 173
Havancsak, K 13
Henrist, C 207
Hongpinyo, V 134
Huhn, G 13
Ilcheva, V 118
Illy, S 188
Iordanova, S 41
Ivanov, A 45
Ivanov, K 192
Ivanov, T 134
Ivanova, P 32, 37
Ivanova, T 71, 75, 173
Ivanova, V 62
Karashanova, D 156
Karatodorov, S 184
Kasarova, S N 212
Kaschieva, S 23
Kayama, M E 27
Kazakov, R 102, 156
Kern, S 188
Kiss'Ovski, Z 41, 45
Kitova, S 102
Kolev, S 45, 207
Koleva, I 41
Koleva, M E 9
Koleva, N K 54
Konstantinova, M 122
Koprinarov, N 122
Kostov, K G 27
Kotseva, I 32
Koutzarova, T 207
Kovacheva, D 207
Krezhov, K 207
Lazarov, J 58
Lazovic, S 1
Lovchinov, K 106, 110
Lutov, L 75
Maciel, H 166
Maletic, D 1
Malovic, G 1
Massi, M 166
Mateev, E 87, 91, 95
Mihailescu, I N 13, 161
Mihailova, D 184
Mikli, V 106
Minkov, I P 161
Mitov, M 37
Mladenova, D 169

AUTHOR INDEX

Mocioiu, O C 201
Monchev, B 118
Mota, R P 27
Nascente, P A P 27
Nedkov, I 166, 207
Nedyalkov, N N 9
Nichev, H 106
Nicolescu, M 79, 83
Nikolov, I D 212
Nizhankovskii, V 207
Nurgaliev, T 87, 91, 95
Oishi, S S 142
Ooi, B S 134
Ouzzane, I 169
Pamukchieva, V 83, 201
Petkov, P 62, 118
Petkova, T 118
Petrov, P 146
Petrovic, Z L 1
Podolesheva, I 138
Popov, T K 32, 37
Predoana, L 79
Puac, N 1
Radeva, E 58
Rassokha, I 197
Rauschenbach, B 66
Ristoscu, C 13, 161
Rogovska, R 197
Ruskov, T 122
Sabchevski, S 188
Safonov, V 197
Sendova-Vassileva, M 114
Silva, L L G 142
Simeonov, S 161
Sinigersky, V 169
Sismanoglu, B 166
Slaveeva, S I 50
Šmatko, V 87, 91
Smolik, J 197
Socol, G 13, 161
Spasic, K 1
Spasov, G 18, 138
Spassova, E 156
Stankova, M 150
Stefanov, P 114, 150
Stepanek, P 169
Stöckel, J 32, 37
Stoica, M 79, 83
Štrbík, V 87, 91
Sultanova, N G 212
Szekeres, A 13, 79, 83, 130, 161, 201
Taskova, E 126
Tellaleva, T 134
Temelkov, K A 50, 54
Thelander, E 66
Thumm, M 188
Todorov, G 126
Todorov, R 98
Todorova, K 83, 201
Toth, A 142
Tutulkov, K 37
Tzacheva, T 122
Ueda, M 142
Vachkov, V 41
Vala, M 169
Van Der Mullen, J 184
Van Dijk, J 184
Vasileva, E 188
Velchev, N B 166
Vergov, L 58
Vertruyen, B 207
Vines, P 134
Vitanov, P 173
Vlaikova, E 13, 130
Vuchkov, N K 50, 54
Weiter, M 169
Zaharescu, M 79, 201
Zaidan, A 62
Zaleski, A 207
Zhelyazkov, I 188
Zhikharev, V A 177
Zhivkov, I 169
Zykova, A 197

Institute of Physics
Dirac House, Temple Back
Bristol BS1 6BE UK

CURRAN ASSOCIATES INC
proceedings
.com